Quipu
**Decorated Permutation
Representations of Finite Groups**

SERIES ON KNOTS AND EVERYTHING

ISSN: 0219-9769

Editor-in-charge: Louis H. Kauffman *(Univ. of Illinois, Chicago)*

The Series on Knots and Everything: is a book series polarized around the theory of knots. Volume 1 in the series is Louis H Kauffman's Knots and Physics.

One purpose of this series is to continue the exploration of many of the themes indicated in Volume 1. These themes reach out beyond knot theory into physics, mathematics, logic, linguistics, philosophy, biology and practical experience. All of these outreaches have relations with knot theory when knot theory is regarded as a pivot or meeting place for apparently separate ideas. Knots act as such a pivotal place. We do not fully understand why this is so. The series represents stages in the exploration of this nexus.

Details of the titles in this series to date give a picture of the enterprise.

Published:

Vol. 77: *Quipu: Decorated Permutation Representations of Finite Groups*
by Y. Bae, J. S. Carter & B. Kim

Vol. 76: *Combinatorial Knot Theory*
by R. A. Fenn

Vol. 75: *Scientific Legacy of Professor Zbigniew Oziewicz: Selected Papers from the International Conference "Applied Category Theory Graph-Operad-Logic"*
edited by H. M. C. García, José de Jesús Cruz Guzmán, L. H. Kauffman & H. Makaruk

Vol. 74: *Seeing Four-Dimensional Space and Beyond: Using Knots!*
by E. Ogasa

Vol. 73: *One-Cocycles and Knot Invariants*
by T. Fiedler

Vol. 72: *Laws of Form: A Fiftieth Anniversary*
edited by L. H. Kauffman, F. Cummins, R. Dible, L. Conrad, G. Elisbury, A. Crompton & F. Grote

Vol. 71: *The Geometry of the Universe*
by C. Rourke

Vol. 70: *Geometric Foundations of Design: Old and New*
by J. Kappraff

More information on this series can also be found at http://www.worldscientific.com/series/skae

KE Series on Knots and Everything — Vol. 77

Quipu
Decorated Permutation Representations of Finite Groups

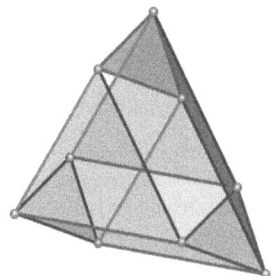

Yongju Bae
Kyungpook National University, South Korea

J Scott Carter
University of South Alabama, USA

Byeorhi Kim
Pohang University of Science and Technology, South Korea

World Scientific

NEW JERSEY · LONDON · SINGAPORE · BEIJING · SHANGHAI · HONG KONG · TAIPEI · CHENNAI · TOKYO

Published by

World Scientific Publishing Co. Pte. Ltd.
5 Toh Tuck Link, Singapore 596224
USA office: 27 Warren Street, Suite 401-402, Hackensack, NJ 07601
UK office: 57 Shelton Street, Covent Garden, London WC2H 9HE

Library of Congress Control Number: 2024022144

British Library Cataloguing-in-Publication Data
A catalogue record for this book is available from the British Library.

Series on Knots and Everything — Vol. 77
QUIPU
Decorated Permutation Representations of Finite Group

Copyright © 2024 by World Scientific Publishing Co. Pte. Ltd.

All rights reserved. This book, or parts thereof, may not be reproduced in any form or by any means, electronic or mechanical, including photocopying, recording or any information storage and retrieval system now known or to be invented, without written permission from the publisher.

For photocopying of material in this volume, please pay a copying fee through the Copyright Clearance Center, Inc., 222 Rosewood Drive, Danvers, MA 01923, USA. In this case permission to photocopy is not required from the publisher.

ISBN 978-981-12-9275-0 (hardcover)
ISBN 978-981-12-9276-7 (ebook for institutions)
ISBN 978-981-12-9277-4 (ebook for individuals)

For any available supplementary material, please visit
https://www.worldscientific.com/worldscibooks/10.1142/13827#t=suppl

Desk Editors: Nambirajan Karuppiah/Nijia Liu

Typeset by Stallion Press
Email: enquiries@stallionpress.com

About the Authors

Yongju Bae is a faculty member at Kyungpook National University. He has served as Dean of Admissions. He has published 35 research papers with focus on algebraic aspects of knot theory that includes recent work on the structure of quandles. He has lectured in Korea, Japan, Russia, the United States, and the United Kingdom. He enjoys playing tennis. He is a proud father and grandfather.

J. Scott Carter is a Professor Emeritus from the University of South Alabama where he served for 29 years. He has published 84 works including two prior books with World Science Publishing. In addition to developing mathematical illustrations that, he hopes, enhance our understanding of the mathematical world, he enjoys walking, biking, and swimming as well as playing guitar. He is a proud father of three boys.

Byeorhi Kim is one of several recent Ph.D.s who earned her degree under the direction of Professor Bae. She is currently working as a post-doctoral researcher at Pohang University of Science and Technology. Her research interests include 2-dimensional surfaces in 4-manifolds and quandle structures related to knot theory. Even in her short career, she has attended conferences and lectured in England, the United States, Korea, Japan, and China. Her cat enjoys helping her with her mathematical research.

Acknowledgments

We gratefully acknowledge a grant in The Basic Science Research Program through the National Research Foundation of Korea (NRF) funded by the Ministry of Education (2016R1D1A3B01007669). JSC was supported by Simons Foundation Grant 381381 that initiated our conversations. He also enjoyed support from Kyungpook National University in 2016 and 2017. He visited Japan under a Japanese Society for the Promotion of Science grant numbered L18511. B. Kim was supported by National Research Foundation of Korea (NRF) Grant No. 2019R1A3B2067839 and No. 2022R1A6A3A01086872.

Other support for the work was by means of research grants of Seonmi Choi, Mark Siggers, and Seung Yeop Yang.

We would like to give personal thanks to Seiichi Kamada and Sang Youl Lee for several interesting conversations.

Contents

About the Authors   v

Acknowledgments   vii

1. **Introduction**   1

2. **Background and Motivating Examples**   11
   - 2.1. Elements of group theory 11
     - 2.1.1. Permutations 16
     - 2.1.2. Permutation representations 20
     - 2.1.3. Short exact sequences 21
   - 2.2. $(\mathbb{Z}/2)^2$ and $\mathbb{Z}/4$ 23
   - 2.3. (mod n)-quipu 31
     - 2.3.1. Application: $\mathbb{Z}/n \times \mathbb{Z}/n$ and $\mathbb{Z}/(n^2)$ 38
     - 2.3.2. Fanciful application: The circle of fifths . . . 49
     - 2.3.3. Application: Dihedral groups 54

3. **Matrix Descriptions**   73
   - 3.1. Overview 73
   - 3.2. The 8-element dihedral group revisited 74
   - 3.3. (mod 2)-quipu that describe a 24-element group . . . 80
     - 3.3.1. Actions upon the cube and an embedded tetrahedron 83
   - 3.4. Signed permutation 93
     - 3.4.1. Signs of permutations and determinant 94

		3.4.2.	The group and a matrix representation	97
		3.4.3.	Subgroups	104
		3.4.4.	The 4-dimensional cube	113
	3.5.	Semi-direct products		117
	3.6.	Proof of Theorem 1		121
		3.6.1.	An alternative view of the dihedral group D_8	122
		3.6.2.	An alternative view of the alternating group A_4	128

4. The 3-Dimensional Sphere is a Group — 137

	4.1.	Balls and their boundaries	139
		4.1.1. Boundaries of cubes	143
		4.1.2. Boundaries of simplices	148
		4.1.3. Taking the boundary twice	150
	4.2.	The unit vectors i, j, and k	154
		4.2.1. The quaternions: Part 1	157
		4.2.2. Stereograph projection	160
		4.2.3. The quaternions: Part 2	168
		4.2.4. Matrix representations	171
	4.3.	The dicyclic groups	174
		4.3.1. Revisiting the dihedral group	174
		4.3.2. Projection from S^3 to the 3-dimensional rotation group	176
		4.3.3. The dicyclic group as a subset of the 3-sphere	178
		4.3.4. Matrices that correspond to the 2-strings-with-quipu representations	181
		4.3.5. The dicyclic group Dic_2 is the group of quaternions Q_8	185
		4.3.6. Describing the projections from the dicyclic groups to the dihedral groups	186
	4.4.	Culmination and anticipation	188

5. Extensions of the Permutation Group Σ_4 — 189

	5.1.	The symmetric group Σ_4	190
		5.1.1. The cosets of the dihedral group	200

Contents

- 5.2. The group $GL_2(\mathbb{Z}/3)$ 202
  - 5.2.1. 4-strings-with-(mod 2)-quipu 205
  - 5.2.2. The semi-dihedral group 217
  - 5.2.3. A 3-strings representation of $GL_2(\mathbb{Z}/3)$ 221
  - 5.2.4. Peculiar correspondences 231
- 5.3. The group $SL_2(\mathbb{Z}/4)$ 240
  - 5.3.1. A 3-strings representation of $SL_2(\mathbb{Z}/4)$ 252
  - 5.3.2. The 2-Sylow subgroup of $SL_2(\mathbb{Z}/4)$ 267
  - 5.3.3. An alternative projection to Σ_4 271
- 5.4. The binary octahedral group 279
  - 5.4.1. Cosets of $A = (1, a, a^2, -1, -a, -a^2)$ 282
  - 5.4.2. Cosets of the dicyclic group Dic_4 of order 16 294
  - 5.4.3. Section summary 306

6. The Binary Tetrahedral Group — 309

- 6.1. The binary tetrahedral group as a subgroup of S^3 310
- 6.2. Correspondences among representations 313
  - 6.2.1. Cosets of $A = \langle a : a^6 = 1 \rangle$ 314
  - 6.2.2. The subgroup Q_8 of $\widetilde{A_4}$ 317

7. The Binary Icosahedral Group — 323

- 7.1. Powers of t 326
  - 7.1.1. Cosets of T 328
- 7.2. Cosets of A 341

8. Computing Group 2-cocycles — 353

- 8.1. Set-theoretic sections 355
  - 8.1.1. Classifying seseqs 363
  - 8.1.2. A geometric interpretation 369
- 8.2. Example 1. The symmetric group Σ_4 378
- 8.3. Example 2. The group $SL_2(\mathbb{Z}/4)$ 385
  - 8.3.1. Computing inverses 394
- 8.4. Example 3. The binary tetrahedral group 397
  - 8.4.1. The cocycle η_1 401
  - 8.4.2. The cocycle η_2 404

	8.4.3. The cocycle η_3	407
	8.4.4. The cocycle η_4	408
8.5.	Example 4. The group $\mathrm{GL}_2(\mathbb{Z}/3)$	420
8.6.	Example 5. The binary octahedral group	425
8.7.	Epilogue	432

References   435

Index   437

Chapter 1

Introduction

This book is written by novices for novices. We beg the indulgence of the experts. Please, let us play and share our joy of discovery while forgiving our transgressions.

There is a problem in writing about mathematics that will be apparent within the next paragraph. It is specialized jargon. So, while we have struggled to make the text as elementary as we can, often we need to write in a way that uses technical language. Whenever it is feasible, we provide metaphors and analogies, and we often provide side comments such as these in the current paragraph to say, "Watch out, there's some heavy stuff coming, but we'll try to make it easier within a few sentences."

Herein, we develop a methodology that uses permutation representations of finite groups to diagrammatically or visually represent the elements. Multiplication is achieved by juxtaposing diagrams vertically with a calculus of diagrammatic manipulations. We believe our methods are novel applications of some theorems that are known among the experts. Furthermore, they lead to methods to explicitly compute 2-dimensional group cocycles.

Permutations commonly can be represented by string diagrams that are analogues of braid diagrams. In general, the number of strings can be quite large, and the diagrams are not visually appealing since it is a daunting task to follow an individual string. On the other hand, elements of cyclic subgroups, in particular, can be presented in a manner that we call "quipu." A given group acts upon its quotient by cyclic and related subgroups. So, a smaller set of strings that are decorated with quipu can often be used to represent the group elements. Using these techniques, subgroups, quotients, and group extensions become more apparent.

While we are assuming that the reader is familiar with the ideas of a group, a subgroup, and the quotient set of a group by a subgroup, these ideas will be briefly reviewed in Section 2.1 of Chapter 2. The quotient may, or may not, be a group depending on whether or not the subgroup that defines the quotient is normal. The word "normal" is used here in a technical sense that will be reviewed in the aforementioned section.

However, our techniques are derived from the topological realm. Equivalences between diagrams are motivated by kinematic considerations. Specifically, the quipu that represent elements in the cyclic subgroups are thought of as sliding upward along the strings on which they appear. Once at the top, they are combined using modular arithmetic.

The most unfamiliar idea in this introduction is that of "quipu." According to a google search on the word, a *quipu* is "an ancient Inca device for recording information, consisting of variously colored threads knotted in different ways." The noun "quipu" sounds to us as if it should be its own plural in a way similar to "fish" and "sheep." In our usage, we will introduce a (mod n)-quipu that will record the number of twists in a fiber of n threads. Other more complicated quipu will also be constructed. These are compositions of permutations and (mod n)-quipu. More specifics will follow the introduction.

The material herein will be developed from the ground up by means of examples. We attempt to proceed linearly from the most simple examples toward the most complicated. The examples, then, will point toward a higher theory with which the experts might already be familiar. As the discussion continues, the theory will be developed incrementally.

The time in which this work is being written is the end of the first quarter of the 21st century, but much of our own training (at least for the senior members of the the research team) occurred in the last half of the 20th century — a time in which abstraction ruled.

We were taught, for example, that $(a-b)(a+b) = a^2 - b^2$, but it seldom occurred to our own instructors to condescend and remark that the arithmetic fact $19 \times 21 = 399$ was a consequence of the general algebraic relationship. Or indeed that one could deduce the product is 399 as an application of the difference of squares.

Not that abstraction in itself is bad. Sometimes, to get from one city to the next, one drives along a super highway or takes a high-speed train. But an equally valid experience is to wander about a village or countryside examining the local shops or exploring the regional fauna and flora. Mathematics, too frequently, neglects the sublime satisfaction that is derived from direct and intricate calculation. We envision these intricate calculations to be as comforting as wandering through the village.

Many colleagues — and ourselves at times — relegate those detailed computations to a computer. In this way, we often substitute the sense of wonder that happens when patterns emerge for the gratification of having written an elegant program. Even then, an awkward program that finds a result may wind up being the primary tool of inquiry. Where is the elegance in that?

One of us (Carter), in collaboration [2], proposed that there is a subject called "Diagrammatic Algebra." The matters addressed in this text could be, and should be, included in that realm, for we are doing algebraic computations by means of manipulating diagrams. But this work stands in the shadow of that work. There are categorical ramifications that can be derived. However, our focus here is much more limited.

We are interested in a few families of finite groups. We develop permutation representations by examining specific cyclic subgroups. Permutation representations are presented as string diagrams. Subgroups, in general, and cyclic subgroups, in particular, are collected into bundles or spools of threads. As we remarked earlier, the elements of the cyclic groups will be called "quipu." Rules for diagrammatic compositions will be established, and simple calculi will be applied in various contexts.

The successors to the current paragraph contain an obligatory outline of the context of the book. Following this, we present a guide to our expectations of the reader's mathematical ability. It is safe for the reader to skip the outline now and come back to it at a later time should the narrative seem muddled or as a checklist of topics covered. The material following the outline, however, is important.

The book is organized into eight chapters.

This first chapter is an introduction and overview. The most important part of the chapter is the section on the authors' expectations of you, the reader. Fear not. Very few readers will meet the expectations of a mathematical author. But enter into the text with your eyes open. Even though our intentions are good, we cannot give all the background material that you will need to understand the things about which we write. We certainly try, but we fall short.

Chapter 2 gives a rapid review of group theory. After defining and exemplifying subgroups and quotient groups in the case of the integers, we discuss permutations in general and in very specific cases. It is very helpful to you if you take the time to understand the string notations for the permutation groups on four objects. The main idea of quipu is introduced by means of several examples. We have a little fun with an associated notion from music theory. Then we discuss the dihedral groups that have a fairly few number of elements.

Chapter 3 begins by playing with some groups that have 2-strings-with-(mod 2)-quipu representations for their elements. As it builds general structures, it points out that the quipu depictions of group elements can be thought of by means of matrix representations.

Chapter 4 begins with a short digression about topological spaces in general. The chapter introduces the 3-dimensional sphere which is a quite remarkable space because it is also a group. That group structure will be studied in the rest of the text by means of the finite subgroups that can be found within it. Many more topological discussions occur in this chapter. Some are not so central to the main theme of the book, but will recur in the last chapter. Namely, we introduce the boundaries of cubes and simplices. The latter are higher-dimensional generalizations of triangles, tetrahedra and the like. Quipu representations of the elements in the unit quaternions and their generalizations, the dicyclic groups, are presented. A nice application of the techniques in the chapter is the group isomorphism between the 3-dimensional sphere and a particular group[1] of (2×2)-matrices whose entries are complex numbers.

Chapter 5 might be the most important section of the book. It contains descriptions of three of the four groups, which have 48 elements, and which map surjectively to the permutation group of four letters. It gives differing quipu descriptions of these groups and also of the symmetric group. Despite its importance, a further description of the material of this chapter at this time is difficult. The difficulty arises because the groups themselves have very specialized names. Two, in particular, are matrix groups, and their common names are usually presented notationally as opposed to being verbal or word-like. Another difficulty that we have here is that the descriptions that we present of the group elements are via quipu. In this introduction, we don't yet have the tools of quipu representations available.

Chapter 6 summarizes many of the things that were uncovered in both Chapters 4 and 5 about the binary tetrahedral group which is also one of the finite subgroups of the 3-dimensional sphere. It serves as a short summary of many of the important ideas found in the previous chapter, and it provides a segue into the next.

Chapter 7 wraps up the discussion of the finite subgroups of the 3-dimensional sphere by giving two quipu representations of the generators for the binary icosahedral group. This group has 120 elements.

The finite subgroups of the 3-dimensional sphere are important in mathematics (and perhaps physics) for reasons that we will not discuss. One of our hopes in writing this text is that we provide a less scary introduction to these groups.

Chapter 8 contains general material about 2-dimensional group cohomology. It provides computations that can be used to distinguish the groups that were discussed in Chapter 5. This chapter appears to give additional purpose to the quipu representations of group elements.

We have attempted to make this book as self-contained, readable, and pleasant as possible. Our attempts require us to recall that

---

[1] Specifically, the special unitary group.

many ideas with which we are familiar might be unfamiliar to the reader. For example, in Section 2.1 of Chapter 2, the notion of a group, a subgroup, a normal subgroup, and a quotient group are reviewed.

On the other hand, we doubt that a reader who has never encountered these ideas will be fully comfortable with that discussion. If you become confused, please consult external sources. Certainly, a Wikipedia entry might be helpful, or a more classical text should be used as a reference. Alternatively, be patient with the authors. Our text usually exemplifies the notions that have been introduced.

We assume that you have at least one technical skill that will make certain sections easier to understand. For example, we hope that you know how to multiply matrices. It is easy for us to write the following:

"Suppose that A is an $(m \times k)$-matrix (m rows and k columns) where the (i, ℓ) entry of A is a_ℓ^i. Suppose that B is a $(k \times n)$-matrix whose (ℓ, j) entry is b_j^ℓ. The entry a_ℓ^i lies in the ith row, ℓth column of A, and the entry b_j^ℓ lies in the ℓth row, jth column of B. Then the (i, j) entry c_j^i of the product $C = AB$ is

$$c_j^i = \sum_{\ell=1}^{k} a_\ell^i b_j^\ell.$$

The matrix product C is an $(m \times n)$-matrix. In general, superscripts indicate row indices, and subscripts indicate column indices."

However, that paragraph, while technically correct, falls into that abyss of abstraction from which we hope the reader can escape. We know that it is devoid of context when you have never gone through the arithmetical agony of having computed matrix products — aligning two pens: one along a row of A, another along a column of B, multiplying the corresponding entries and adding the results, jotting down partial calculations, and keeping track of the signs of the products, all the while muttering "first row, first column; first row, second column, ..."

To allay your fears, most of the matrix entries that we encounter herein are 0s, 1s, or symbolic expressions. Moreover, the matrices are seldom of a size greater than (3×3), and more often than not, we are interested in square matrices.

To reiterate, we hope that you know how to multiply matrices. In Chapter 8, row reduction of some matrices will be used. Row reduction is a standard method that is used in solving linear systems of equations. We hope that you have encountered it within your studies.

We also hope that you are comfortable with the juxtaposition of one diagram above another as a binary operation. In addition to these juxtapositions, there are some kinematic symbol gymnastics that are used to simplify the resulting diagram. At the very least, modular arithmetic should be performed on pairs of quipu that lie adjacent at the top of the strings. Moreover, the quipu themselves are allowed to float upward along strings and through the crossings that otherwise might appear to obstruct them. These rules for manipulating quipu will be reviewed, exemplified, and justified via typical rigorous discussions (see Section 2.3 of Chapter 2).

Many of the groups that will be considered are not commutative. The word "commutative" indicates that $ab = ba$ for elements $a, b \in G$. A synonym, among mathematicians, of commutivity is "abelian." This is an adjectival form of the proper noun "Abel." Please consult standard sources on the pronunciations of these words. Since ab may not be equal to ba, it matters which diagram appears above the other in their vertical juxtaposition. The product ab is juxtaposed as $\frac{D(a)}{D(b)}$, where $D(a)$ indicates the diagram that is associated to a. The text will often re-emphasize these conventions.

Finally, we expect that you have some comfort with the standard mathematical notations of set, element containment, union, intersection, real numbers $\mathbb{R}$, complex numbers $\mathbb{C}$, and so forth. At several times, we mention relations that are equivalences. These are reflexive, symmetric, and transitive relations that are defined upon sets. We expect fluency with calculations that involve sines and cosines, and we move between those things that are within a standard secondary school curriculum and the representation of trigonometry via complex numbers.

The *real numbers* is a totally ordered Archimedian field[2] that is denoted by $\mathbb{R}$. It can safely be confused with the more familiar

---

[2] A *field* is set with an associative, commutative, unital (0), addition, a notion of additive inverse (negative), a commutative multiplicative group structure on the non-zero elements, and distributive laws for multiplication over addition.

concept of numbers that are represented as infinite decimals. It has the usual operations of addition, subtraction, multiplication, and division appropriately defined. Addition and multiplication are related by means of distributive laws. The real number system does not have much to do with reality since it is the realm of infinite precision. Any measurement in reality has an error that is associated to the accuracy of the measuring device. In brief, the real numbers form the number system with which we imagine you are familiar.

The *complex numbers* are denoted by $\mathbb{C}$. We think of it as a set

$$\mathbb{C} = \{a + bi : a, b \in \mathbb{R}\}$$

with an addition defined by

$$(a + bi) + (c + di) = (a + c) + (b + d)i,$$

and a multiplication that is defined by

$$(a + bi)(c + di) = (ac - bd) + (ad + bc)i.$$

Exercises have been interspersed throughout the text. Some are explicitly listed as exercises. Others are mentioned in passing. There are other points at which we make various assertions without adorning them as lemmas, propositions, theorems, or corollaries. To do so would interrupt the narrative flow. Often within a paragraph or two, a justification of the assertion appears. If not, then please try to figure out why what we are writing is true. If you are having trouble, again consult a standard source. We urge you, when a notion is introduced, try to understand and exemplify it for yourself before reading further.

This work draws on knowledge that the authors have gained over the years. Some standard sources deserve mention. For algebraic background, see, e.g., [5, 8, 9]. For topological background, see [3] or [12]. While we were writing this, we often looked through several Wikipedia articles. We have often been told that doing so is "sloppy scholarship." Yet, the mathematical materials that we used from there were easily verified. More often than not, we included the verifications that we found within the text. The bibliography will be quite short because of our desire for the text to be self-contained.

Finally, the ideas about quipu that we give here are our own. With the exception of Kauffman's paper [10], we know of no other source.

Chapter 2
Background and Motivating Examples

2.1. Elements of group theory

A *group* is defined to be a set G upon which a binary operation $\cdot : G \times G \to G$ is defined (and written in in-fix notation):

- *associative*:
$$(a \cdot b) \cdot c = a \cdot (b \cdot c)$$
for all $a, b, c \in G$;
- *unitary*: there is an element $1 \in G$ such that
$$1 \cdot a = a \cdot 1 = a$$
for all $a \in G$;
- *invertible*: for any $a \in G$, there is an element $a^{-1} \in G$ such that
$$a \cdot a^{-1} = a^{-1} \cdot a = 1.$$

A *subgroup* is a non-empty subset $H \subset G$ that is closed under multiplication ($a, b \in H$ implies that $a \cdot b \in H$) and inversion (if $a \in H$, then $a^{-1} \in H$). If G is finite and H is a subgroup, then H is also finite. The *order* of a group $|G|$ is the number of elements in its underlying set. It is considered to be an elementary fact that the order of H divides the order of G. So, if $|H| = k$, then $|G| = nk$ for some $n \in \mathbb{N} = \{1, 2, \dots, \}$. A sketch of this fact follows.

An equivalence relation upon G, that is called *congruence modulo* H and that is written $\equiv_R$, is defined by the relationship

$$[a \equiv_R b \mod H] \Leftrightarrow [ba^{-1} \in H].$$

It is not too difficult to demonstrate that this is a symmetric, reflexive, and transitive relation. Thus, the equivalence classes $\{Ha : a \in G\}$ partition the underlying set of G. Here,

$$Ha = \{ha : h \in H\}$$

is the set of multiples of the elements of H with the fixed element $a \in G$. Each of the *cosets* Ha has the same number of elements, k, as the subgroup H. So, as a set, G is decomposed as a disjoint union

$$G = Ha_1 \cup Ha_2 \cup \cdots \cup Ha_n.$$

According to Wikipedia, the elements of $\{Ha : a \in G\}$ are called *right cosets*.

An analogous equivalence relation $\equiv_L$ is defined by the relationship

$$[a \equiv_L b \mod H] \Leftrightarrow [a^{-1}b \in H].$$

The equivalence classes $\{aH : a \in G\}$ are called *left cosets*.

On a day-to-day basis, it may be difficult to recall which of Ha or aH is a right coset. In general, the text will usually use cosets of the form aH, and it will neglect to call the coset a left coset. The collection of cosets will be denoted by G/H.

Some authors adopt a clever notation that $H\backslash G$ denotes the collection of right cosets, while G/H denotes the collection of left cosets. We will not yield to that temptation.

A subgroup H is said to be a *normal subgroup* if every left coset is a right coset. Equivalently, the subgroup is fixed by *conjugation* by elements of G. That is, $g^{-1}Hg = H$ for all $g \in G$. There are many criteria that characterize normal subgroups. We do not list these here. Normal subgroups are related to group homomorphisms.

A function $G' \xleftarrow{\phi} G$ between groups G and G' is said to be a *homomorphism* if and only if

$$\phi(a \cdot b) = \phi(a) \cdot \phi(b).$$

(As is customary, henceforth, juxtaposition will often be used to denote group multiplication.) The *kernel of ϕ* is the set

$$\ker(\phi) = \{g \in G : \phi(g) = 1\}.$$

(Our arrows point from right to left. The domain of a function is on the right; the codomain is written to the left. In this way, composition — $C \xleftarrow{g} B \xleftarrow{f} A$ — is written as $g(f(x))$ for an element $x \in A$. Similarly, we imagine the domain of a string diagram to be written at the bottom of the diagram. These conventions may not be usual, and they may take some time before you are accustomed to them.)

Standard exercises[1] in an elementary course in group theory include showing that the kernel is a normal subgroup of G, showing that the function

$$G/H \xleftarrow{\phi} G$$

that sends a group element g to the coset gH is a homomorphism from G to G/H and that the image of a surjective homomorphism is isomorphic to G/H. Here, $H = \ker(\phi)$. The group operation on G/H is given by $aHbH = abH$. Generally, one spends some time demonstrating this formula is well defined in the sense that neither side of the equation depends upon the choice of a coset representative. Note that an *isomorphism* is a bijective[2] homomorphism. In this case, $\ker(\phi) = \{1\}$.

In a few lines, we will give examples of groups, subgroups, and group homomorphisms.

---

[1] See the exercises listed in the following.
[2] One-to-one (or injective): $\phi(a) = \phi(b) \Rightarrow a = b$; onto (or surjective): for all $g' \in G'$, there is an element $g \in G$ such that $\phi(g) = g'$.

Exercise 1.

1. Verify that the relations $\equiv_R$ and $\equiv_L$ are equivalence relations (symmetric, reflexive, and transitive relations).
2. Verify that if every left coset of a subgroup H is a right coset if and only if $g^{-1}Hg = H$ for every $g \in G$. The equality of sets means that for every $h \in H$, there is a unique $h' \in H$ such that $g^{-1}hg = h'$.
3. Verify that the kernel of a homomorphism is a normal subgroup.
4. Verify that the operation $aHbH = abH$ is a well-defined group operation on G/H when H is a normal subgroup.
5. Suppose that $G' \xleftarrow{\phi} G$ is a surjective group homomorphism. Show that there is an isomorphism (bijective homomorphism) between G' and G/H, where $H = \ker(\phi)$.
6. If there are other claims that the authors made in the prior text that you still don't fully understand, try to prove them for yourself. When the authors write phrases such as, "It is not too difficult to show that [...]," take the time to try and show the result, and take pleasure in cursing us if you found it difficult.

A group G is said to be *abelian* or *commutative* if and only if for all $a, b \in G$, we have

$$ab = ba.$$

Often, but not always, the operation in an abelian group is denoted by $+$. So, the commutative law is written as

$$a + b = b + a.$$

The group structure with which many people are familiar is the following. The set of integers

$$\mathbb{Z} = \{0, \pm 1, \pm 2, \pm 3, \ldots\}$$

forms an abelian group under the operation of $+$. Addition of integers is associative, the identity element is 0, and the additive inverse of an element a is $-a$. Note also that $-(-a) = a$.

In the case of an abelian group, all subgroups are normal.

Let n denote a positive integer. Then

$$n\mathbb{Z} = \{0, \pm n, \pm 2n, \pm 3n, \ldots\}$$

is a normal subgroup of $\mathbb{Z}$. It is non-empty since $0 \in n\mathbb{Z}$. It is closed under addition since $an + bn = (a+b)n$. It is normal because $\mathbb{Z}$ is abelian. The associative property for addition is inherited from the integers $\mathbb{Z}$, and $-(an) = (-a)n$, so the additive inverse of any element in $n\mathbb{Z}$ is also in $n\mathbb{Z}$.

The set of cosets $\mathbb{Z}/(n\mathbb{Z})$ is also denoted by the more brief expression $\mathbb{Z}/n$. A list of cosets follows:

$$n\mathbb{Z}, 1 + n\mathbb{Z}, 2 + n\mathbb{Z}, \ldots, (n-1) + n\mathbb{Z}.$$

The elements $0, 1, \ldots, n-1 \in \mathbb{Z}$ are called the *canonical (or standard) set of coset representatives*. For a temporary emphasis, we write $[k]$ to indicate the coset $k+n\mathbb{Z}$. The representative k may not be canonical. In general, a *coset representative* is any element in the coset. So, for example, $k - 2n$, $k - n$, k, $k + n$, and $k + 2n$ are a few of the representatives of the coset $k + n\mathbb{Z}$.

Modular arithmetic is performed on coset representatives as follows:

$$[a] + [b] = [a+b],$$

where if the value of the sum $a + b \in \mathbb{Z}$ is greater than n, then (repeatedly) subtract n from the result to compute the canonical representative in $\{0, 1, \ldots, n-1\}$. If that sum happens to be less than zero, repeatedly add n until the sum is non-negative and bounded above by $n - 1$.

In addition to the integers $\mathbb{Z}$, perhaps the most familiar abelian group is the integers modulo 2, $\mathbb{Z}/2 = \mathbb{Z}/(2\mathbb{Z})$. The canonical representatives are 0 and 1 (respectively, even and odd) and a table of values for addition is presented as follows:

+	0	1
0	0	1
1	1	0

It might be familiar because it expresses the addition of parity: even + even is even, even + odd is odd, odd + even is odd, and odd + odd is even.

If G and G' are both groups, then their *cartesian product*
$$G \times G' = \{(g, g') : g \in G, \ \& \ g' \in G'\}$$
is a group under component-wise multiplication:
$$(a, a') \cdot (b, b') = (ab, a'b').$$

In the following section, the cartesian product $\mathbb{Z}/2 \times \mathbb{Z}/2 = (\mathbb{Z}/2)^2$ will be contrasted with $\mathbb{Z}/4$. Note that both groups have four elements
$$(\mathbb{Z}/2)^2 = \{(0,0), (0,1), (1,0), (1,1)\}$$
and
$$\mathbb{Z}/4 = \{0, 1, 2, 3\}.$$
Observe further that if the integers modulo 4 were written in binary notation, then $0 \leftrightarrow [00]$, $1 \leftrightarrow [01]$, $2 \leftrightarrow [10]$, and $3 \leftrightarrow [11]$. So, the elements of the two groups in question will have similar looking names.

2.1.1. Permutations

Let $n \in \mathbb{N}$ denote a positive integer. Let
$$[n] = \{1, 2, \ldots, n\}$$
denote the set of integers from 1 to n inclusive. Since $1 < 2 < \cdots < n$, this set is tacitly ordered. On the other hand,
$$\Sigma_n = \{[n] \xleftarrow{s} [n] : s \text{ is bijective}\},$$
the *set of permutations* on n-elements, plays mischief with any preassigned ordering. This set is a group under composition of functions. The identity function $j \mapsto j$ is the group identity. Composition of functions is associative. If a given function, s, maps i to j, that is, $s : i \mapsto j$, then the inverse maps j to i, that is, $s^{-1} : j \mapsto i$.

It is convenient to represent such a function as follows. Arrange n dots equally spaced along a line. For brevity, we will consider the case $n = 3$.

A permutation will be represented as a set of three upward pointing arrows. The six permutations in Σ_3 are presented in Fig. 2.1.

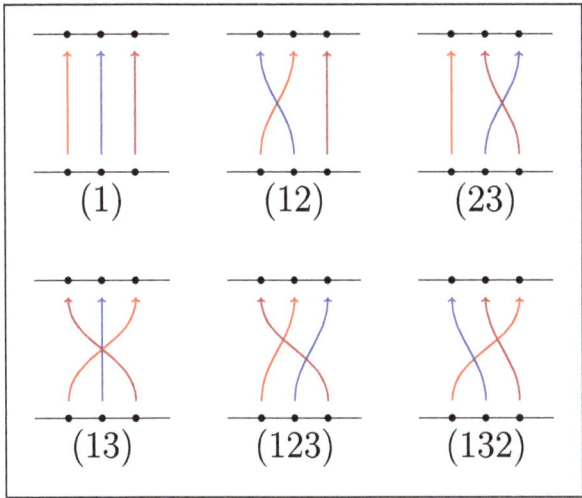

Figure 2.1. The permutations in Σ_3 drawn as arrows.

Since the diagrams are to be read from bottom to top, the arrows are redundant. Similarly, the dots along the line are not necessary (Figure 2.2).

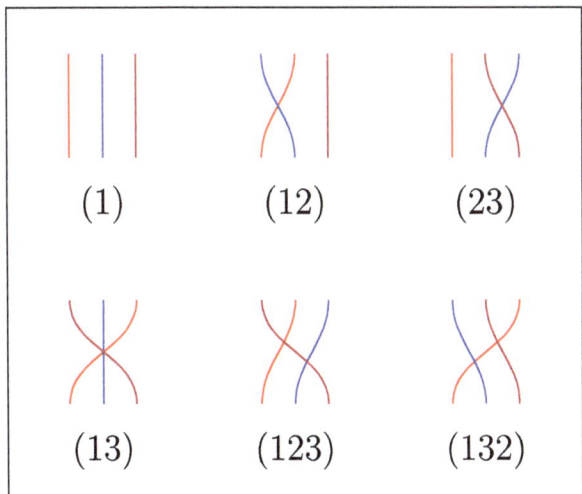

Figure 2.2. The permutations in Σ_3 drawn as string diagrams.

On the other hand, these figures indicate the standard names for the permutations as given in "cycle" notation that we exemplify. The permutation that is labeled (123) indicates the composition of arrows

$$[3] \xleftarrow{(12)} [3] \xleftarrow{(23)} [3]$$

that is alternatively expressed as

$$(12) \circ (23)(\{1,2,3\}).$$

In this way, compositions are written from left to right with the domain variable written on the far right, and they are drawn from top to bottom with the domain variables fed in from the bottom. The cycle (123) indicates that $1 \mapsto 2$, $2 \mapsto 3$ and $3 \mapsto 1$. Similarly, the cycle (12) indicates that $2 \mapsto 1$ and $1 \mapsto 2$.

In our discussion, it will be rare to consider a permutation upon a larger order set, and so commas are usually removed from the cycle notation.

The permutation group Σ_n is usually written in terms of generators and relations as follows:

$$\left\langle t_1, t_2, \ldots t_{n-1} \left| \begin{array}{ll} t_i^2 = 1, & \text{for } i = 1, \ldots, n-1, \\ t_i t_{i+1} t_i = t_{i+1} t_i t_{i+1} & \text{for } i = 1, \ldots, n-2, \\ t_i t_j = t_j t_i & \text{for } 1 < |i-j| \end{array} \right. \right\rangle.$$

The above conglomerate of symbols is known as *a group presentation*. The presentation indicates that any permutation can be written in terms of the (adjacent) transpositions $t_i = (i, i+1)$. The relations that hold among these are as follows: each is of order 2, that is, $t_i^2 = 1$; distant transpositions commute, that is, $t_i t_j = t_j t_i$ when $1 < |i-j|$; and the transposition $(i, i+2)$ can be achieved in two different ways: $t_i t_{i+1} t_i$ or $t_{i+1} t_i t_{i+1}$. All other relationships that hold among expressions that are written in terms of the generators can be determined as consequences of these three relationships. The relation $t_i t_{i+1} t_i = t_{i+1} t_i t_{i+1}$. can be visualized as follows:

The permutation group Σ_n is also called the *symmetric group*. The symmetric group, Σ_4, of permutations on $[\mathbf{4}] = \{1, 2, 3, 4\}$ will be a main topic of interest throughout the narrative. Consequently, we will often refer to the "string diagram" representations of its elements. These are cataloged in Fig. 2.3.

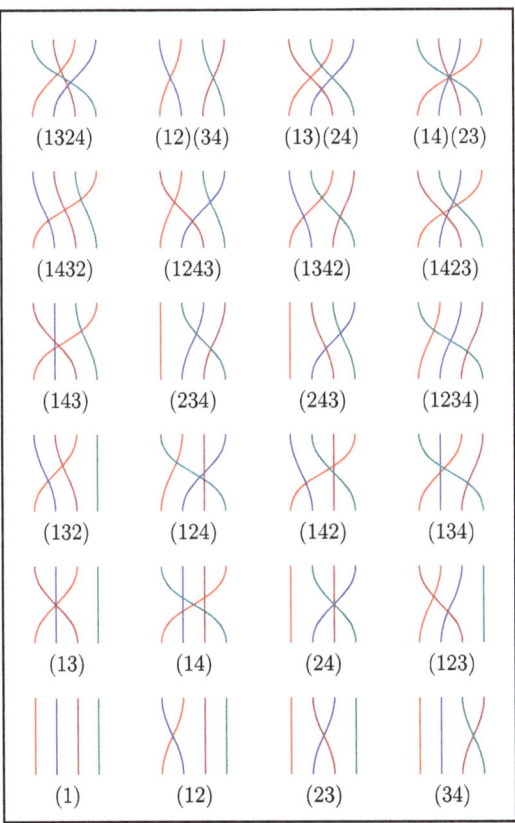

Figure 2.3. The permutations in Σ_4 drawn as string diagrams.

A sample diagrammatic computation of the product $(123)(143) = (14)(23)$ is presented in Fig. 2.4.

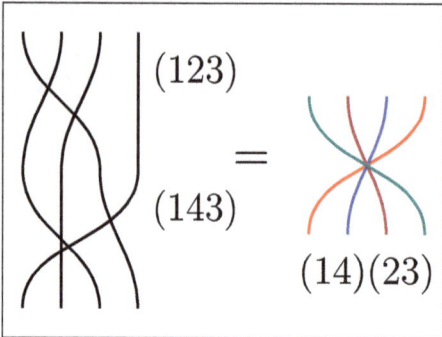

Figure 2.4. A computation with string diagrams in Σ_4.

We know of verbal gymnastics that allow such products to be computed. One starts at the rightmost cycle, (143), in the expression (123)(143) and observes that "1 goes to 4, and 4 doesn't move in the left cycle. But 4 goes to 3 which goes to 1. So, there is a closed cycle (14) in the product. Where does 3 go? Let's see 3 goes to 1 which goes to 2. So, there is an open[3] cycle (32. The element 2 is fixed in the right cycle, but 2 goes to 3 in the left cycle. So, the product must be (14)(23)." These verbalizations might be enhanced by imagining the corresponding string diagrams.

2.1.2. Permutation representations

Let G denote a group. Let $1 \in G$ denote its identity: $1g = g1 = g$ for all $g \in G$. For greater intuition, you may suppose that G is finite, but that assumption is not necessary for the discussion that follows. Choose a fixed element $g \in G$. The set $\{gx : x \in G\}$ of multiples of g is, in fact, the set of elements of G because g is invertible. So, any $y \in G$ can be written as $g(g^{-1}y)$. Therefore, for each $g \in G$, the function $\sigma_g : x \mapsto gx$ is a permutation of the underlying set of G. In this way, σ_1 is the identity permutation, and the composition $\sigma_g \circ \sigma_h = \sigma_{gh}$ of two such permutations is the permutation that is associated to the group product gh. To see this, compute

$$\sigma_g \circ \sigma_h(x) = \sigma_g(hx) = g(hx) = (gh)x = \sigma_{gh}(x).$$

---

[3] Since it is open, we have yet to close the parentheses upon it.

In this way, there is a representation of the group G onto the permutation group $\Sigma_{|G|}$ where $|G|$ denotes the order of G.

In general, when one group is *represented* into another group, this indicates an injective homomorphism as exemplified above. But more specifically, the image of the representation has some geometric context. In the case of permutation representations, we are representing any (finite) group to act as if it is a set of permutations. In the act of "representing" the group, we pretend the group (which is given abstractly) is a set of symmetries of some underlying object. In this case, it is the set of symmetries of the underlying set of the group.

In general, we will choose an ordering (left-to-right) upon the set of cosets of a specific subgroup H of a group G and also an ordering the subgroup H. Then we'll develop permutation diagrams for the cosets. The strings that represent the permutation actions on the cosets will be adorned by quipu that represent the permutation action upon the subgroup.

2.1.3. Short exact sequences

Suppose that G is a group, H is a normal subgroup, and $Q = G/H$ is the quotient group.[4] We often say that there is a short exact sequence

$$1 \longleftarrow Q \xleftarrow{q} G \xleftarrow{i} H \longleftarrow 1.$$

The homomorphism $H \longleftarrow 1$ indicates the inclusion of the identity element into the subgroup H. The homomorphism $G \xleftarrow{i} H$ is the inclusion of the subgroup H into the group G. At times, the element $h \in H$ might need to be distinguished from its image $i(h) \in G$. The homomorphism $Q \xleftarrow{q} G$ is the quotient map that sends an element $g \in G$ to the coset gH that it represents. The homomorphism $1 \longleftarrow Q$ is the function that sends every coset to the identity. Saying that the sequence is *exact* is to say that the image of a homomorphism on the right is the kernel of the homomorphism that appears immediately to the left. In particular, $\text{Im}(i) = \ker(q)$. The subgroup H is included

---

[4]Here and elsewhere, we usually don't make distinctions between a group G and other groups G' that are isomorphic to G. Consequently, we say that a given group "is" this or that. The more appropriate phrasing is that G is isomorphic to this or that.

into G since the image of 1 in H is just 1, and the homomorphism from G to Q is a surjection since its image is in the kernel of the function $1 \longleftarrow Q$, and every element of Q maps to the identity 1 of the trivial group.

We adopt a neologism *seseq* to indicate that a sequence of five such groups that begin and end at the identity is a short exact sequence.

The rest of the text will be replete with short exact sequences, so we only give a few examples here.

This first example of a seseq is the following:

$$0 \longleftarrow \mathbb{Z}/2 \xleftarrow{q} \mathbb{Z} \xleftarrow{2\cdot} \mathbb{Z} \longleftarrow 0.$$

Since the groups in question are abelian, the trivial groups will be denoted additively as 0 rather than 1. The inclusion map $\mathbb{Z} \xleftarrow{2\cdot} \mathbb{Z}$ maps the integer k to the even integer $2k$.

This sequence has an analogue:

$$0 \longleftarrow \mathbb{Z}/n \xleftarrow{q} \mathbb{Z} \xleftarrow{n\cdot} \mathbb{Z} \longleftarrow 0.$$

The inclusion $\mathbb{Z} \xleftarrow{n\cdot} \mathbb{Z}$ maps the integer k to the integer $n \cdot k$ that is divisible by n.

Let G denote the cartesian product $\mathbb{Z}/2 \times \mathbb{Z}/2$. Then there is an inclusion $\mathbb{Z}/2 \times \mathbb{Z}/2 \xleftarrow{(0,i)} \mathbb{Z}/2$ in which $y \in \mathbb{Z}/2$ is mapped to $(0,i)(y) = (0,y)$ in $\mathbb{Z}/2 \times \mathbb{Z}/2$. The image $\{0\} \times \mathbb{Z}/2$ is the kernel of the projection $\mathbb{Z}/2 \xleftarrow{p_1} \mathbb{Z}/2 \times \mathbb{Z}/2$ onto the first factor: $p_1(x,y) = x$. Thus, there is a seseq

$$0 \longleftarrow \mathbb{Z}/2 \xleftarrow{p_1} \mathbb{Z}/2 \times \mathbb{Z}/2 \xleftarrow{(0,i)} \mathbb{Z}/2 \longleftarrow 0.$$

Consider the canonical coset representatives $\{0,1,2,3\}$ of the elements of $\mathbb{Z}/4$. There is an inclusion $\mathbb{Z}/4 \xleftarrow{2\cdot} \mathbb{Z}/2$ that is defined by $2\cdot(y) = 2y$. So, $0 \mapsto 0$ and $1 \mapsto 2$. The image of this map (the even numbers modulo 4) is the kernel of the parity map $\mathbb{Z}/2 \xleftarrow{p} \mathbb{Z}/4$ for which $p(1) = p(3) = 1 \in \mathbb{Z}/2$ and $p(0) = p(2) = 0 \in \mathbb{Z}/2$. In this way, there is a seseq

$$0 \longleftarrow \mathbb{Z}/2 \xleftarrow{p} \mathbb{Z}/4 \xleftarrow{2\cdot\_} \mathbb{Z}/2 \longleftarrow 0.$$

The goal of the following section is to compare and contrast these two seseqs and establish diagrams that represent the elements of these two groups. In this way, we will establish motivating examples. In the section thereafter, these examples will be generalized.

2.2. $(\mathbb{Z}/2)^2$ and $\mathbb{Z}/4$

One of the beautiful aspects of group theory is that one can start with a fairly arbitrary description of elements, define a way of composing them, and with a bit of luck, or ingenuity, inverses will become apparent, and a group will arise. On the other hand, in the case of finite groups that have a relatively small number of elements, the isomorphism classes of these are catalogued. So, by following the creative path, don't expect to find something new. Instead, imagine that you are finding your own description of something that is known by others.

We will begin abstractly with three nice examples. Then we will identify these as among the groups that have already been discussed and give rationale for the isomorphisms to the more standard descriptions. In Section 2.3.1, which follows this one, generalizations will be made.

Example 1. Consider the two-element set,

$$T = \left\{ \,\big|\, , \,\big\vert\!\!\bullet\, \right\}.$$

The elements are combined via stacking them vertically and rescaling. Additional rules are imposed:

The multiplication table, for which the element in the row is stacked above that in the column, appears in Fig. 2.5.

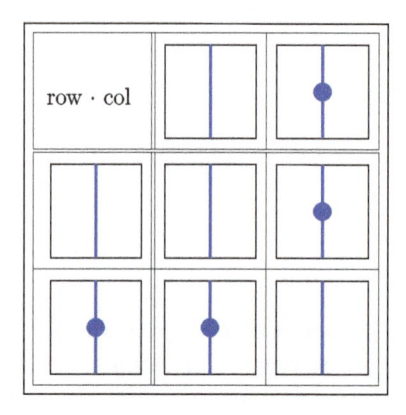

Figure 2.5. A multiplication table for (mod 2)-quipu.

The resulting group apparently is $\mathbb{Z}/2$. The blue dot along a string is called a *(mod 2)-quipu*. The apparent isomorphism to $\mathbb{Z}/2$ occurs by stacking a larger number of dots upon a string and counting the result having canceled successive pairs. This is a parity count.

Before presenting the following two examples, two rules for manipulating (mod 2)-quipu will be presented:

While the rules are declared to be equalities, in practice they are used to slide the quipu upward along the strings and beyond any interfering crossings.

Example 2.

$$A = \left\{ \vphantom{\Big|} \big| \big| \, , \, \times\!\!\!\bullet \, , \, \big|\bullet\big| \, , \, \bullet\!\!\!\times \right\}.$$

One additional rule will be imposed:

$$\vcenter{\hbox{⧖}} = \vcenter{\hbox{||}}.$$

We use the diagrammatic substitution rules that were given above to compute the following:

$$\vcenter{\hbox{⧖}} = \vcenter{\hbox{|• •|}}; \quad \vcenter{\hbox{)(}} = \vcenter{\hbox{)(}};$$

$$\vcenter{\hbox{⧖}} = \vcenter{\hbox{||}}.$$

The *order of an element a* in a group G is the minimum exponent n such that $a^n = a \cdot a^{n-1} = 1$. The element ╳, therefore, has order 4 in the group A. A similar computation that is the vertical reflection of the one above gives that the element ╳ also has order 4. The group A that has been described via (mod 2)-quipu is apparently $\mathbb{Z}/4$.

Example 3. The set B that is listed in the following is also a group:

$$B = \{ \,||\,,\,╳\,,\,\overset{\bullet}{|}\overset{\bullet}{|}\,,\,╳\, \}.$$

Each of the non-identity elements has order 2:

$$\vcenter{\hbox{⧖}} = \vcenter{\hbox{||}}\,;\, \vcenter{\hbox{⧖}} = \vcenter{\hbox{||}}\,;\, \vcenter{\hbox{||}} = \vcenter{\hbox{||}}.$$

The product of any one of these with another is the third:

$$\text{(diagrams showing crossing-quipu identities)}$$

In $\mathbb{Z}/2 \times \mathbb{Z}/2$, we have

$$(1,0) + (0,1) = (0,1) + (1,0) = (1,1);$$

$$(1,1) + (0,1) = (0,1) + (1,1) = (1,0);$$

$$(1,0) + (1,1) = (1,1) + (1,0) = (0,1),$$

and each is of order 2:

$$(1,0) + (1,0) = (0,1) + (0,1) = (1,1) + (1,1) = (0,0).$$

Thus, B is isomorphic to $\mathbb{Z}/2 \times \mathbb{Z}/2$.

In both cases, there are seseqs that can be understood and imagined from the visual point of view. The homomorphism from the topmost group $T \cong \mathbb{Z}/2$ replicates the strings. Its image consist of those two elements that do not involve crossings. The lower map ignores the quipu upon the strings. Those diagrams in the middle group (A or B) that do not involve crossings map to the identity elements when the quipu are ignored. After all, they represent a pair of straight strings (Fig. 2.6).

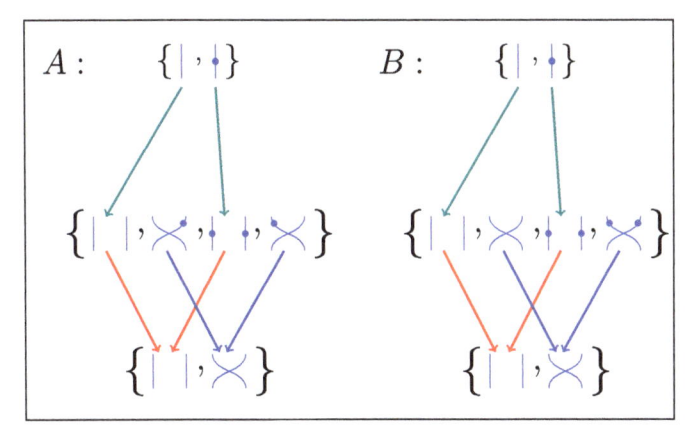

Figure 2.6. Two seseqs with $\mathbb{Z}/2$ as kernel and as image.

Why do these examples work? The short answer is that the strings-with-quipu which are used to represent the group elements are a short-hand notation for specific permutations in the group Σ_4. Furthermore, both $\mathbb{Z}/2 \times \mathbb{Z}/2$ and $\mathbb{Z}/4$ are isomorphic to subgroups of Σ_4. The longer answer will specify these elements, and it will specify the elements in Σ_4 that the strings-with-quipu correspond. Furthermore, the quipu are schematics for some transpositions.

To facilitate the reader's understanding of the longer answer, the text digresses into a brief discussion of the set of permutations, Σ_2, upon $\{1, 2\}$. Let us analyze the presentation

$$\left\langle t_1, t_2, \ldots t_{n-1} \;\middle|\; \begin{array}{ll} t_i^2 = 1, & \text{for } i = 1, \ldots, n-1, \\ t_i t_{i+1} t_i = t_{i+1} t_i t_{i+1} & \text{for } i = 1, \ldots, n-2, \\ t_i t_j = t_j t_i & \text{for } 1 < |i - j| \end{array} \right\rangle$$

in the case that $n = 2$. The set of *generators* $\{t_1, t_2, \ldots t_{n-1}\}$ when $n = 2$ consists of the unique generator t_1. As before, we remark that t corresponds to the transposition (12) in which the elements 1 and 2 switch positions. Among the relations, only the first one, $t_1^2 = 1$, has any content because of the ranges upon the indices i and j in the second and third set of relations. The interchange $t_1 \leftrightarrow (12)$ of the positions of 1 and 2 is represented vertically as ⨯.

The vertical stacking

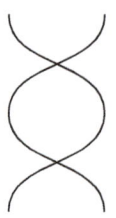

corresponds to the composition of t_1 with itself. Read as a permutation $1 \mapsto 1$ and $2 \mapsto 2$. So, this composition is the identity permutation. Figure 2.7 indicates this equality and the correspondence of Σ_2 with the group $T = \{\,\vert\vert\, ,\, \vert\!\cdot\!\vert\,\}$, namely, the transposition (12) corresponds to a single quipu upon a thicker blue string, as indicated in the upper rectangle, and an unadorned thick blue string corresponds to a pair of straight black strings as indicated on the right side of the lower rectangle.

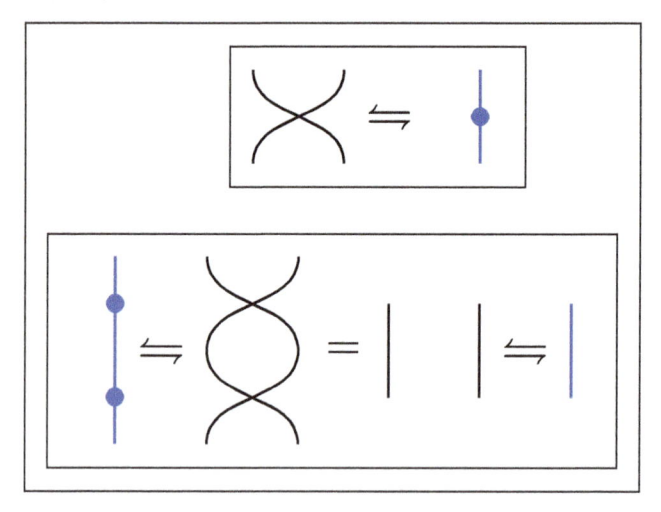

Figure 2.7. The correspondence between (mod 2)-quipu and transpositions.

The harpoons ($\rightleftharpoons$) in these pictures indicate the correspondences between elements in the isomorphism between the groups Σ_2 and T that was defined above. In both cases, when a change happens twice, then no change occurs. That involutive phenomenon is the essence of the group $\mathbb{Z}/2$.

Consider the elements (1), (12)(34), (13)(24), (14)(23), (1423), and (1324) in Σ_4. Slightly modified string-diagram depictions of these six elements are depicted in Fig. 2.8. The modifications are as follows. First, the strings that start on the bottom at 1 and 2 have been colored blue while those that start at 3 and 4 have been colored red. The colorations run to the top. Second, a horizontal line segment joins 1 to 2 at the bottom, and a similar segment joins 3 to 4. These segments suggest a ribbon, so there are horizontal segments at the top that inherit their color from below that also join 1 to 2 or 3 to 4.

The modified diagrams suggest the fact that each of these six permutations can be thought of as a permutation between the sets $\{1, 2\}$ and $\{3, 4\}$. The color is meant to emphasize this fact.

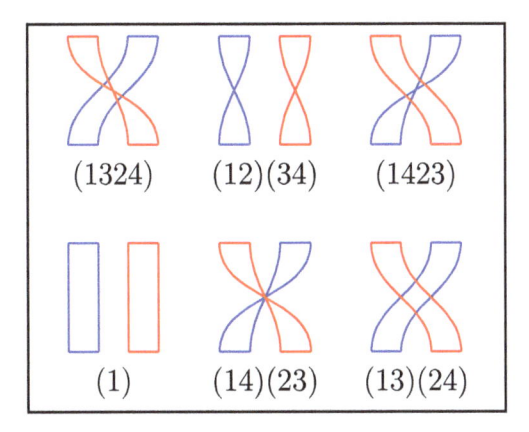

Figure 2.8. Grouping $\{1, 2\}$ and $\{3, 4\}$ to form ribbons.

The diagrams that are depicted above are *ribbon diagrams*. Some ribbons remain planar and others have half-twists in them. The half-twists are represented by quipu.

Thus, as a set of permutations on four elements,

$$A = \{(1), (1423), (12)(34), (1324)\}$$

and

$$B = \{(1), (13)(24), (12)(34), (14)(23)\}.$$

For emphasis, the correspondence between the quipu pictures and the ribbon pictures are compiled in Fig. 2.9.

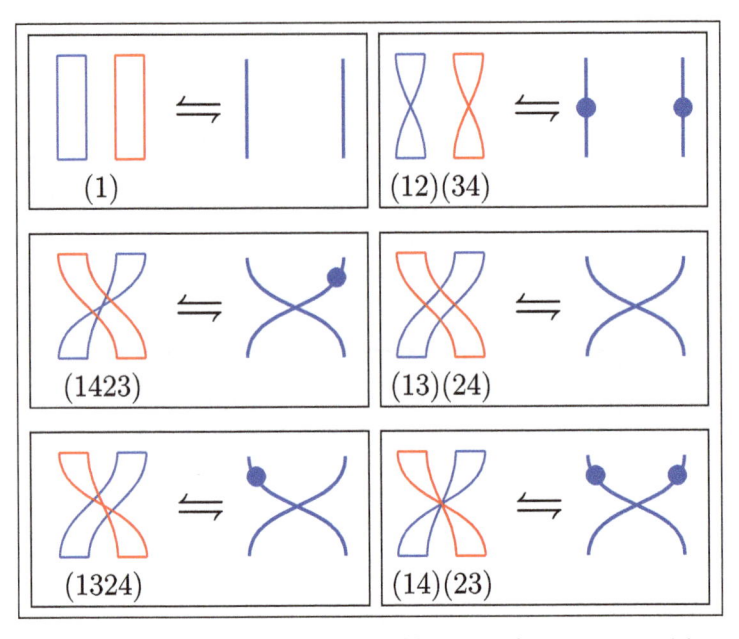

Figure 2.9. Some permutations as ribbons and as transpositions with (mod 2)-quipu.

The subgroup $B = \{(1), (13)(24), (12)(34), (14)(23)\} \cong \mathbb{Z}/2 \times \mathbb{Z}/2$ of the permutation group Σ_4 plays a special role in considerations about the structure of the larger group. It is often called the *Klein 4-group*.

While it should be clear that $\mathbb{Z}/2 \times \mathbb{Z}/2$ and $\mathbb{Z}/4$ are not isomorphic (since the latter group has two elements of order 4), it may not be so clear from the point of view of the seseqs (Fig. 2.6) that contain these groups. We will return to this point later.

From the point of view of (mod 2)-quipu, the choice of $A = \{(1), (1423), (12)(34), (1324)\}$ seems fairly natural. However, the more obvious cyclic subgroup of order 4 in the permutation group is $\{(1), (1234), (13)(24), (1432)\}$. In a moment, we will use this subgroup and analogues in Σ_n to define (mod n)-quipu.

2.3. (mod n)-quipu

There is an inductive definition of the generating mod n-quipu. We define the (mod 3)-quipu that are denoted $\triangleright\!\!1$ and $\triangleright\!\!2$.

$$\triangleright\!\!1 \;=\; \text{\Large{$\times\!\!\times$}} \;;\quad \triangleright\!\!2 \;=\; \text{\Large{$\times\!\!\times$}}.$$

By default,

$$\triangleright\!\!0 \;=\; | \;=\; |||.$$

The quipu are composed by stacking. They are merely a short-hand notation for the corresponding permutations. The composition rules are indicated in the following. To put it briefly, the indices inside the (mod 3)-quipu are added using modular arithmetic. The blue dots at the various vertices of the triangles are a redundant indicator in case the integral indices become unreadable under miniaturization.

Figure 2.10 contains some sample computations.

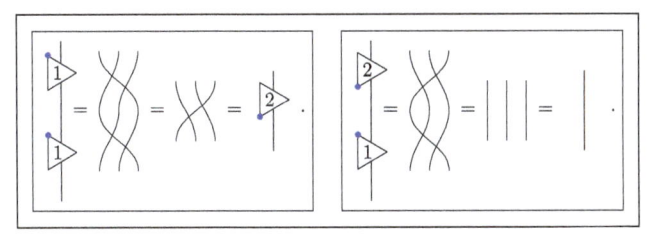

Figure 2.10. Sample calculations with (mod 3)-quipu and their associated permutations.

The graphic depictions for the (mod 4)-quipu appear in Fig. 2.11. These are also composed by stacking them vertically. The integral indices are added modulo 4 when one (mod 4)-quipu appears over the other.

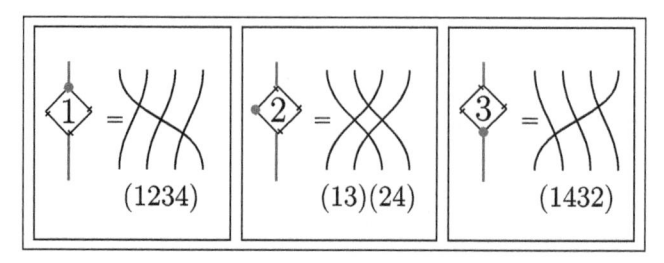

Figure 2.11. The correspondence between (mod 4)-quipu and permutations in Σ_4.

Since ⟨1⟩ corresponds to the 4-cycle $(1234) \in \Sigma_4$, the group of (mod 4)-quipu is isomorphic to $\mathbb{Z}/4$. The glypography that we use for these quipu includes drawing the squares that enclose the integer indices with a symmetric tick marks. These appear to remind the user that they respect cyclic, rather than dihedral,[5] symmetry. However, for higher modulus quipu, the symmetry breaking is visually inconvenient.

In the symmetric group, Σ_n consider the n-cycle $(1, 2, \ldots, n)$. In terms of the standard generators, this cycle is written as the product $t_1 t_2 \cdots t_{n-1}$. In cycle notation,

$$t_1 t_2 \cdots t_{n-1} = (1,2) \cdot (2,3) \cdot \cdots \cdot (n, n-1).$$

The *generating* (mod n)-*quipu* is a short-hand notation for this n-cycle. It is denoted in Fig. 2.12.

---

[5]The notion of dihedral symmetry will be described in detail in Section 2.3.3.

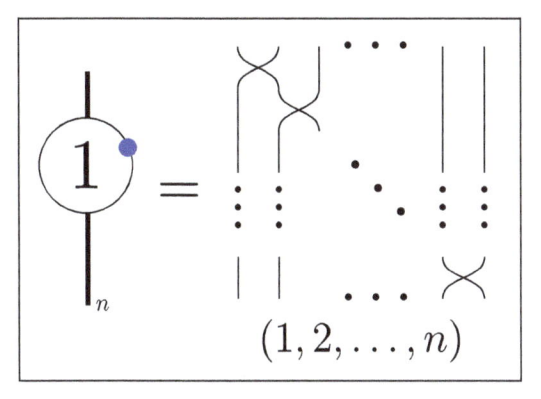

Figure 2.12. The correspondence between the generating (mod n)-quipu and the cyclic permutation $(1, 2, \ldots, n)$.

At times, the circle will be replaced by a regular polygon (often a hexagon or an octagon). Also, the index n at the bottom of the glyph will be omitted. The blue dot is meant to be located at an angle of $2\pi/n$. There are also (mod n)-quipu that represent the powers of $(1, 2, \ldots, n)$. These have indices $1, \ldots, n-1$ encircled, and the blue dot, which is a secondary indicator, is incremented around the circle.

These quipu are composed by stacking them vertically. When one appears immediately below another, internal indices are added modulo n. In addition, a (mod n)-quipu can pass through a crossing, as indicated in Fig. 2.13. As in the case of the (mod-2)-quipu, we imagine that the quipu are migrating toward the top of the diagrams in these relations. Then while at the tops of the strings, the indices within the quipu that are stacked vertically along a given string are added modulo n. The action is reminiscent of computations upon an abacus or the sliding of beads along wires in a popular infant's toy.

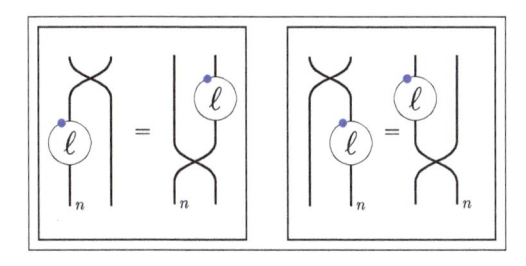

Figure 2.13. (Mod n)-quipu can pass through crossings of strings.

In practice, the unlabeled strings through which the quipu pass, say on the bottom right of the first boxed illustration (Fig. 2.13), consist of n parallel threads or strings. But the key property is more general. **The quipu are allowed to pass through any number of strings.** The reason for these diagrammatic moves come down to repeated applications of the relations $t_i t_{i+1} t_i = t_{i+1} t_i t_{i+1}$ and $t_i t_j = t_j t_i$ for $1 < |i - j|$ in the symmetric group Σ_k, where the index $k > n$. The idea of the proof is to move the quipu across one string at a time.

It suffices to prove the result for one string crossing the generating quipu ①⃗ by the application of two inductive arguments.

These two arguments are sketched as follows. In the following, we will demonstrate how to move the generating quipu through one crossing. If multiple strings are crossing the quipu, then move the quipu above one string at a time. If the quipu represents a modular integer $i > 1$, then write the quipu as a repeated sum of the generating quipu, and move the generating quipu above the crossing string, one quipu at a time.

Now, we demonstrate how to move a generating quipu above one string that crosses it.

If there is only one string crossing the other n threads, then there are $(n+1)$ strings that are involved, and the crossing that is above and to the right of the quipu is expressed as $t_1 t_2 \cdots t_n$. The quipu ①⃗, in this case, is expressed as the product $t_1 t_2 \cdots t_{n-1}$. The claim to be proven in the symmetric group Σ_{n+1} is that

$$t_1 t_2 \cdots t_n \, [t_1 t_2 \cdots t_{n-1}] = [t_2 t_3 \cdots t_n] \, t_1 t_2 \cdots t_n.$$

The product in the braces ("[" and "]") represents the quipu. Since the notation and the steps in even this simplified case are cumbersome, we first present a proof that

$$t_1 t_2 t_3 t_4 \, [t_1 t_2 t_3] = [t_2 t_3 t_4] \, t_1 t_2 t_3 t_4.$$

The proof of the general rule will be sketched after the specific case.

Background and Motivating Examples

First, let us establish some glyphs that express the equalities $t_i t_{i+1} t_i = t_{i+1} t_i t_{i+1}$ and $t_i t_j = t_j t_i$ for $1 < |i - j|$. The glyphs will be drawn in black, but the permutation generators will be colored, and the colors will extend to the glyphs.

1. The relation $t_i t_{i+1} t_i = t_{i+1} t_i t_{i+1}$ will be drawn (and read from top to bottom) as

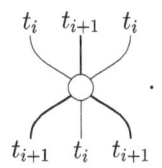

$$t_i \quad t_{i+1} \quad t_i$$
$$t_{i+1} \quad t_i \quad t_{i+1}$$

This relationship may be called a *triple point*.

2. The relation $t_i t_j = t_j t_i$ in case $i < j + 1$ will be read from top to bottom and drawn as

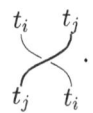

$$t_i \quad t_j$$
$$t_j \quad t_i$$

3. The relation $t_j t_i = t_i t_j$ in case $i < j + 1$ will be read from top to bottom and drawn as

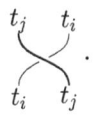

$$t_j \quad t_i$$
$$t_i \quad t_j$$

These last two relations will be called *commutations*.

Figure 2.14 indicates the string of equalities that start from $t_1 t_2 t_3 t_4 [t_1 t_2 t_3]$ at the top and proceed to $[t_2 t_3 t_4] t_1 t_2 t_3 t_4$ at the bottom.

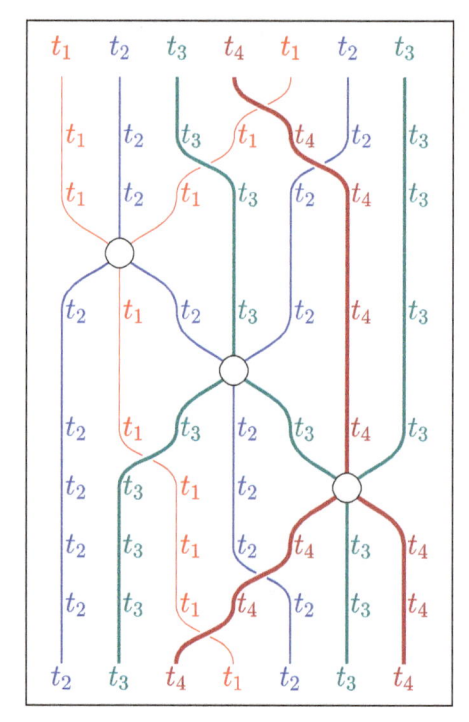

Figure 2.14. The calculation of an equality between two sequences of transpositions.

Each horizontal row of elements represents a product in Σ_4. The equivalences of these expressions is demonstrated by means of the glyphs that were introduced and itemized above. Thus, all the expressions in the rows are equal. Of course, there is a truncation at the fourth row from the bottom. Thus, the equality holds.

It may not be too difficult now to augment this picture with ellipses and demonstrate that

$$t_1 t_2 \cdots t_n \, [t_1 \cdots t_{n-1}] = [t_2 \cdots t_n] \, t_1 t_2 \cdots t_n.$$

Instead, we opt for a more algebraic proof. We believe that the reader will follow the steps more easily having attempted the case when $n = 4$, or equivalently when five strings are involved.

For $i = 1, \ldots, n$, we let the index i stand in for the generator t_1. In this revised notation, we aim to demonstrate that

$$12 \cdots n \; [12 \cdots (n-1)] = [23 \cdots n] \; 12 \; \cdots (n-1)n.$$

First, regroup and move the 1 that is to the right of "[" adjacent to the leftmost 2:

$$12\cdots n\ [12\cdots(n-1)] = [121]\ 3\cdots(n-1)n\ [2\cdots(n-2)(n-1)].$$

Apply the triple point relation $\underset{2\ 1\ 2}{\overset{1\ 2\ 1}{\bigvee}}$:

$$[121]\ 3\cdots(n-1)n\ [2\cdots(n-2)(n-1)]$$
$$= [212]\ 3\cdots(n-1)n\ [2\cdots(n-2)(n-1)].$$

Regroup and move the 2 that is to the right of "[" to the left by means of commutations:

$$[212]\ 3\cdots(n-1)n\ [2\cdots(n-2)(n-1)]$$
$$= 21\ [232]\cdots(n-1)n\ [3\cdots(n-2)(n-1)].$$

Apply the triple point relation $\underset{3\ 2\ 3}{\overset{2\ 3\ 2}{\bigvee}}$:

$$21\ [232]\cdots(n-1)n\ [3\cdots(n-2)(n-1)]$$
$$= 21\ [323]\cdots(n-1)n\ [3\cdots(n-2)(n-1)].$$

Then apply a commutation and regroup.

$$21\ [323]\cdots(n-1)n\ [3\cdots(n-2)(n-1)]$$
$$= [23][12]34\cdots(n-1)n]\ [3\cdots(n-2)(n-1)].$$

Continue in this fashion. Commute the entry just to the right of the "[" on the right of the expression toward the left. Then apply the triple point relation $\underset{(i+1)\ i\ (i+1)}{\overset{i\ (i+1)\ i}{\bigvee}}$ and move the left value $(i+1)$ further left. We obtain

$$[23][12]34\cdots(n-1)n]\ [3\cdots(n-2)(n-1)]$$
$$= \cdots = [23\cdots(n-1)]\ [12\cdots(n-1)n(n-1)]$$
$$= [23\cdots(n-1)]\ [12\cdots n(n-1)n] = [23\cdots n]\ 12\cdots(n-1)n.$$

In this way, quipu can move upward above crossing strings. This completes the proof of the relationship indicated in Fig. 2.13.

2.3.1. Application: $\mathbb{Z}/n \times \mathbb{Z}/n$ and $\mathbb{Z}/(n^2)$

Let n denote an integer with $2 < n$. The case of $n = 2$ was covered above in Section 2.2. The groups $\mathbb{Z}/n \times \mathbb{Z}/n$ and $\mathbb{Z}/(n^2)$ fit into seseqs

$$0 \longleftarrow \mathbb{Z}/n \xleftarrow{q} \mathbb{Z}/(n^2) \xleftarrow{n\cdot} \mathbb{Z}/n \longleftarrow 0,$$

and

$$0 \longleftarrow \mathbb{Z}/n \xleftarrow{p_1} \mathbb{Z}/n \times \mathbb{Z}/n \xleftarrow{y} \mathbb{Z}/n \longleftarrow 0.$$

In the first case, a coset representative $[a] \in \mathbb{Z}/n$ is mapped to $[n \cdot a]$, and the homomorphism q reduces the residue class $[b] \in \mathbb{Z}/(n^2)$ to its equivalence class in $\mathbb{Z}/n$, that is, $q[b] = [b] \mod n$. In the second case, $y(b) = (0, b) \in \mathbb{Z}/n \times \mathbb{Z}/n$ while $p_1(a, b) = a$.

The quipu and their rules for visual manipulation allow the elements of these two groups to be represented in a concise diagrammatic manner. To begin, the elements in $\mathbb{Z}/9$ are tabulated in Fig. 2.15, and those in $\mathbb{Z}/3 \times \mathbb{Z}/3$ are tabulated in Fig. 2.16.

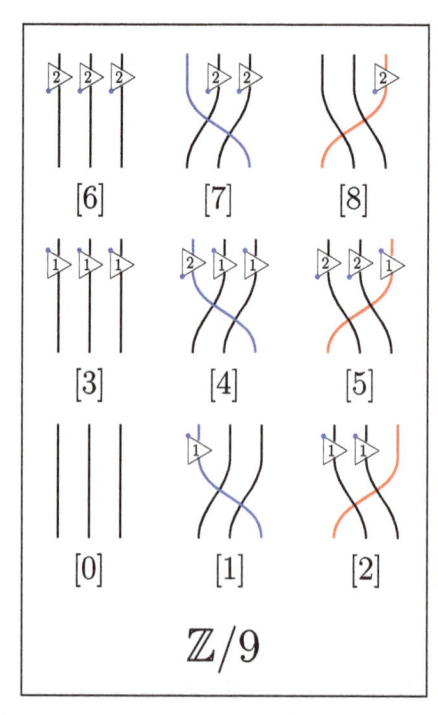

Figure 2.15. (Mod 3)-quipu representation of the elements in $\mathbb{Z}/9$.

Here is the method by which the above table in Fig. 2.15 was made. The subgroup $H = \{0, 3, 6\} \subset \mathbb{Z}/9$ is the image of the injective homomorphism $a \mapsto 3 \cdot a$. It is isomorphic to $\mathbb{Z}/3$. The larger group $\mathbb{Z}/9$ is partitioned into cosets $H, 1 + H$, and $2 + H$. Both H and the set of cosets are ordered. So, write $H = (0, 3, 6)$, and write

$$G = \mathbb{Z}/9 = (H, 1+H, 2+H) = ((0,3,6), (1,4,7), (2,5,8)).$$

Then analyze the action of 1 both on the ordered cosets and upon the ordering of the cosets.

It should be clear that by adding 1 to each coset, the coset is incremented to the (cyclically) next most coset. So, $+1 \leftrightarrows (012)$ on the collection of cosets. Also, $1 + (0, 3, 6) = (1, 4, 7)$, $1 + (1, 4, 7) = (2, 5, 8)$, and $1 + (2, 5, 8) = (3, 6, 0)$, where the integers are, of course, reduced modulo 9. The last relation $1 + (2, 5, 8) = (3, 6, 0)$ can be written as $1 + (2 + H) = [H, 1]$ to indicate that the coset $[2 + H]$ increments to H, but there is a cyclic transposition in the order of the elements of H. Thus, the correspondence

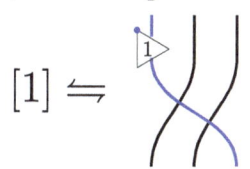

$$[1] \leftrightarrows$$

arises.

Please note that the color for the longer arc (in this case the blue arc) is added to add some visual emphasis to distinguish the cycles (012) and (021).

Since the group $G = \mathbb{Z}/9$ is generated by the element $[1]$, the remaining elements are obtained by vertically stacking the generator above the previously defined elements and manipulating the diagrams according to the rules for quipu. So, for example,

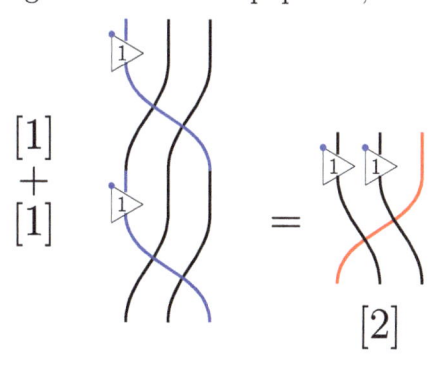

$$[1] + [1] = [2]$$

Here, the triple point rule,

$$\bowtie\!\!\bowtie = \bowtie\!\!\bowtie$$

was also applied, but in this case to three bundles of three strings each.

There are two permutation actions. The elements of G act to permute the elements in a coset, and they act to permute the cosets. The quipu account for the permutation actions within the cosets. Since the collection of cosets are also being permuted, there are string-diagram representations of those actions. These are generated by adjacent transpositions, and the $t_i t_{i+1} t_i = t_{i+1} t_i t_{i+1}$ relation holds.

$\mathbb{Z}/3 \times \mathbb{Z}/3$: Figure 2.16 lists the elements of $\mathbb{Z}/3 \times \mathbb{Z}/3$ using their quipu form. Please recall the seseq

$$0 \longleftarrow \mathbb{Z}/3 \xleftarrow{p_1} \mathbb{Z}/3 \times \mathbb{Z}/3 \xleftarrow{y} \mathbb{Z}/3 \longleftarrow 0.$$

The value of the injection y upon $b \in \mathbb{Z}/3$ is $y(b) = (0,b)$. So, the subgroup $H = \{(0,0),(0,1),(0,2)\}$ is the kernel of the projection p_1 onto the first factor. The subgroup H is ordered as $H = ((0,0), (0,1),(0,2))$. The remaining cosets are $H_1 = ((1,0),(1,1),(1,2))$ and $H_2 = ((2,0),(2,1),(2,2))$. The action of $(0,1)$ upon any of these cosets is to cyclically permute the elements therein. On the other hand, $(1,0)$ cyclically permutes the (ordered) set (H, H_1, H_2). Each column in Fig. 2.16 represents a coset. The group $G = \mathbb{Z}/3 \times \mathbb{Z}/3$ is generated by the elements $(1,0)$ and $(0,1)$. We leave it as an exercise to verify that if these two elements have the stated form, then their various compositions agree with the entries within the figure.

Background and Motivating Examples   41

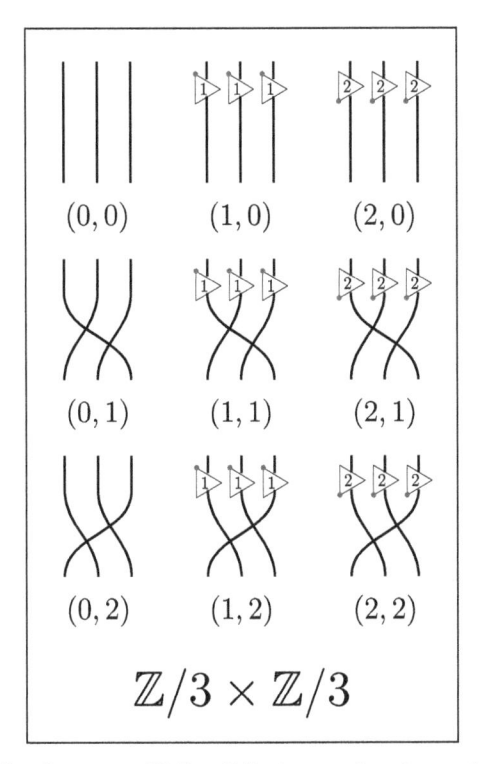

Figure 2.16. The elements of $\mathbb{Z}/3 \times \mathbb{Z}/3$ given as 3-strings-with-(mod 3)-quipu.

These visual representations of the groups $\mathbb{Z}/9$ and $\mathbb{Z}/3 \times \mathbb{Z}/3$ generalize to $\mathbb{Z}/(n^2)$ and $\mathbb{Z}/n \times \mathbb{Z}/n$. Moreover, they are engendered by means of permutation representations of the respective groups. To begin the discussions of both the generalization and the permutation representation, we present the following.

Proposition 1.

1. *An injective (one-to-one) permutation representation*

$$\Sigma_{n^2} \xleftarrow{f} \mathbb{Z}/n^2$$

is given by the following formula:

$$1 \mapsto (1, 2, \ldots, n)$$
$$\cdot [(1, n+1, \ldots, (n-1)n+1)$$
$$\cdot (2, n+2, \ldots, (n-1)n+2)$$
$$\cdot \ldots$$
$$\cdot (n, 2n, \ldots, n^2)].$$

2. There is also an injective permutation representation

$$\Sigma_{n^2} \xleftarrow{f} \mathbb{Z}/n \times \mathbb{Z}/n$$

that is given by

$$(1, 0) \mapsto [(1, 2, \ldots, n)$$
$$\cdot (n+1, n+2, \ldots, 2n)$$
$$\cdot \ldots$$
$$\cdot (n(n-1)+1, \ldots, n^2)],$$

while

$$(0, 1) \mapsto [(1, n+1, \ldots, (n-1)n+1)$$
$$\cdot (2, n+2, \ldots, (n-1)n+2)$$
$$\cdot \ldots$$
$$\cdot (n, 2n, \ldots, n^2)].$$

To help parse these expressions, let us begin by examining item 1 in the case that $n = 3$. Then $\Sigma_9 \xleftarrow{f} \mathbb{Z}/(9)$ is given by the following formula:

$$1 \mapsto (1, 2, 3) \cdot [(1, 4, 7)(2, 5, 8)(3, 6, 9)].$$

This is depicted as follows:

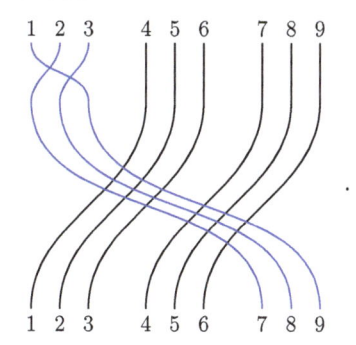

We observe directly that

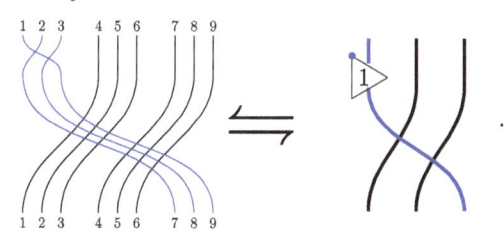

To observe that this element has order 9, either successively stack the permutation diagram on the left above itself or stack the diagrams that contain the quipu. Manipulate either by means of the relations that define the permutation group.

Caution. The set $\{1, 2, \ldots, 9\}$ is not apparently related to the elements $0, 1, \ldots, 8 \in \mathbb{Z}/9$. In fact, it reflects the ordering that is determined by the subgroup $H = (0, 3, 6)$ and its cosets $1 + H = (1, 4, 7)$ and $2 + H = (2, 5, 8)$.

Here is a clever way to achieve that correspondence. Start by arranging the standard representatives in $\mathbb{Z}/9$ into a square array as indicated:
$$\begin{bmatrix} 0 & 1 & 2 \\ 3 & 4 & 5 \\ 6 & 7 & 8 \end{bmatrix}.$$

Pretend, for a moment, that this is a matrix and take its transpose:
$$\begin{bmatrix} 0 & 1 & 2 \\ 3 & 4 & 5 \\ 6 & 7 & 8 \end{bmatrix}^t = \begin{bmatrix} 0 & 3 & 6 \\ 1 & 4 & 7 \\ 2 & 5 & 8 \end{bmatrix}.$$

Then write down the resulting entries in the order that they appear left-to-right along the first row, left-to-right on the second row, and left-to-right on the third row. This articulates the ordering $((0, 3, 6), (1, 4, 7), (2, 5, 8))$ that is induced from ordering the subgroup and subsequently ordering the cosets.

Thus, the indices in the permutation illustration correspond to the elements in $\mathbb{Z}/9$ in this fashion

$$\begin{array}{ccc ccc ccc}
1 & 2 & 3 & 4 & 5 & 6 & 7 & 8 & 9 \\
\updownarrow & \updownarrow & \updownarrow & \updownarrow & \updownarrow & \updownarrow & \updownarrow & \updownarrow & \updownarrow \\
0 & 3 & 6 & 1 & 4 & 7 & 2 & 5 & 8
\end{array}.$$

Next, we generalize the diagram and the ordering to $\mathbb{Z}/n$. We often think of this process of generalization as a "strategic insertion of ellipses."[6] Let's begin with the ordering. The transpose

$$\begin{bmatrix}
0 & 1 & \ldots & n-1 \\
n & n+1 & \ldots & 2n-1 \\
\vdots & \vdots & & \vdots \\
n(n-1) & n(n-1)+1 & \ldots & n^2-1
\end{bmatrix}^t$$

$$= \begin{bmatrix}
0 & n \cdot 1 & \ldots & n \cdot (n-1) \\
1 & n+1 & \ldots & (n-1) \cdot n + 1 \\
\vdots & \vdots & & \vdots \\
n-1 & 2n-1 & \ldots & n^2-1
\end{bmatrix}.$$

The subgroup $H \subset \mathbb{Z}/(n^2)$ that is the image of the homomorphism $\mathbb{Z}/(n^2) \xleftarrow{n\cdot} \mathbb{Z}/n$ consists of the standard representatives that are multiples of n. The first row of the transposed matrix $(0, n, \ldots, n \cdot (n-1))$ gives the ordering upon this set. The subsequent rows articulate the orderings of the other cosets. These cosets correspond to the visual groupings of the strings in Fig. 2.17.

---

[6] *An ellipsis* is a sequence of three dots (...) that is usually pronounced "and so forth." The plural, ellipses, should not be confused with a collection of ovals.

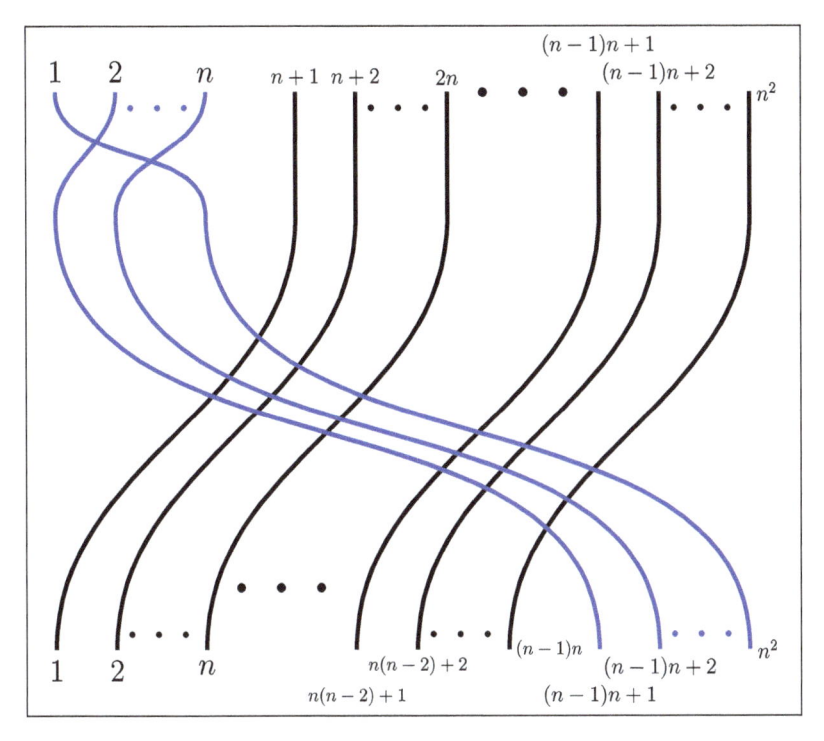

Figure 2.17. Adding 1 and the resulting permutation representation.

Carefully follow the strings to see that the permutation representation that is written above, in the statement of the proposition, corresponds to this diagram.

Next, we describe this in terms of the decomposition of the group $\mathbb{Z}/(n^2)$ into cosets of H that is the set of multiples of n in $\mathbb{Z}/(n^2)$. The cosets are $H, 1+H, \ldots, i+H, \ldots, (n-1)+H$. For $0 \leq i \leq (n-2)$, the action of adding 1 to $i+H$ is to increment it in the same order to $(i+1)+H$. The coset $(n-1)+H$ is ordered as the last row of the transposed matrix above, namely,

$$(n-1)+H = (n-1, 2n-1, \ldots, n^2-1).$$

Add 1 to each of these elements

$$1+(n-1, 2n-1, \ldots, n^2-1) = (n, 2n, \ldots, 0).$$

That is the subgroup H, but with a cyclic rotation among the elements. Some notation above is mimicked. We write

$$1 + ((n-1) + H) = [H, 1]$$

to indicate that the subgroup H is the result of incrementing the last coset, but its elements have been cyclically reordered. The action upon the ordering of the group $\mathbb{Z}/(n^2) = (H, 1+H, \ldots, (n-1)+H)$ can also be written in terms of the generating quipu (Fig. 2.18).

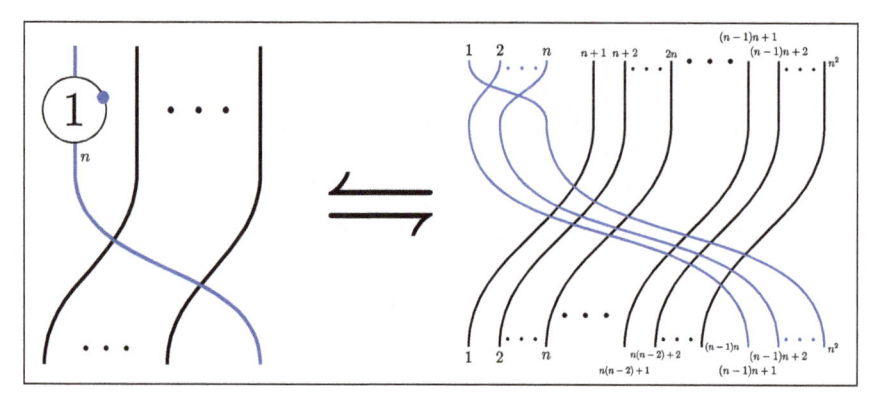

Figure 2.18. Permutation representation and n-strings-with-(mod n)-quipu.

We think that the quipu notation expeditiously demonstrates that this permutation has order n^2.

Exercise 2.

1. Compile a table of the 16 quipu representatives of the elements in $\mathbb{Z}/16$.
2. Demonstrate that the illustrations that are compiled above both for $\mathbb{Z}/9$ and for the group $A \cong \mathbb{Z}/4$ are early cases for this quipu notation.

$\mathbb{Z}/n \times \mathbb{Z}/n$: Let us examine the second statement of Proposition 1. In the case that $n = 3$, $(1,0) \mapsto (123)(456)(789)$ and

$(0,1) \mapsto (147)(258)(369)$. As string diagrams of permutations, these are presented in Fig. 2.19 with their correspondences, the "three-strings-with-quipu" illustrations presented.

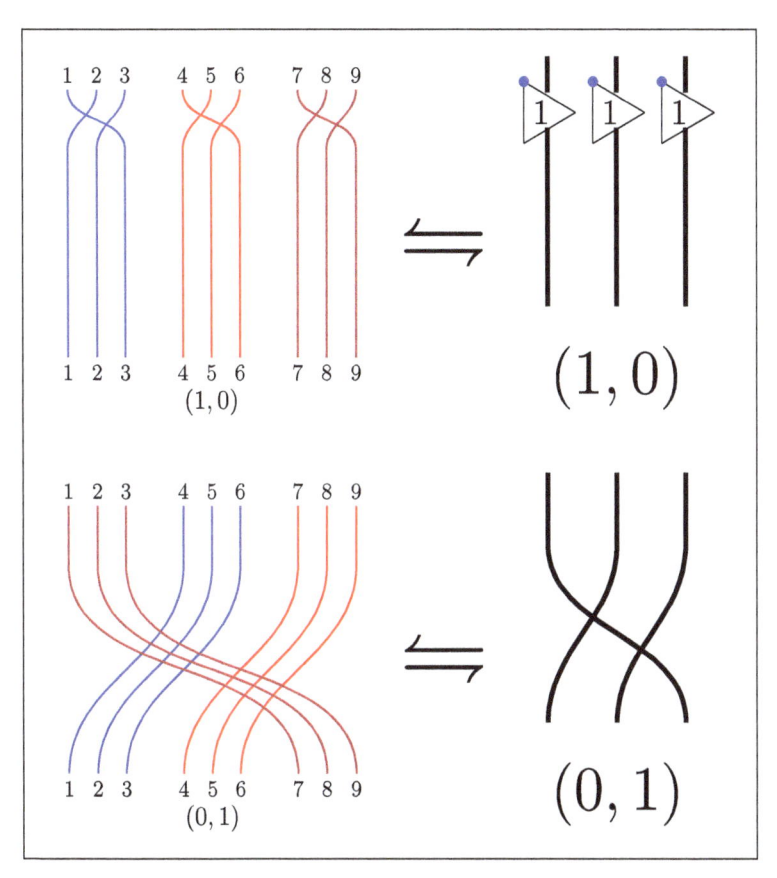

Figure 2.19. Permutation representations of $\mathbb{Z}/3 \times \mathbb{Z}/3$ and 3-strings-with-(mod 3)-quipu.

The table in Fig. 2.20 is abstracted from the table of elements of $\mathbb{Z}/3 \times \mathbb{Z}/3$. Instead of three strings, n-strings are imagined, and the quipu are (mod n)-quipu.

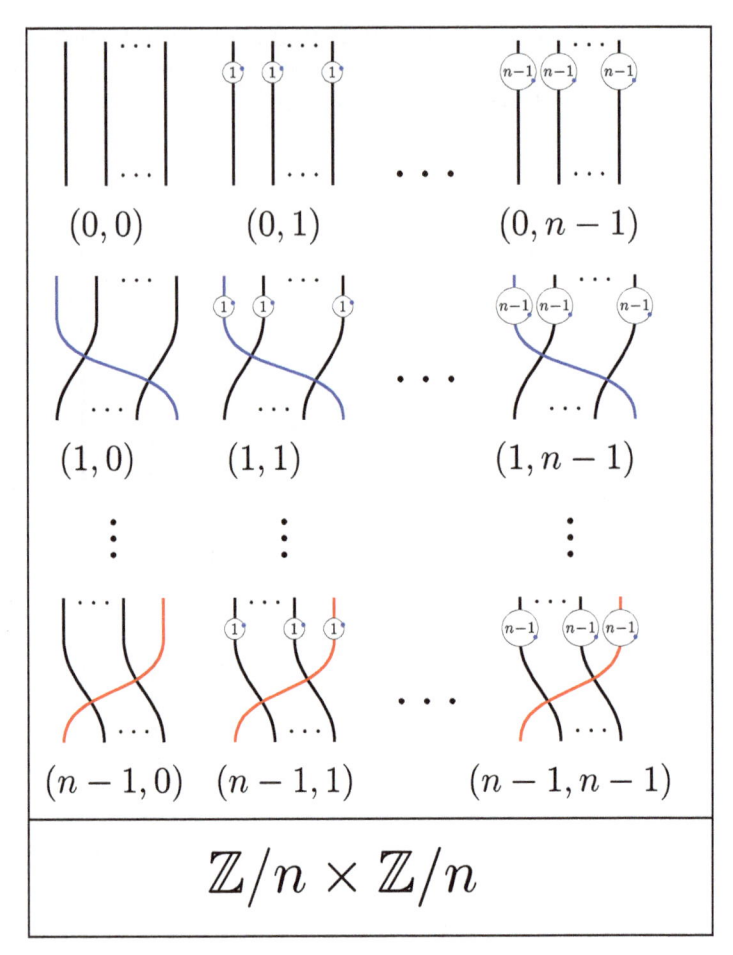

Figure 2.20. Permutation representations of $\mathbb{Z}/n \times \mathbb{Z}/n$ and n-strings-with-(mod n)-quipu.

To complete the proof of part 2 of Proposition 1, we abstract from the case of $n = 3$ to the general case by strategically inserting ellipses and carefully modifying the notation of the string diagrams that are indicated. Figure 2.21 does this for the generators $(1, 0)$ and $(0, 1)$ of $\mathbb{Z}/n \times \mathbb{Z}/n$.

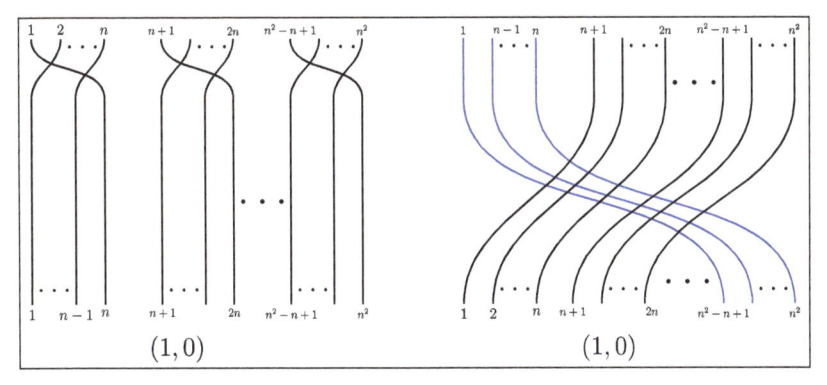

Figure 2.21. Permutation representations of the standard generators of $\mathbb{Z}/n \times \mathbb{Z}/n$.

In contrast to associating the diagram to denote the generator of $\mathbb{Z}/n^2$, we observe that the figures that are assigned to $(1,0)$ and $(0,1)$ in $\mathbb{Z}/n \times \mathbb{Z}/n$ cause the quipu to behave, more or less, independently from the strings upon which they sit. This is primarily due to the fact that the strings are saturated with quipu. This completes the proof of part 2 of Proposition 1.

2.3.2. Fanciful application: The circle of fifths

Those with advanced understanding of music theory may wish to skip this section. It merely describes the locality of added sharps and flats within scales in an evenly tempered instrument as the scales progress around the circle of fifths. Similarly, those who have no desire to acquire this knowledge may safely skip the section.

On the other hand, if you wish to learn why the sharps that are added in the order $F\sharp$, $C\sharp$, $G\sharp$, $D\sharp$, $A\sharp$, $E\sharp$, $B\sharp$ wind up in peculiar places in the scales to be played, then read on. This section does not neglect flats, but relies on the tonal equality between $C\sharp$ and $D\flat$. We acknowledge that these are approximations, and we encourage learning musicians to practice and understand scales (in particular minor scales, which will not be covered here) from the point of view of your own music instructor.

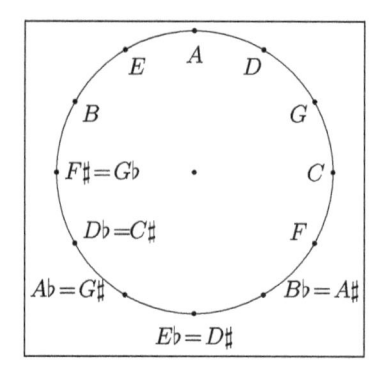

Figure 2.22. The circle of fifths.

The *circle of fifths* is depicted in Fig. 2.22. It is used as a device to indicate how to add sharps (or flats) in order to change keys. The key of C contains all naturals, and these are listed in the order C,D,E,F,G,A,B. For example, the key of G has one sharp, $F\sharp$. The key of D has two sharps $F\sharp$ and $C\sharp$. To move from one key to the next that is located one step counter-clockwise along the circle, add an additional sharp that is antipodal to the key from which you are incrementing.

For example, the key of A has three sharps: $F\sharp$, $C\sharp$, and $G\sharp$. To move to the key of E, observe that $D\sharp$ is antipodal to A. So, the key of E has the sharps $F\sharp$, $C\sharp$, and $G\sharp$ that were present in A and the additional sharp $D\sharp$.

There are better ways of expressing this same idea. The last note that is sharped is alphabetically before the name of the scale. The notes that "belong to the scale" are cyclic permutations of (C, D, E, F, G, A, B) that start at the name of the scale and continue with the appropriate sharps added. So, in this section, when we write "alphabetically," we mean in this cyclic order.

In general, there are 12 notes to consider. A major scale skips various *accidentals or half steps*. So, the notes of a scale are found at the localities of the asterisks ($*$) in the sequence

$$(*, \smile, *, \smile, *, *, \smile, *, \smile, *, \smile, *).$$

For example, the seven unique tones of the C major scale are represented by the white keys in a standard keyboard (Fig. 2.23).

Background and Motivating Examples   51

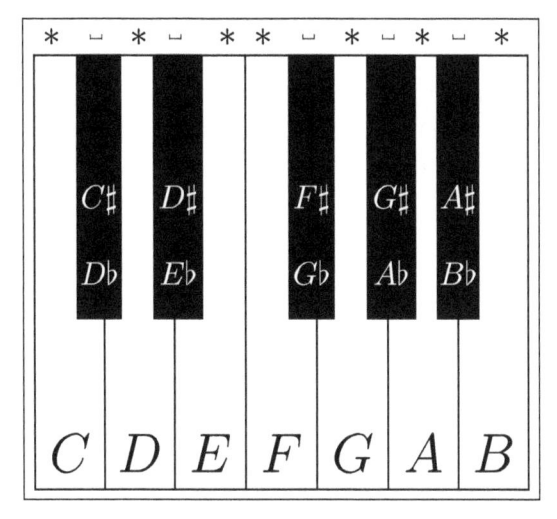

Figure 2.23. An octave on a standard piano keyboard with note names indicated.

The key of G contains the notes in order $(G, A, B, C, D, E, F\sharp)$. The change of key from C to G is schematized by the string diagram with $\sharp$-quipu in Fig. 2.24. Moreover, this same diagram can be used to move from one spot along the circle of fifths to the next spot counter-clockwise around.

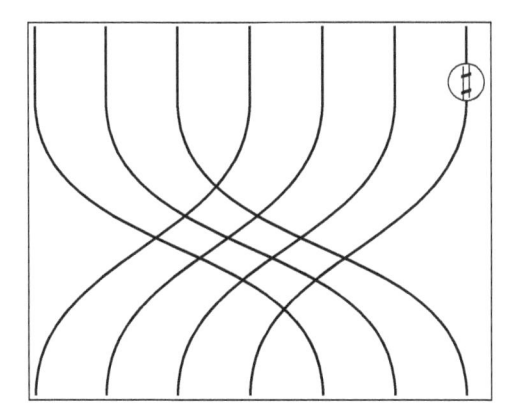

Figure 2.24. The change of key permutation as a string diagram with $\sharp$-quipu.

The *change of key permutation* adds one sharp successively, and it can be composed by vertical juxtaposition to indicate subsequent key changes as in Figs. 2.25 and 2.26. In Fig. 2.25, the so-called sharp keys are indicated as transformations from the key of C from G to $F\sharp$. The note names of the various keys are indicated to the left of the compositions. Note that from a layperson's point of view, $E\sharp$ may be confused with F, but since the key is $F\sharp$, the letter F is already spoken for.

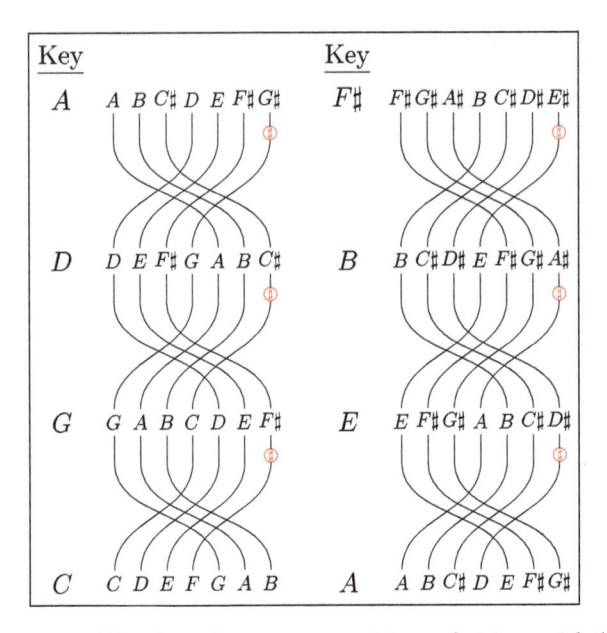

Figure 2.25. The sharp keys as compositions of strings-with-$\sharp$-quipu.

A reasonable objective might be to have at most six accidentals. The key of $F\sharp$ and its enharmonic $G\flat$ both have six accidentals. The former includes the peculiar $E\sharp$, and the peculiarity of the latter is $C\flat$. Specifically,

$$F\sharp = (F\sharp, G\sharp, A\sharp, B, C\sharp, D\sharp, E\sharp),$$

while

$$G\flat = (G\flat, A\flat, B\flat, C\flat, D\flat, E\flat, F).$$

The addition of a $\sharp$ to a flatted note is to naturalize it. The chart that follows continues successively sharping the seventh note of the scale and cyclically permuting the note names. But the names of the keys are the more customary "flatted keys."

Background and Motivating Examples 53

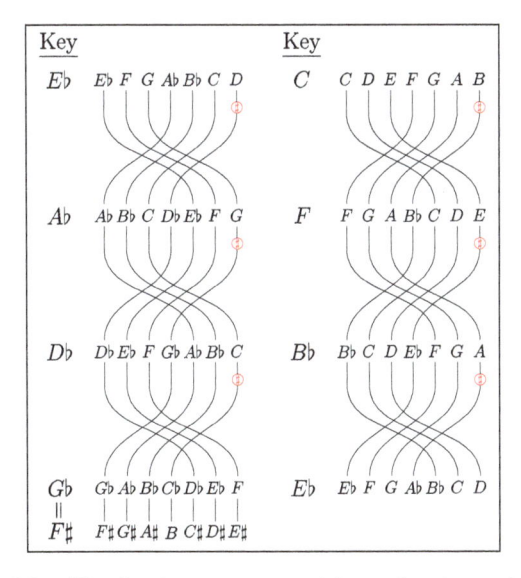

Key		Key	
E♭	E♭ F G A♭ B♭ C D	C	C D E F G A B
A♭	A♭ B♭ C D♭ E♭ F G	F	F G A B♭ C D E
D♭	D♭ E♭ F G♭ A♭ B♭ C	B♭	B♭ C D E♭ F G A
G♭ = F♯	G♭ A♭ B♭ C♭ D♭ E♭ F / F♯ G♯ A♯ B C♯ D♯ E♯	E♭	E♭ F G A♭ B♭ C D

Figure 2.26. The flat keys as compositions of strings-with-♯-quipu.

Exercise 3. This section was written from the point of view of a string player. A horn player may prefer to travel clockwise around the circle of fifths. In that case, it becomes a "circle of fourths."

Compute that the flattening permutation-with-quipu is given by means of the diagram in Fig. 2.27. Under the operation that $\sharp^{-1} = \flat$, observe in the figure that the flattening permutation-with-quipu is the inverse of the sharpening permutation-with-quipu.

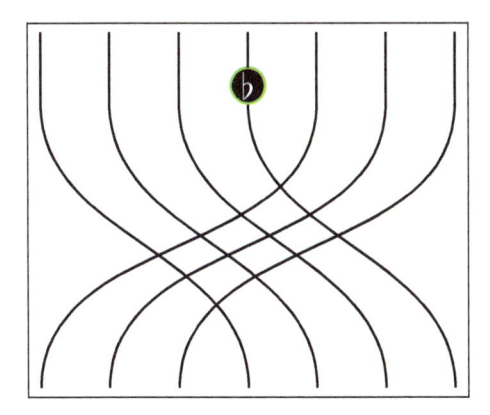

Figure 2.27. The permutation with ♭-quipu that flattens a key.

Create tables that proceed in the fashion $C \mapsto F \mapsto B\flat \mapsto \cdots \mapsto G$ analogous to those above.

2.3.3. Application: Dihedral groups

A myth that we would like to perpetuate is that group theory grew out of a systematic study of symmetry. We use the word "myth" here because none of the authors has done any extensive research in the history of mathematics and science, nor have we consulted with an expert to verify this perspective.

We have been told that Felix Klein established group theory as a perspective to classify geometries: the collection of possible geometries were related to the collection of symmetries that described the geometry. We also have been told that Évariste Galois coined the name "group" to describe the set of symmetries of field extensions that are associated to the solutions of polynomial equations. However, we expect that Galois used the French word *groupe*. To continue on this theme, Emmy Nöther observed that the conservation laws in physics were an expression of symmetry: conserved quantities are conserved under transformations of spacetime. That conservation is an expression of the symmetries among the quantities.

All of nature, art, music, and architecture involves ideas of symmetry and balance to express those things that we call harmony whether it be literal, visual, or metaphysical.

Let us move from these platitudes to more specificity. The collection of transformations of a symmetric object, that preserve its symmetry, forms a group. When an object is maintained, this is the identity transformation; two transformations successively applied to the object represents a third transformation — the composition of the two; the composition of three transformations is associative; and the inverse of a transformation is to undo it.

For more context, imagine an equilateral triangle whose vertices have been labeled 1, 2, and 3.

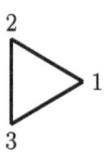

The transformations of the figure are as follows:

1. Do nothing:

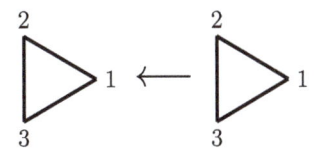

2. Rotate counter-clockwise by 120°:

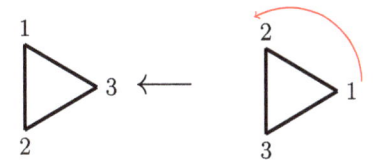

3. Rotate counter-clockwise by 240°:

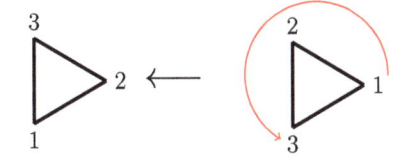

4. Flip about the horizontal axis:

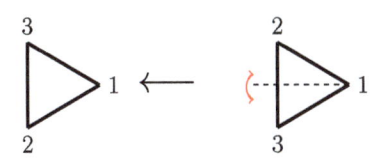

5. Flip about the perpendicular to the segment 13 that passes through the vertex 2:

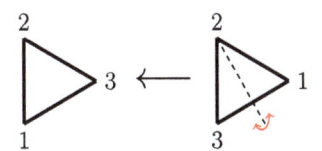

6. Flip about the perpendicular to the segment 12 that passes through the vertex 3:

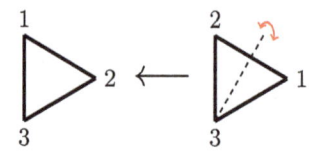

Any rigid transformation of the triangle is a composition of one of these. Any composition of any two of these is one of the others. Both statements require proof. The second will be exemplified as we proceed.

The astute, or previously informed, reader might recognize that the group of symmetries of the equilateral triangle is isomorphic to the group of permutations Σ_3 on three elements.

This example will be generalized in two different directions.

It seems a bit contrived to claim the set, Σ_n, of permutations of the set $\{1, 2, \ldots, n\}$ is its set of symmetries when the set of n dots, $\bullet \bullet \cdots \bullet$, is arranged along a line. Instead, let us consider these points as being arranged in a higher-dimensional space.

In $(n+1)$-dimensional space

$$\mathbb{R}^{n+1} = \{(x_1, x_2, \ldots, x_{n+1}) : x_j \in \mathbb{R}\},$$

consider the *standard unit vectors*

$$e_1 = [1, 0, \ldots, 0]^t, e_2 = [0, 1, \ldots, 0]^t, \ldots,$$
$$e_\ell = [0, 0, \ldots, \underbrace{1}_{\ell}, \ldots, 0]^t, \ \ldots \ , e_{n+1} = [0, 0, \ldots, 1]^t.$$

Here, t denotes the transpose, so these vectors are, in fact, column vectors. The n-*dimensional simplex*, Δ_n, consist of the convex hull of this set of points. That is, line segments are included between every pair of these vertices, triangles are included among every triple of these points, tetrahedra are included among every quadruple, and so forth.

A point $p \in \Delta_n$ has coordinates in $\mathbb{R}^n$

$$p = \sum_{\ell=1}^{n+1} \lambda_\ell e_\ell,$$

when $\lambda_1, \lambda_2, \ldots, \lambda_{n+1}$ are non-negative real numbers so that

$$0 \leq \lambda_\ell \quad \text{for } \ell = 1, 2, \ldots, n+1,$$

$$\sum_{\ell=1}^{n+1} \lambda_\ell = \lambda_1 + \lambda_2 + \cdots + \lambda_{n+1} = 1.$$

The 1-dimensional simplex is the line segment $\{(x, y) \in \mathbb{R}^2 : 0 \leq x, y,\ x + y = 1\}$. Here, x is used instead of λ_1 and y instead of λ_2. An alternative expression is $\vec{v}(t) = t[0, 1] + (1 - t)[1, 0]$, where $t \in [0, 1]$. Its set of symmetries is Σ_2 because Σ_2 interchanges $e_1 = [1, 0]$ and $e_2 = [0, 1]$.

Consider the vertices of the equilateral triangle to be $e_1 = [1, 0, 0]^t$, $e_2 = [0, 1, 0]^t$, and $e_3 = [0, 0, 1]^t$. The group Σ_3 acts upon the triangle by permuting its vertices.

The parameter space

$$\{(\lambda_1, \lambda_2, \lambda_3) : 0 \leq \lambda_1, \lambda_2, \lambda_3,\ \lambda_1 + \lambda_2 + \lambda_3 = 1\}$$

appears more friendly when the λs are replaced by x, y, and z. Furthermore, the variable z is not needed since $z = 1 - (x + y)$. The inequality $0 \leq z$, translates to $(x + y) \leq 1$, and since $0 \leq x, y$, the parameter space can be identified with the isosceles right triangle in the plane that is bounded by the three lines $x = 0$, $y = 0$, and $x + y = 1$. The symmetries of this figure are generated by the reflections through the line $x = y$. In Fig. 2.28, the parameter space is indicated on the upper left side of the illustration. The equilateral triangle is drawn as it appears in three dimensions with the parameter space indicated on the floor ($z = 0$) of that space. Note that the other walls, $x = 0$ and $y = 0$, could just as easily be used to contain the isosceles triangles that parameterize the equilateral triangle.

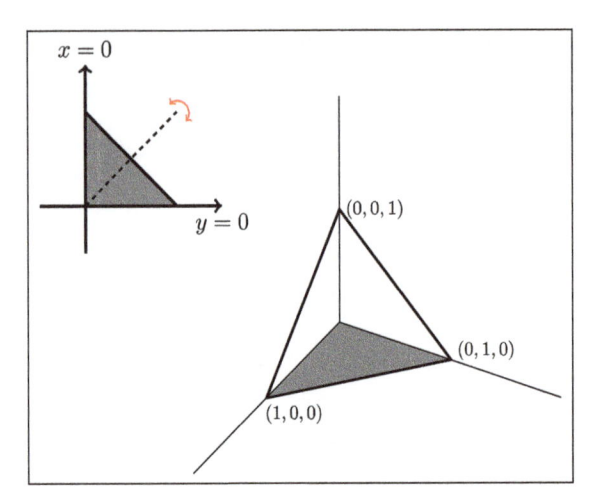

Figure 2.28. The parameter space for the equilateral triangle that is formed by the unit coordinate vectors.

The transformation of exchanging two vertices of an equilateral triangle cannot be achieved as motion of the triangle in the plane; 3-dimensional space is needed.

More generally and indeed inductively, the group Σ_n is the set of symmetries of the $(n-1)$-dimensional simplex Δ_{n-1}. However, we are obliged to insert a word of caution here. The group Σ_4 is the set of symmetries of the tetrahedron Δ_3 *as it sits in 4-dimensional space.*

When you think about that tetrahedral game die that is buried among the unused games on your toy shelf, you might note that the symmetries that interchange two vertices of it cannot be realized in 3-dimensional space. To interchange two vertices of a 3-dimensional tetrahedron requires a reflection. So, to interchange vertices, imagine the physical object in a mirror. Mirrors simulate reflections in 3-space. Moreover, a reflection in 3-dimensional space can be achieved by a rotation in 4-dimensional space.

Since points in the n-dimensional simplex Δ_n are linear combinations of the basis vectors e_ℓ, the symmetries thereof can be expressed as matrices. The basis vector e_ℓ has been chosen as a column vector,

$$e_\ell = [0, 0, \ldots, \underbrace{1}_{\ell}, \ldots, 0]^t,$$

since we prefer matrices to be written to the left of the vectors upon which they act.

The (2×2)-matrix $T = \begin{bmatrix} 0 & 1 \\ 1 & 0 \end{bmatrix}$ corresponds to the transposition $t_1 = (12) \in \Sigma_2$. If $x + y = 1$, then

$$T \begin{pmatrix} x \\ y \end{pmatrix} = \begin{bmatrix} 0 & 1 \\ 1 & 0 \end{bmatrix} \begin{pmatrix} x \\ y \end{pmatrix} = \begin{pmatrix} y \\ x \end{pmatrix}.$$

So, T interchanges the coordinates of the line segment that is Δ_1. Let I_ℓ denote the $(\ell \times \ell)$ identity matrix whose off-diagonal entries are 0 and the diagonal entries of I_ℓ all are 1. In particular, $I_1 = [1]$, $I_2 = \begin{bmatrix} 1 & 0 \\ 0 & 1 \end{bmatrix}$, and I_0 is empty. Then the $(n \times n)$-matrix

$$T_{\ell+1} = \begin{bmatrix} I_\ell & 0 & 0 \\ 0 & T & 0 \\ 0 & 0 & I_{n-\ell-2} \end{bmatrix}$$

corresponds to the transposition $t_{\ell+1} = (\ell+1, \ell+2)$. Here, the 0 entries are appropriately sized rectangular blocks of 0s.

When $\ell = 0$, there is no identity block that appears above the T. The maximum value of ℓ is $n - 2$, and in this case, there is no identity block that appears below the T.

Suppose that $0 \leq x, y, z$ and $x + y + z = 1$. Then $(x, y, z)^t$ represents a point on the standard triangle Δ_2. We compute

$$\begin{bmatrix} 0 & 1 & 0 \\ 1 & 0 & 0 \\ 0 & 0 & 1 \end{bmatrix} \cdot \begin{bmatrix} 1 & 0 & 0 \\ 0 & 0 & 1 \\ 0 & 1 & 0 \end{bmatrix} \begin{pmatrix} x \\ y \\ z \end{pmatrix} = \begin{bmatrix} 0 & 0 & 1 \\ 1 & 0 & 0 \\ 0 & 1 & 0 \end{bmatrix} \begin{pmatrix} x \\ y \\ z \end{pmatrix} = \begin{pmatrix} z \\ x \\ y \end{pmatrix}.$$

To illustrate how these matrices act, to establish conventions about the order of multiplication, and the correspondences left/right $\leftrightarrow$ above/below, the calculation is replicated, step by step, as a pair of actions on Δ_2 in Fig. 2.29.

$$\begin{bmatrix} 0 & 1 & 0 \\ 1 & 0 & 0 \\ 0 & 0 & 1 \end{bmatrix} \qquad \begin{bmatrix} 1 & 0 & 0 \\ 0 & 0 & 1 \\ 0 & 1 & 0 \end{bmatrix}$$

$$\begin{bmatrix} 0 & 1 & 0 \\ 1 & 0 & 0 \\ 0 & 0 & 1 \end{bmatrix} \cdot \begin{bmatrix} 1 & 0 & 0 \\ 0 & 0 & 1 \\ 0 & 1 & 0 \end{bmatrix} = \begin{bmatrix} 0 & 0 & 1 \\ 1 & 0 & 0 \\ 0 & 1 & 0 \end{bmatrix} \leftrightharpoons (123)$$

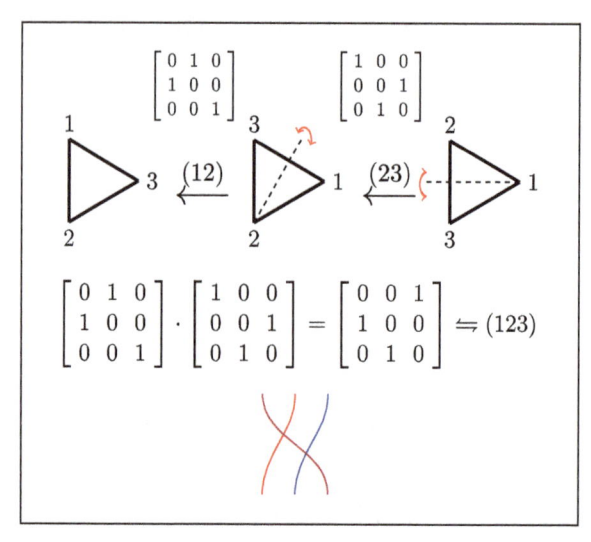

Figure 2.29. The order of matrix multiplication and multiplication of permutations.

The group of symmetries of an equilateral triangle can be generalized in another way that does not involve a higher-dimensional perspective. Consider a regular planar polygon. For definiteness, let $\zeta = e^{2\pi i/n} = \cos(2\pi/n) + i \sin(2\pi/n)$ denote an nth root of unity. It is customary to use i to indicate the complex number $i \in \mathbb{C}$, such that $i^2 = -1$.

The set $V_n = \{\zeta^\ell : \ell = 0, 1, \ldots, n-1\}$ is the set of vertices of a regular polygon, P_n. There are two types of symmetry for such a figure: (1) rotations through an angle that is a multiple of $2\pi/n$, and (2) reflections. An axis of the reflection is a diameter of the unit circle. When n is even, there are two types of these. Those that pass through antipodal vertices, and those that pass through opposite edges. For example, in a hexagon, whose vertices are cyclically labeled $\{1, 2, 3, 4, 5, 6\}$, the reflection $(14)(23)(56)$ has as its axis the diameter that passes through the segments $[2, 3]$ and $[5, 6]$. If n is odd, then an axis of reflection passes through a vertex and an edge opposite to it.

The group of symmetries of the polygon P_n (which is also called an n-gon) is given by the presentation

$$D_{2n} = \langle x, r : x^n = r^2 = 1, xr = rx^{-1} \rangle.$$

This is called *the dihedral group of order $2n$*. For definiteness, and with a slight abuse of notation, we let x denote the transformation $\zeta^\ell \mapsto \zeta^{\ell+1}$ on the set of vertices V_n, where the exponents are read modulo n. Let $\overline{\zeta^\ell} = \cos 2\pi\ell/n - i\sin 2\pi\ell/n = \zeta^{-\ell}$ denote the complex conjugate of ζ^ℓ for $\ell = 0, 1, \ldots, n-1$. Then r is the transformation $r : \zeta^\ell \mapsto \zeta^{-\ell}$. In this choice for r, the element $\zeta^0 = (1,0)$ is fixed. So is $\zeta^{n/2}$ when n is even. The polygon P_n is determined by its vertices, and so these actions on the set of vertices determine an action on the polygon.

Since the exponents are read modulo n, then $x^n(\zeta^\ell) = \zeta^{\ell+n} = \zeta^\ell$. Negation is involutory, so $r(r(\zeta^\ell)) = r(\zeta^{-\ell}) = \zeta^\ell$. Finally, $xr(\zeta^\ell) = \zeta^{1-\ell}$, while $rx^{-1}(\zeta^\ell) = r(\zeta^{\ell-1}) = \zeta^{1-\ell}$. So, the purported relations hold for these choices of x and r.

Before we described the quipu representation of the dihedral group D_{2n}, we'd like to point out that the edges in the $(n-1)$-simplex Δ_{n-1} can map to chords of the polygon P_n by means of the matrix

$$\begin{bmatrix} 1 & \cos(2\pi/n) & \ldots & \cos(2\pi\ell/n) & \ldots & \cos(2\pi(n-1)/n) \\ 0 & \sin(2\pi/n) & \ldots & \sin(2\pi\ell/n) & \ldots & \sin(2\pi(n-1)/n) \end{bmatrix}.$$

The basis vector e_1 maps to $(1,0) = \zeta^0$ and more generally $e_{\ell+1}$ maps to ζ^ℓ. The edge $te_a + (1-t)e_b$ of the simplex maps to the chord that connects the corresponding points on P_n. By ignoring the intersections among the chords in the interior of the n-gon P_n, a visualization of the *complete graph on n-vertices*, which is defined in the following paragraph, is presented.

If you were unhappy with the description of the group Σ_n acting upon the $(n-1)$-dimensional simplex Δ_{n-1} because of some difficulty in imagining objects in n-space, then you may alternatively think of Σ_n as the group of symmetries of this complete graph. The graph is an inherently 1-dimensional object that has n-vertices, and between any pair of vertices, there is a unique edge. Be aware that if $n > 4$, then this graph does not embed in the plane. Still you can imagine this graph in the "diamond-like" configuration of all lines that join any pair of vertices of the n-gon.

The presentation
$$D_{2n} = \langle x, r : x^n = r^2 = 1, xr = rx^{-1} \rangle$$
can be used directly to take any word that involves r and x, and reduce it to a word in the list
$$\{1, x, x^2, \ldots, x^{n-1}, r, rx, rx^2, \ldots, rx^{n-1}\}.$$
The elements of the form rx^ℓ for $\ell = 0, \ldots, n-1$ are the reflections. The method of reducing a word is to first cancel any power of r that is 2 or greater. Similarly, any exponent of x should be reduced modulo n. If an r is to the right of a power of x, move it left at the expense of negating the exponent on x. So, $x^s r = r x^{-s}$. Keep moving factors of r to the left, and cancel expressions of the form r^2. Multiply the powers of x that appear to the right, and reduce the resulting exponent modulo n.

The set $\{1, x, x^2, \ldots, x^{n-1}\} = H$ forms a subgroup of D_{2n}. It is isomorphic to the cyclic group $\mathbb{Z}/n$. Moreover, H is normal. This gives a seseq
$$0 \longleftarrow \mathbb{Z}/2 \xleftarrow{p} D_{2n} \xleftarrow{i} \mathbb{Z}/n \longleftarrow 0.$$

We now turn to the main idea of this treatise.

Put the elements of H in their obvious cyclic order:
$$H = (1, x, x^2, \ldots, x^{n-1}).$$
Consider the coset rH to have an induced order:
$$rH = (r, rx, rx^2, \ldots, rx^{n-1}).$$
Then compute
$$xH = (x, x^2, \ldots, x^{n-1}, 1)$$

and
$$xrH = rx^{-1}H = (rx^{n-1}, r, \ldots, rx^{n-2}).$$

We write $xH = [H, 1]$ to indicate that as cosets, $xH = H$, but as cyclically ordered sets, the elements have moved one step forward. Similarly, $xrH = [rH, n-1]$ since the elements have moved one step backward. That motion is the same as $(n-1)$ steps forward. Since $r(H) = rH$, and $r(rH) = H$, the orders are not affected by the action of r, and the reflection switches the cosets.

Quipu representatives of the generators appear in Fig. 2.30.

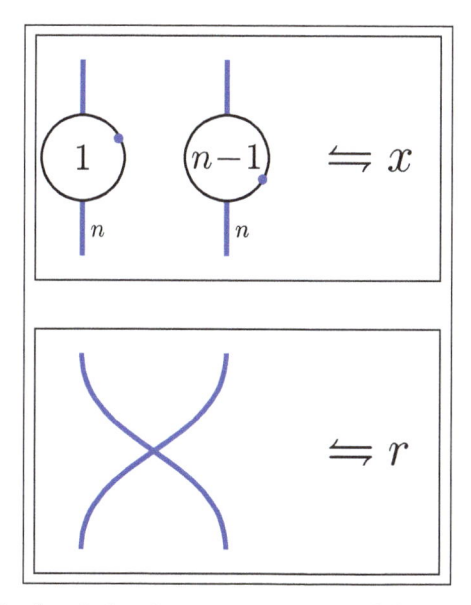

Figure 2.30. The (mod n)-quipu representations of the generators of the dihedral group D_{2n}.

Examples will be presented when $n = 3$ and 4. In these cases, all the elements in the dihedral groups D_{2n} will be presented as transpositions with quipu. In case $n = 3$, we'll demonstrate that the relations $xr = rx^{-1}$ hold. Furthermore, we will illustrate that the relationship $aba = bab$ holds when a is the quipu picture of r and b is that of rx. Finally, the case $n = 2$ will be revisited.

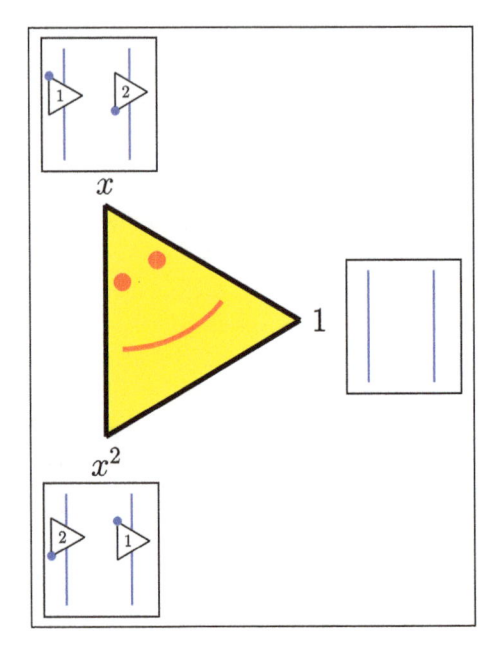

Figure 2.31. The rotations in D_6.

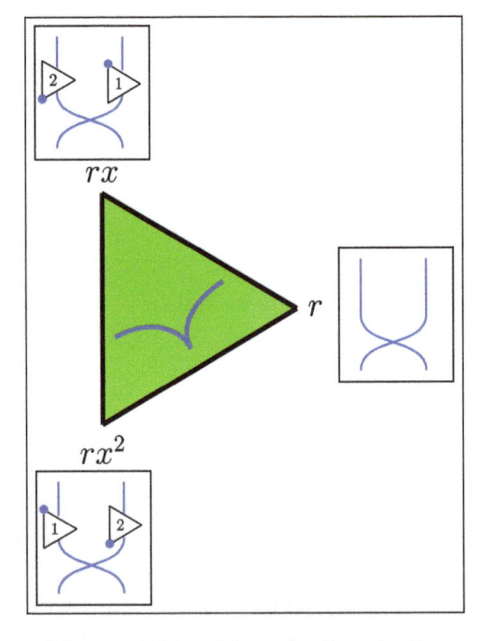

Figure 2.32. The reflections in D_6.

The central triangles in Figs. 2.31 and 2.32 represent the same triangle, with the "heads" side appearing in yellow with a smiley face, and the "tails" side appearing green with a bird flying away. Each side represents a coset of the cyclically ordered subgroup $H = (1, x, x^2)$. The six elements of the dihedral group D_6 adorn the vertices. These are presented in their transpositions-with-quipu forms.

The identities, $x^3 = r^2 = 1$ and $xr = rx^2$, are demonstrated in Fig. 2.33.

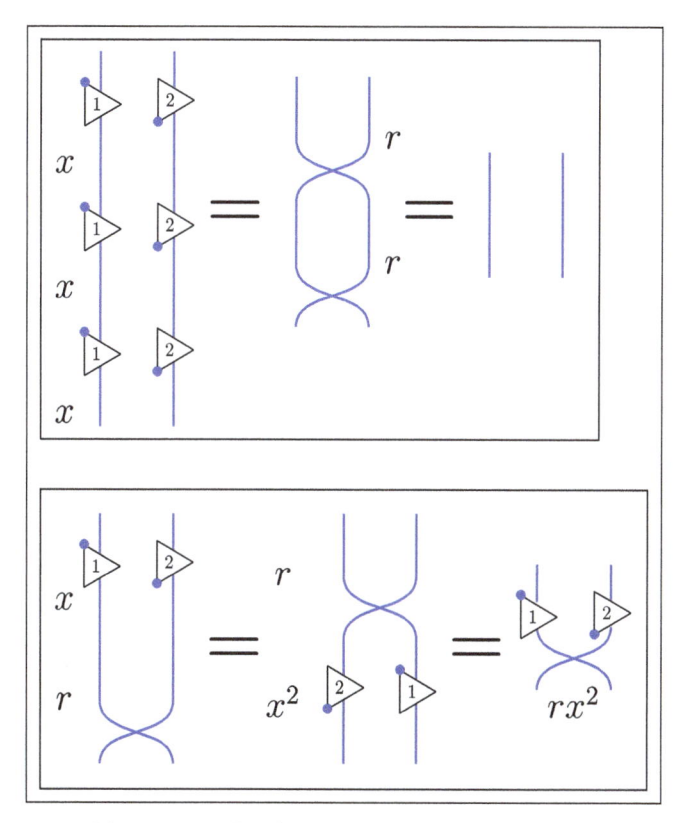

Figure 2.33. Identities in D_6 demonstrated by means of transpositions with (mod 3)-quipu.

Figure 2.34 demonstrates that $aba = bab$ by using transpositions-with-quipu when a represents r and b represents rx.

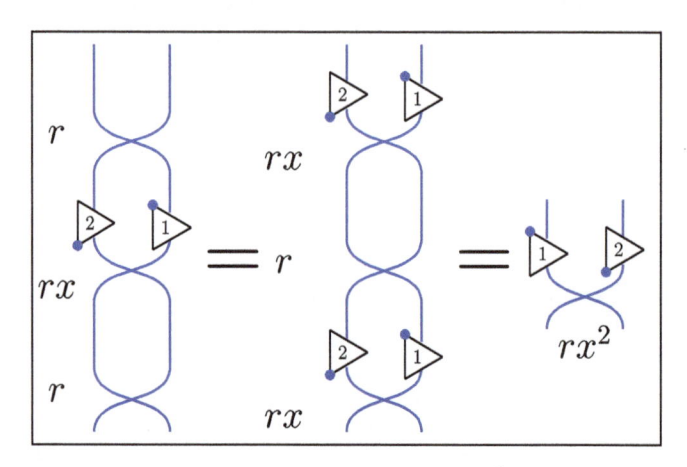

Figure 2.34. The $r(rx)r = (rx)r(rx) = rx^2$ identity in D_6.

Both sets of identities hold because (1) quipu can be slid upward through crossings, (2) the indices of adjacent quipu upon a single string are added modulo 3, and (3) a pair of crossings that happen between a pair of blue strings that are unobstructed by quipu cancel. That is to say, the pair of blue strings represent elements of the group $\langle t : t^2 = 1 \rangle$ of transpositions.

The elements of the 8-element dihedral group D_8 are illustrated in Fig. 2.35 as transpositions-with-quipu.

Exercise 4. Consider the transposition-with-quipu representation that corresponds to the elements x and r in D_8. Mimic the calculations above that juxtaposing the diagrams to demonstrate that $x^4 = r^2 = 1$ and that $xr = rx^3$.

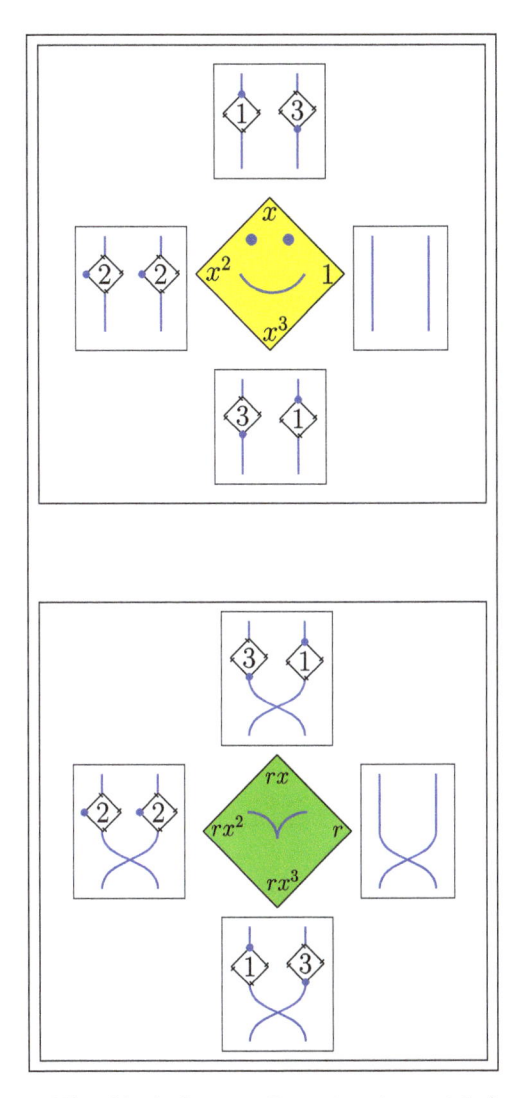

Figure 2.35. The dihedral group D_8 as 2-strings-with-(mod 4)-quipu.

Consider the extreme case of $D_4 = \langle x, r : x^2 = r^2 = 1, xr = rx^{-1} \rangle$. Since $x^2 = 1$, we have $x^{-1} = x$. So, the relation $xr = rx^{-1} = rx$ indicates that x and r commute. Since x and r are otherwise independent, then $D_4 = \mathbb{Z}/2 \times \mathbb{Z}/2$. We turn to schematizing this group as a set of symmetries of a *bigon* — this is a degenerate polygon that only has two vertices and two edges. Those edges are drawn as arcs. The symmetry x rotates the figure 180°, and the symmetry r is a reflection. Both transformations are involutory. Figure 2.36 illustrates that $xr = rx$ by way of labeling the vertices and edges of the bigon and demonstrating the transformation.

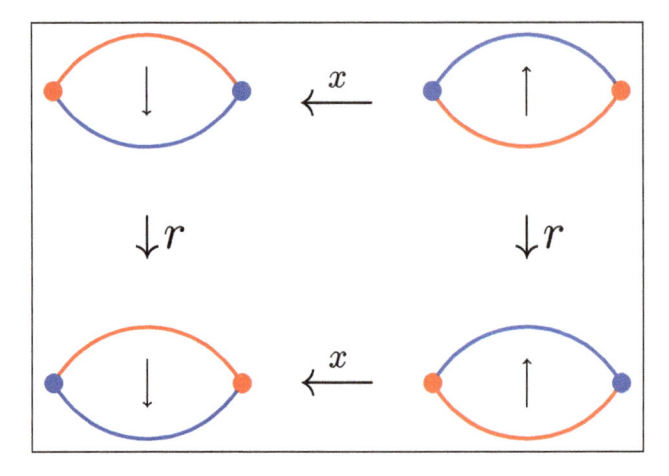

Figure 2.36. In the 4-element dihedral group D_4, x and r commute.

The dihedral group D_4 is singular in that it is the only dihedral group that is commutative. When $n = 2$, both the quipu in the description of x are (mod 2)-quipu, and so x can be represented as . In Fig. 2.37, the analogue of the depictions of D_6 and D_8 is illustrated for D_4. For technical reasons, the bigons are not colored.

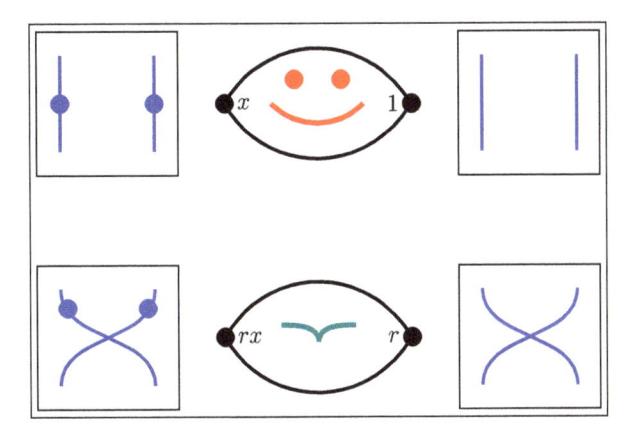

Figure 2.37. The 4-element dihedral group D_4 is isomorphic to $\mathbb{Z}/2 \times \mathbb{Z}/2$.

When n is a relatively small number, the number of groups (when considered up to isomorphism) that are of order n is also relatively small. Recall that the order of a group is the number of elements in the underlying set of the group. For example, there is only one group of order 1. This is known as the *trivial group*. It consists of the element 1 with the multiplication $1 \cdot 1 = 1$. There is only one group of order 2. It is $\mathbb{Z}/2$. There is only one group of order 3: $\mathbb{Z}/3$. But there are two groups of order 4: $\mathbb{Z}/4$ and $Z/2 \times \mathbb{Z}/2$.

In general, if p denotes a prime number, any group of order p is isomorphic to the cyclic group $\mathbb{Z}/p$. That is not too difficult to show. Recall that the order of an element $a \in G$ is the smallest exponent n such that $a^n = 1$. In a finite group, every element has finite order. The subset $H = \langle a^j : j = 0, 1, 2, \ldots \rangle$, that consists of the powers of a non-trivial ($a \neq 1$) element of G, is a subgroup of G. Then the order of H, $|H|$ divides the order of G. But since the order of G is the prime p, the order $|H|$ must be either 1 or p. But $a \neq 1$. So, $H = G$.

When $n = 6$, there are precisely two different groups of order 6.[7] One of these is the cyclic group $\mathbb{Z}/6$. The other is non-commutative,

---

[7] More precisely, any group of order 6 is isomorphic to one of the groups mentioned in this paragraph.

and it is isomorphic to the permutation group Σ_3. A curious reader might be thinking about the product $\mathbb{Z}/2 \times \mathbb{Z}/3$, whose elements consist of ordered pairs,

$$\mathbb{Z}/2 \times \mathbb{Z}/3 = \{(a,b) : a \in \mathbb{Z}/2, \ \& \ b \in \mathbb{Z}/3\}.$$

In this group, the element $(1,1)$ has order 6 (verify), and so $\mathbb{Z}/2 \times \mathbb{Z}/3$ is cyclically generated by $(1,1)$.

Exercise 5. Consider the subgroups of orders 2 and 3 in $\mathbb{Z}/6$ and also in $\mathbb{Z}/2 \times \mathbb{Z}/3$. Let H denote such a subgroup. Order it cyclically, and order the cosets of H. If $|H| = 3$, there are two such. While if $|H| = 2$, there are 3 cosets.

1. Give transpositions-with-quipu presentations for both $\mathbb{Z}/6$ and $\mathbb{Z}/2 \times \mathbb{Z}/3$. In these cases, the quipu will be (mod 3)-quipu.
2. Give presentations that involve 3-strings-with-(mod 2)-quipu for both $\mathbb{Z}/6$ and $\mathbb{Z}/2 \times \mathbb{Z}/3$.
3. Rewrite the generating diagram (either $1 \in \mathbb{Z}/6$ or $(1,1) \in \mathbb{Z}/2 \times \mathbb{Z}/6$) as a 6-cycle in the permutation group Σ_6.
4. (Answer unknown to us) Can you discover other quipu representations by examining the 120 various 6-cycles in Σ_6? (Known) Why are there 120 such 6-cycles?

You ask your phone or type the question: "How many groups of order 8 are there?" The answer will be 5. Either that or the internet is broken. One of these is the cyclic group $\mathbb{Z}/8$. Another is $\mathbb{Z}/2 \times \mathbb{Z}/4$. Another is $\mathbb{Z}/2 \times \mathbb{Z}/2 \times \mathbb{Z}/2$. All those are commutative.

There are two non-commutative groups of order 8. One of these is the dihedral group D_8. The other can be thought of as a dicyclic group. This is a 2-fold extension of the dihedral group $D_4 = \mathbb{Z}/2 \times \mathbb{Z}/2$. But that description is arcane. Alternatively, it is called the group of quaternions $Q_8 = \{\pm 1, \pm i, \pm j, \pm k\}$. It has two very nice descriptions as strings-with-quipu that will be discussed in Section 4.2 of Chapter 4.

Background and Motivating Examples 71

Our goal, herein, is to give permutation-with-quipu representations of many important finite groups. Some of these groups (the finite subgroups of the matrix group SU(2)) are considered to be important because of their relationships with other branches of mathematics.

In general, human-kind's understanding of finite groups is facilitated by two important ideas. The first idea is that of the Sylow theorems. Given a finite group, determine how many elements that it may have. Let n denote $|G|$ — the order of G. Then suppose that $n = p^q k$, where p is a prime that does not divide the remaining factor k. One of the Sylow theorems states that there is a subgroup of G that is of order p^q. That subgroup often is a product of cyclic groups that are also of prime power orders. More can be said. The Sylow theorems were discovered in the last half of the 19th century.

The second idea is far more encompassing. It involves the classification of the finite simple groups. A group is *simple* if it does not contain a non-trivial normal subgroup. The trivial normal subgroups are the identity subgroup and the group itself. In the above discussion, we indicated that a group whose order is prime is simple. A less trivial fact is that the alternating groups A_n for $5 \leq n$ are simple. A much broader fact is that the remaining simple groups are of Lie type or found in some so-called sporadic families. This proof is found (according to Wikipedia) in "tens of thousands of pages in several hundred journal articles written by about 100 authors, published mostly between 1955 and 2004."

Clearly, any investigation of the classification of finite simple groups would lead us far astray. On the other hand, we may encounter the Sylow theorems here and there throughout the text.

It should be of no surprise to a beginning graduate student that a question on the comprehensive exam in group theory usually involves an application of the Sylow theorems. Often, the calendar year, if it is easily factorable and if it does not contain a large number of prime power factors, becomes food for thought. For example, the number $2023 = 7 \times 17^2$. A typical exam question in that year might be: "Classify all the groups of order 2023." Beware.

Chapter 3
Matrix Descriptions

3.1. Overview

This chapter starts with one foot grounded in examples and gingerly steps into some abstraction. First, the examples

$$A = \{\,||\,,\,\times\,,\,|\!\!\uparrow\,\,\uparrow\!\!|\,,\,\times\,\}$$

and

$$B = \{\,||\,,\,\times\,,\,|\!\!\uparrow\,\,\uparrow\!\!|\,,\,\times\,\}$$

will be combined and augmented by a pair of additional diagrams to give an alternate description of the dihedral group D_8. That description can also be thought of as set of non-singular (2×2)-matrices whose non-zero entries are ± 1. An outline of the process that goes from a permutation with (mod 2)-quipu to a matrix is initiated.

In the next example, we look back at two 3-strings descriptions of $\mathbb{Z}/6$, combine these and determine the matrix representation of the resulting group. By doing so, we will be better able to identify the group that was constructed by quipu methods.

Following these two examples, some remarks are made about the signed permutation matrices that were engendered. Generalizations will be made. These generalizations are constructions of groups that are common within the group-theoretic literature. The construction is known as a semi-direct product.

More examples will be presented both as permutations with quipu and in matrix form.

The quipu descriptions of semi-direct products are un-maskings of a folktale: permutation matrices in which the entries are elements of a finite group fully describe semi-direct products. That folktale, if it is a folktale, can be more fully codified into a special case of the Krasner–Kaloujine theorem.[1] To our knowledge, its use in the current context is novel. A proof of Theorem 1 is provided in Section 3.6.

Theorem 1. *Let G denote a finite group of order nk. Let H denote a subgroup of order k. Then there is an inclusion $G \subset H^n \rtimes \Sigma_n$, where the second factor permutes the coordinates of H^n.*

We beg the forgiveness of the reader for stepping into the more abstract setting here. Before walking with reckless haste into that abstract world, we will provide signposts and preview the journey with the following outline.

Section 3.2 discusses the 8-element dihedral group as transpositions with (mod 2)-quipu and as matrices. Section 3.3 creates a group of order 24 from 3-cycles and (mod 2)-quipu. The group is identified. Section 3.4 defines the signed permutation matrices, presents these via generating transpositions with (mod 2)-quipu, and describes some associated geometric configurations, for which these are the symmetries. Section 3.5 presents semi-direct products as seseqs, permutations with quipu, and as generalized permutation matrices. More examples are given. Section 3.6 presents the proof of Theorem 1.

3.2. The 8-element dihedral group revisited

We gave you these six toys

$$| \; | \; , \; \bowtie \; , \; \dot{|} \; \dot{|} \; , \; \dot{|} \; \bowtie \; , \; \dot{\bowtie} \; , \; \dot{\bowtie} \; ,$$

---

[1]Thanks to Roger Alperin, David Benson, and Greg Kuperberg.

and we told you to play with them by stacking one on top of the other. Now, to be fair to us, we did give you the toys in separate packages (the sets A and B), but now we have mixed those packages up and removed some duplicates. We hope that while you were playing, you formed the products indicated in Fig. 3.1, and we hope that you wondered why we didn't give you all the pieces.

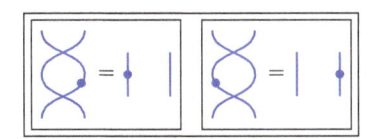

Figure 3.1. Some new products.

With these additional two pieces, one can verify (and you should) that this set

$$D = \{\,|\,|\,, \times, \,\dagger\,\,\dagger, \dagger\,\times\,\dagger, \,|\,\times\,|, \,\dagger\,\times\,\}$$

is closed under the multiplication when the quipu and the crossings are subject to the following relations:

and

As you play with these elements, you should note that this is a non-commutative group. To verify that the set is indeed a group, you should determine the inverses of the elements. All but two of the elements are involutive. The involutive elements are their own inverses. The remaining elements are inverse to each other. A little experimentation finds a pair of elements a and b such that $ab \neq ba$.

We pointed out earlier that there are exactly five groups of order 8. Only two of these are non-commutative. So, the group represented here is either the unit quaternions (that we have yet to discuss), or it is the dihedral group D_8 of order 8. Since we haven't yet talked about the former, the group here must be the latter. Let's see why.

First, the (mod 2)-quipu are better suited as multiplicative elements. Thus, ♦ indicates multiplication by (-1). Next, we insert variables x and y at the bottom of the strings, and imagine how they could be transformed up the strings as indicated with the following example. Here, we have indicated that the transposition-with-quipu corresponds to a matrix action on a column vector $\begin{bmatrix} x \\ y \end{bmatrix}$.

$$\begin{bmatrix} -y \\ x \end{bmatrix} = \begin{bmatrix} 0 & -1 \\ 1 & 0 \end{bmatrix} \cdot \begin{bmatrix} x \\ y \end{bmatrix}.$$

Yes, it is unfortunate that the permutations-with-quipu play well when vectors are arranged as rows, but the corresponding matrices act on columns. It's always a bit off-putting, but we must get used to it.

The set $\{(x, y) : |x| + |y| \leq 1\}$ corresponds to a square in the plane. Then each of the transpositions-with-quipu in the set D has a matrix representation. For completeness, these are listed in Fig. 3.2.[2]

---

[2] We regret the notational transgression. The variable x is used in two different contexts, both as if in the (x, y)-plane and as a generator of the dihedral group.

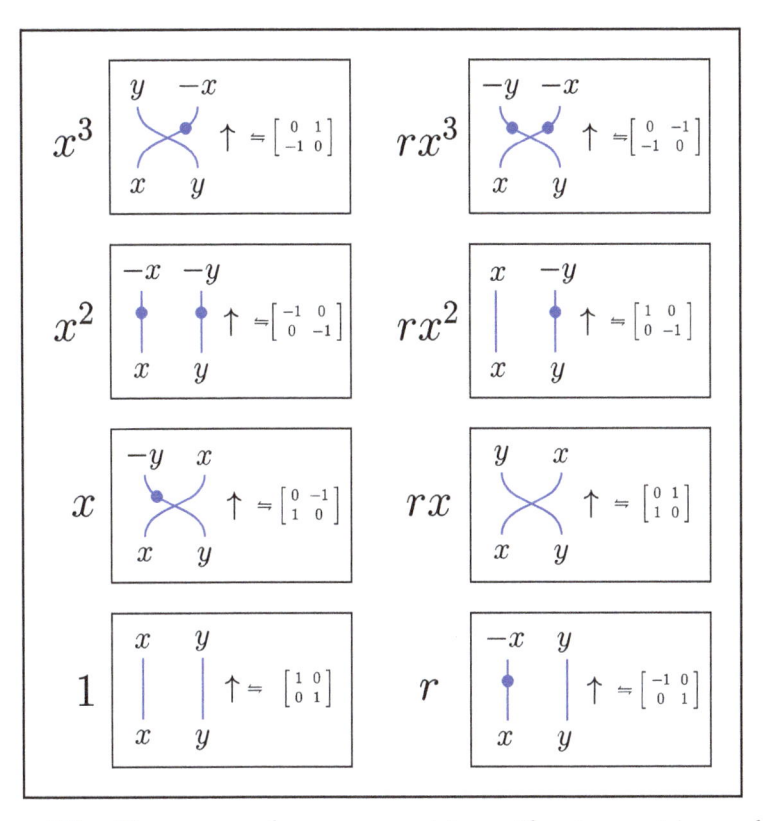

Figure 3.2. The correspondences among strings-with-quipu, matrices and the elements of the dihedral group.

Within the figure, the correspondences with the powers of the usual rotation x and the reflection r in the presentation $D_8 = \langle x, r : x^4 = r^2 = 1, \ rx = x^3 r \rangle$ are given. However, it is not enough to establish these correspondences without further checking both (1) the diagrams transform according to the rules $x^4 = r^2 = 1, \ rx = x^3 r$, and (2) the corresponding matrices do as well.

In the case of the matrices, observe that

$$\begin{bmatrix} 0 & -1 \\ 1 & 0 \end{bmatrix} \cdot \begin{bmatrix} 0 & -1 \\ 1 & 0 \end{bmatrix} = \begin{bmatrix} -1 & 0 \\ 0 & -1 \end{bmatrix} \quad (x^2),$$

$$\begin{bmatrix} -1 & 0 \\ 0 & -1 \end{bmatrix} \cdot \begin{bmatrix} -1 & 0 \\ 0 & -1 \end{bmatrix} = \begin{bmatrix} 1 & 0 \\ 0 & 1 \end{bmatrix} \quad (x^4 = 1),$$

$$\begin{bmatrix} -1 & 0 \\ 0 & 1 \end{bmatrix} \cdot \begin{bmatrix} -1 & 0 \\ 0 & 1 \end{bmatrix} = \begin{bmatrix} 1 & 0 \\ 0 & 1 \end{bmatrix} \quad (r^2 = 1),$$

and

$$\begin{bmatrix} -1 & 0 \\ 0 & 1 \end{bmatrix} \cdot \begin{bmatrix} 0 & -1 \\ 1 & 0 \end{bmatrix} = \begin{bmatrix} 0 & 1 \\ 1 & 0 \end{bmatrix} = \begin{bmatrix} 0 & 1 \\ -1 & 0 \end{bmatrix} \cdot \begin{bmatrix} -1 & 0 \\ 0 & 1 \end{bmatrix}.$$

This last relation corresponds to the relation $rx = x^3r$ in the standard generators. In the transpositions-with-quipu realm, Fig. 3.3 gives the corresponding identities. In the top box, $x^2 = -1$ is illustrated. In the middle $x^4 = 1 = r^2$, and at the bottom, $rx = x^3r$.

There are two types of permutations represented by the strings in these diagrams: the identity and the transposition (12). For both of these, there are four possible distributions of quipu $(+,+)$, $(+,-)$, $(-,+)$, and $(-,-)$. These observations lead to a more concise method of describing the group D_8. It fits into the following seseq:

$$0 \longleftarrow \Sigma_2 \overset{p}{\longleftarrow} D_8 \overset{i}{\longleftarrow} \mathbb{Z}/2 \times \mathbb{Z}/2 \longleftarrow 0$$

that is said to be *split* in the sense that there is an injective group homomorphism $\Sigma_2 \overset{s}{\longrightarrow} D_8$ such that the composition $p \circ s = \mathrm{Id}_{\Sigma_2}$ is the identity on Σ_2. The homomorphism s is said to be a spitting, and $s(12) = rx^3$. Thus, $D_8 = (\mathbb{Z}/2 \times \mathbb{Z}/2) \rtimes \mathbb{Z}/2$ is a *semi-direct product*.

We'll take a few sentences to further describe this phenomenon. It is helpful to refer to Fig. 3.2 while reading this paragraph. Elements in D_8 will be described using the rotation x and the reflection r. Let $H = (1, rx^3)$ denote the image of s. It is a subgroup of D_8. Meanwhile, the subgroup $N = (1, r, x^2, rx^2)$ is normal, for conjugation

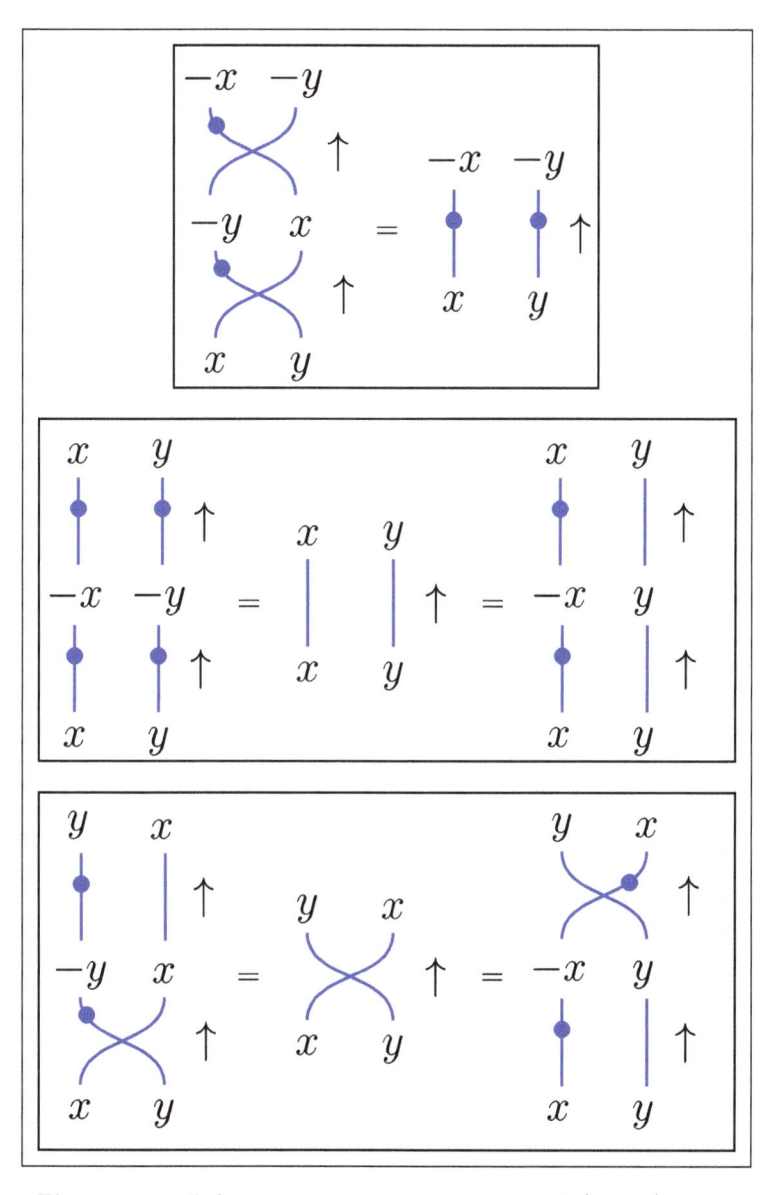

Figure 3.3. Relations among transpositions with (mod 2)-quipu.

by x, x^3, rx, or rx^3 upon an element in N results in another element in N. Specifically, $p(x) = p(x^3) = p(rx) = p(rx^3) = (12)$, and $p(y) = 1$ for $y \in N$. So, conjugation by these transpositions-with-quipu results in elements that project to the identity — the elements in N. Now, every element in D_8 is either in N or in $rx^3 N$. So, every element is in

$$HN = \{g \in D_8 : g = hk, \text{ for some } h \in H, k \in N\}.$$

That $N \cap H = \{1\}$, N is normal, and $D_8 = HN$ is the characterization of D_8 being a *semi-direct product*.

3.3. (mod 2)-quipu that describe a 24-element group

We begin with an exercise.

Exercise 6. The 3-strings-with-(mod 2)-quipu diagrams that are depicted in Fig. 3.4 each generate a cyclic group of order 6.

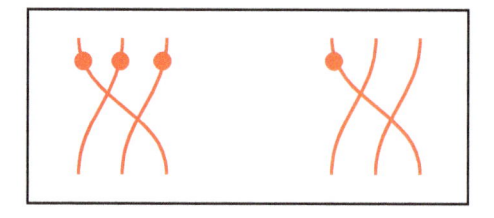

Figure 3.4. These two diagrams both generate a cyclic group of order 6.

Let's compose these two elements in two different orders and see what happens:

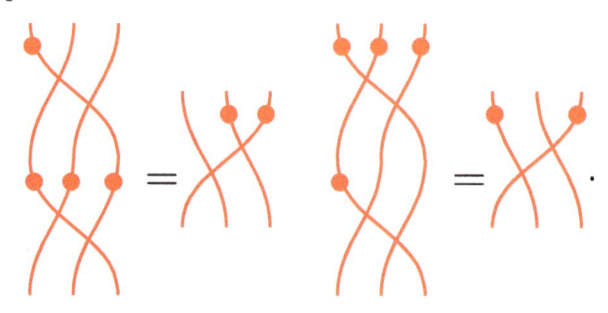

Matrix Descriptions 81

That the reduced diagrams are not equal indicates that the group consisting of the elements in Fig. 3.5 is not a commutative group. Continue multiplying elements together until all 24 elements that are listed in Fig. 3.5 are found. Verify that these form a group.

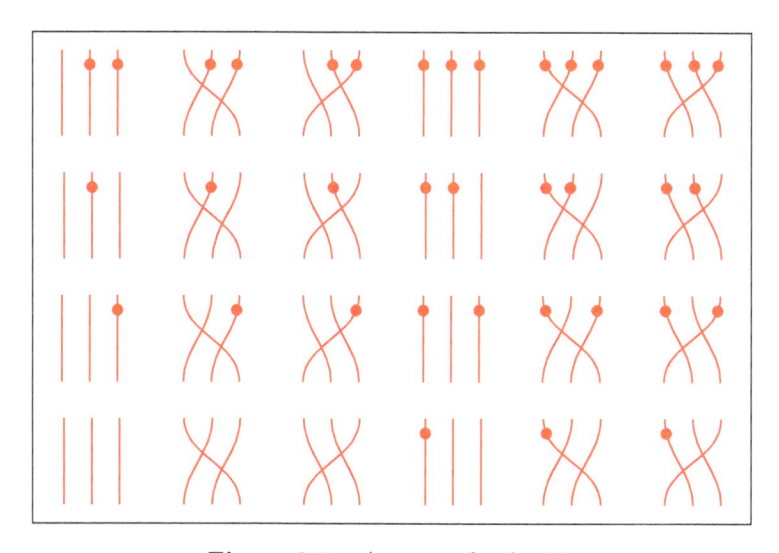

Figure 3.5. A group of order 24.

Mimic the techniques from Section 3.2. Let the quipu represent (-1) in the multiplicative cyclic group of order 2. Insert variables x, y, and z at the bottom of the diagrams and compute the values at the top. In this way, the 3-cycles-with-quipu that are depicted in Fig. 3.5 can be thought of as living in a matrix group. The first and fourth columns of the illustration are diagonal matrices of the form

$$\begin{bmatrix} \epsilon_1 & 0 & 0 \\ 0 & \epsilon_2 & 0 \\ 0 & 0 & \epsilon_3 \end{bmatrix},$$

where $\epsilon_1, \epsilon_2, \epsilon_3 \in \{\pm 1\}$. The second and fifth columns are matrices of the form

$$\begin{bmatrix} 0 & 0 & \delta_1 \\ \delta_2 & 0 & 0 \\ 0 & \delta_3 & 0 \end{bmatrix},$$

where $\delta_1, \delta_2, \delta_3 \in \{\pm 1\}$. The third and sixth columns are matrices of the form

$$\begin{bmatrix} 0 & \eta_1 & 0 \\ 0 & 0 & \eta_2 \\ \eta_3 & 0 & 0 \end{bmatrix},$$

where $\eta_1, \eta_2, \eta_3 \in \{\pm 1\}$. At this reading, you might find it difficult to determine the whereabouts (row versus column) of the signs as compared to the positions of the quipu. So, Fig. 3.6 indicates nine of the strings-with-quipu and their corresponding matrices as well as the method for computing these. The matrices are known as *signed permutation matrices* — matrices that have exactly one non-zero entry in any row or column. The non-zero entry is (± 1). Within this example, not all signed permutation matrices are present.

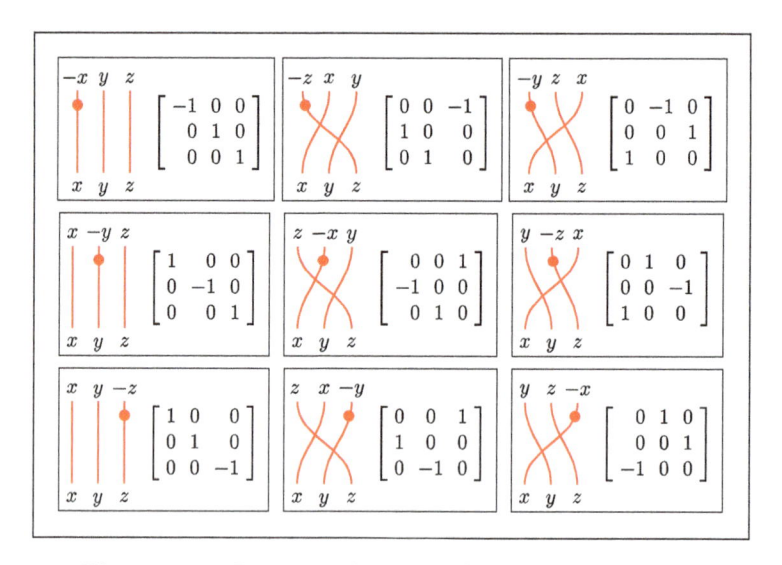

Figure 3.6. Location of quipu and signs in the matrices.

It will be important to understand and recognize that the location left-to-right of a (mod 2)-quipu located at the top of the strings corresponds to a sign top-to-bottom in the associated signed permutation matrix. For this reason, we urge the reader to do the following exercise.

Exercise 7. For the remaining diagrams that are depicted in Fig. 3.5, determine the corresponding signed permutation matrices. In particular, prove by direct calculation: *If a (mod 2)-quipu appears on the ℓth string, then there is a (-1) in the ℓth row of the associated signed permutation matrix.*

Let G denote the group of order 24

$$G = \left\{ \begin{bmatrix} \epsilon_1 & 0 & 0 \\ 0 & \epsilon_2 & 0 \\ 0 & 0 & \epsilon_3 \end{bmatrix}, \begin{bmatrix} 0 & 0 & \delta_1 \\ \delta_2 & 0 & 0 \\ 0 & \delta_3 & 0 \end{bmatrix}, \begin{bmatrix} 0 & \eta_1 & 0 \\ 0 & 0 & \eta_2 \\ \eta_3 & 0 & 0 \end{bmatrix} \right\},$$

where $\epsilon_\ell, \delta_m, \eta_n \in \{\pm 1\}$, for all $\ell, m, n = 1, 2, 3$. There is a seseq

$$0 \longleftarrow \mathbb{Z}/3 \xleftarrow{p} G \xleftarrow{i} (\mathbb{Z}/2 \times \mathbb{Z}/2 \times \mathbb{Z}/2) \longleftarrow 0$$

with an injective splitting homomorphism $\mathbb{Z}/3 \xrightarrow{s} G$ (so that $p \circ s = \text{Id}_{\mathbb{Z}/3}$). In a manner analogous to the preceding section, let $N = (\mathbb{Z}/2 \times \mathbb{Z}/2 \times \mathbb{Z}/2)$. This is a normal subgroup of G. Let H denote the image of s, $H = s(\mathbb{Z}/3)$. Then $N \cap H = \{1\}$, and $G = NH$. So, G is written as a semi-direct product

$$G = (\mathbb{Z}/2)^3 \rtimes \mathbb{Z}/3.$$

At the end of Section 3.3.1, we will demonstrate that the group is isomorphic to the cartesian product $\mathbb{Z}/2 \times A_4$, where $A_4 \subset \Sigma_4$ is the subgroup that consists of the *alternating permutations*. These are the permutations that can be written as a product of an even number of the adjacent transpositions $t_\ell = (\ell, \ell+1)$ for $\ell = 1, 2, 3$.

3.3.1. Actions upon the cube and an embedded tetrahedron

It may have been a little unfair of us to introduce the $(n-1)$-dimensional simplex

$$\Delta_{n-1} = \left\{ \sum_{\ell=1}^{n} \lambda_\ell e_\ell : \sum_\ell \lambda_\ell = 1 \ \& \ 0 \leq \lambda_\ell \text{ for } \ell = 1, \ldots, n \right\}$$

as an early example an object with a plethora of symmetries. Even though the symmetry group of Δ_{n-1} is Σ_n, it is, after all, a higher-dimensional figure. That invisible nature makes it hard to imagine.

Yet, Δ_0 is a point, Δ_1 is a line segment, and Δ_2 is a triangle. Despite being first described as a subset of 4-dimensional space, Δ_3 is a tetrahedron. When it was introduced as having Σ_4 symmetry, we attempted to warn the reader. A standard 3-dimensional tetrahedron only has the alternating group A_4 as its group of symmetries. Non-alternating permutations in Σ_4 can be thought of as reflections.

The situation of the prior paragraph has analogues among the dihedral groups, D_{2n}. These include reflections that, from the perspective of the 2-dimensional embeddings, seem impossible, but are achieved by lifting the polygons into 3-dimensional space to turn them over. In much the same way, but nonetheless conceptually, the reflection of a 3-dimensional solid can be achieved by lifting the object from its surroundings into 4-dimensional space, rotating it therein, and returning it to its space.

For quite some time in the current discussion, we will be interested in elements in the alternating group A_4 that is a subgroup of G. So, we take a few sentences to distinguish it.

The matrix representative of the elements in the larger group G are as follows:

$$(\epsilon_1 e_1, \epsilon_2 e_2, \epsilon_3 e_3) = \begin{bmatrix} \epsilon_1 & 0 & 0 \\ 0 & \epsilon_2 & 0 \\ 0 & 0 & \epsilon_3 \end{bmatrix},$$

$$(\delta_2 e_2, \delta_3 e_3, \delta_1 e_1) = \begin{bmatrix} 0 & 0 & \delta_1 \\ \delta_2 & 0 & 0 \\ 0 & \delta_3 & 0 \end{bmatrix},$$

and

$$(\eta_3 e_3, \eta_1 e_1, \eta_2 e_2) = \begin{bmatrix} 0 & \eta_1 & 0 \\ 0 & 0 & \eta_2 \\ \eta_3 & 0 & 0 \end{bmatrix}.$$

These matrices are of two types: those of determinant[3] 1 and those of determinant -1. Those of determinant 1 correspond to the strings-with-quipu representatives having an even number of quipu.

---

[3]Determinants will be reviewed in general in Section 3.4.1.

Equivalently, either all the ϵs, δs, or ηs are 1 or exactly two are -1. The strings-with-quipu are depicted in Fig. 3.9. If two such strings with quipu are multiplied (juxtaposed vertically), there will still be an even number of quipu present.

While these transformations may be difficult to image, a catalogue of the elements in the alternating group A_4 is easy to provide. Figure 3.7 contains a list of these.

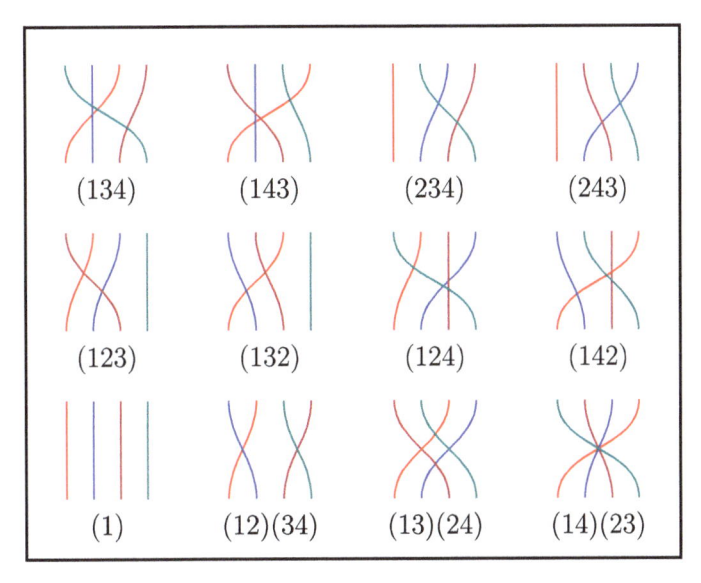

Figure 3.7. The elements of the alternating group A_4.

The alternating group A_4 is of interest here because of the purported isomorphism $(\mathbb{Z}/2)^3 \rtimes \mathbb{Z}/3 \approx \mathbb{Z}/2 \times A_4$. However, Fig. 3.7 contains the standard 4-strings representation, and Section 3.3 is meant to be about a 3-strings picture where there are (mod 2)-quipu that adorn the strings. To resolve these two presentations, we will need to study the actions of the corresponding matrices upon the cube $[-1, 1] \times [-1, 1] \times [-1, 1]$ in which an equilateral tetrahedron lies.

The cube $C_3 = \{(x, y, z) \in \mathbb{R}^3 : x, y, z \in [-1, 1]\}$, after all, is more common place than a simplex Δ_n, for $4 < n$. Some of us had similar cubes among our toys as toddlers. Figure 3.8 indicates such a cube, its coordinates, and the embedded tetrahedron upon which some secondary calculations will be focused.

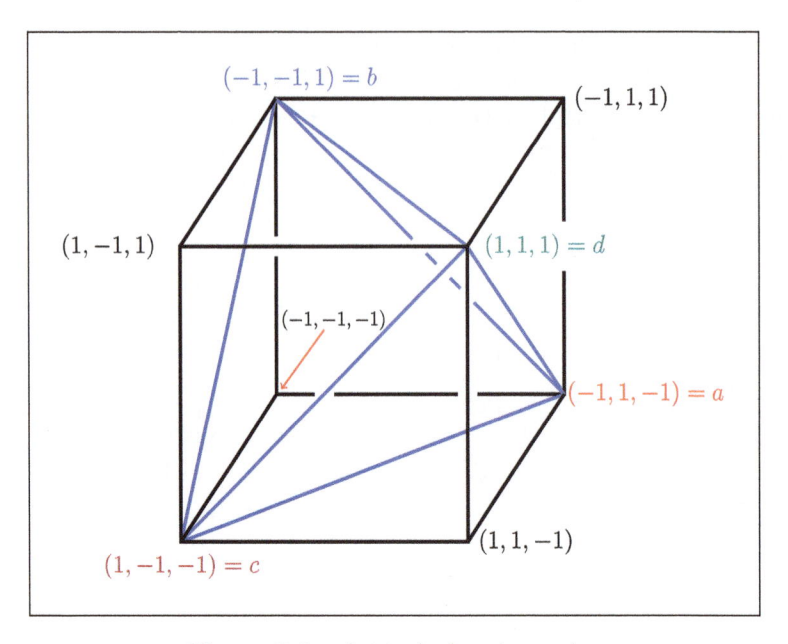

Figure 3.8. A tetrahedron in a cube.

In a few paragraphs, we will track the actions of the matrix representatives of the elements of the 24-element group G by examining their actions upon the vertices $a = (-1, 1, -1)$, $b = (-1, -1, 1)$, $c = (1, -1, -1)$, and $d = (1, 1, 1)$ of the tetrahedron since the remaining vertices are antipodal to these and the matrix action is, after all, linear. We will demonstrate below a method that uses the strings-with-quipu to compute the correspondences between these diagrams and the standard representations.

A meticulous reader may wonder why this particular choice of labels upon the vertices was made. It does seem to be quite arbitrary. The answer to your ponder is as follows. Another more orderly choice was originally made. However, after an extensive analysis with respect to other, more advanced, quipu representations, we found that there were some incongruities between the more advanced presentations and the mod 2 presentations that were, aesthetically, off-putting. So, in brief, this choice causes pretty correspondences later on within the narrative.

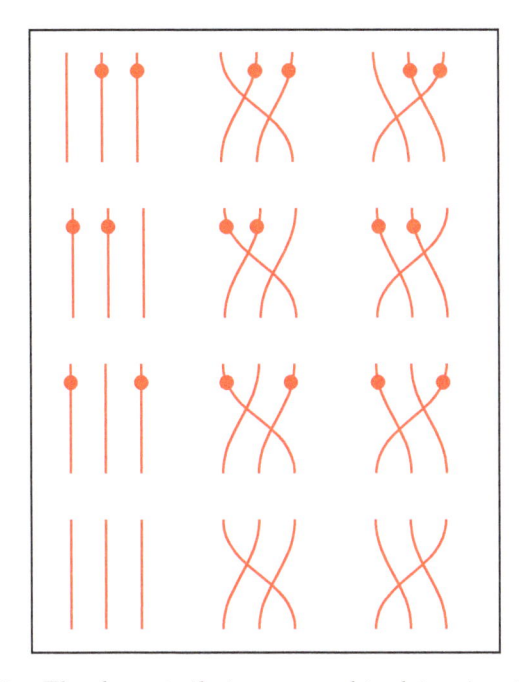

Figure 3.9. The elements that correspond to determinant 1 matrices.

Let us consider Fig. 3.9 to be a (4 × 3) array. The entry ⨯ is in position (1, 2), for example, first row, second column. Its inverse is at position (2, 3). Similarly, the inverse of the (2, 2) entry is in position (3, 3). The inverse of the (3, 2) entry is in position (1, 3). The elements in the first column are their own inverses, and entry (4, 2) is inverse to entry (4, 3). So, this set forms a subgroup of the group G shown in Fig. 3.5.

It is a more direct calculation to observe that the determinant is a group homomorphism onto the multiplicative group $\mathbb{Z}/2 = \{\pm 1\}$, and the elements of Fig. 3.9 are in the kernel of this map.

The correspondences among the signed permutation matrices, the 3-strings-with-quipu diagrams, the 4-strings diagrams, and the cycle structure as rigid[4] symmetries of the tetrahedron depicted in Fig. 3.8 are computed in the following and subsequently tabulated. This information will also be used in Section 4.3.2, where the rigid motions of 3-space SO(3) are lifted to the group SU(2) that coincides with the 3-dimensional sphere.

Each strings-with-quipu diagram (with the exception of the identity diagram) has been adorned below by a vector (x, y, z). Then the location of these elements following the application of diagram appears at the top. The actions upon the vertices a, b, c, and d of the tetrahedron have been indicated.

$$\begin{bmatrix} a \\ b \\ c \\ d \end{bmatrix} = \begin{bmatrix} -1 & 1 & -1 \\ -1 & -1 & 1 \\ 1 & -1 & -1 \\ 1 & 1 & 1 \end{bmatrix} \begin{matrix} -x & -y & z \\ \bullet & \bullet & \\ | & | & | \\ | & | & | \\ x & y & z \end{matrix} \begin{bmatrix} 1 & -1 & -1 \\ 1 & 1 & 1 \\ -1 & 1 & -1 \\ -1 & -1 & 1 \end{bmatrix} \begin{bmatrix} c \\ d \\ a \\ b \end{bmatrix}.$$

$$\xrightarrow{}$$
$$(ac)(bd)$$

$$\begin{bmatrix} a \\ b \\ c \\ d \end{bmatrix} = \begin{bmatrix} -1 & 1 & -1 \\ -1 & -1 & 1 \\ 1 & -1 & -1 \\ 1 & 1 & 1 \end{bmatrix} \begin{matrix} -x & y & -z \\ \bullet & & \bullet \\ | & | & | \\ | & | & | \\ x & y & z \end{matrix} \begin{bmatrix} 1 & 1 & 1 \\ 1 & -1 & -1 \\ -1 & -1 & 1 \\ -1 & 1 & -1 \end{bmatrix} \begin{bmatrix} d \\ c \\ b \\ a \end{bmatrix}.$$

$$\xrightarrow{}$$
$$(ad)(bc)$$

$$\begin{bmatrix} a \\ b \\ c \\ d \end{bmatrix} = \begin{bmatrix} -1 & 1 & -1 \\ -1 & -1 & 1 \\ 1 & -1 & -1 \\ 1 & 1 & 1 \end{bmatrix} \begin{matrix} x & -y & -z \\ & \bullet & \bullet \\ | & | & | \\ | & | & | \\ x & y & z \end{matrix} \begin{bmatrix} -1 & -1 & 1 \\ -1 & 1 & -1 \\ 1 & 1 & 1 \\ 1 & -1 & -1 \end{bmatrix} \begin{bmatrix} b \\ a \\ d \\ c \end{bmatrix}.$$

$$\xrightarrow{}$$
$$(ab)(cd)$$

---

[4]Here, the word *rigid* is used to mean those motions that can be achieved by means of moving a model that might sit upon your desk. It excludes reflections.

Matrix Descriptions

$$\begin{bmatrix} a \\ b \\ c \\ d \end{bmatrix} = \begin{bmatrix} -1 & 1 & -1 \\ -1 & -1 & 1 \\ 1 & -1 & -1 \\ 1 & 1 & 1 \end{bmatrix} \overset{z\ x\ y}{\underset{x\ y\ z}{\bowtie}} \begin{bmatrix} -1 & -1 & 1 \\ 1 & -1 & -1 \\ -1 & 1 & -1 \\ 1 & 1 & 1 \end{bmatrix} = \begin{bmatrix} b \\ c \\ a \\ d \end{bmatrix}.$$

$$\xrightarrow{\quad (abc) \quad}$$

$$\begin{bmatrix} a \\ b \\ c \\ d \end{bmatrix} = \begin{bmatrix} -1 & 1 & -1 \\ -1 & -1 & 1 \\ 1 & -1 & -1 \\ 1 & 1 & 1 \end{bmatrix} \overset{-z\ -x\ y}{\underset{x\ y\ z}{\bowtie}} \begin{bmatrix} 1 & 1 & 1 \\ -1 & 1 & -1 \\ 1 & -1 & -1 \\ -1 & -1 & 1 \end{bmatrix} = \begin{bmatrix} d \\ a \\ c \\ b \end{bmatrix}.$$

$$\xrightarrow{\quad (adb) \quad}$$

$$\begin{bmatrix} a \\ b \\ c \\ d \end{bmatrix} = \begin{bmatrix} -1 & 1 & -1 \\ -1 & -1 & 1 \\ 1 & -1 & -1 \\ 1 & 1 & 1 \end{bmatrix} \overset{-z\ x\ -y}{\underset{x\ y\ z}{\bowtie}} \begin{bmatrix} 1 & -1 & -1 \\ -1 & -1 & 1 \\ 1 & 1 & 1 \\ -1 & 1 & -1 \end{bmatrix} = \begin{bmatrix} c \\ b \\ d \\ a \end{bmatrix}.$$

$$\xrightarrow{\quad (acd) \quad}$$

$$\begin{bmatrix} a \\ b \\ c \\ d \end{bmatrix} = \begin{bmatrix} -1 & 1 & -1 \\ -1 & -1 & 1 \\ 1 & -1 & -1 \\ 1 & 1 & 1 \end{bmatrix} \overset{z\ -x\ -y}{\underset{x\ y\ z}{\bowtie}} \begin{bmatrix} -1 & 1 & -1 \\ 1 & 1 & 1 \\ -1 & -1 & 1 \\ 1 & -1 & -1 \end{bmatrix} = \begin{bmatrix} a \\ d \\ b \\ c \end{bmatrix}.$$

$$\xrightarrow{\quad (bdc) \quad}$$

$$\begin{bmatrix} a \\ b \\ c \\ d \end{bmatrix} = \begin{bmatrix} -1 & 1 & -1 \\ -1 & -1 & 1 \\ 1 & -1 & -1 \\ 1 & 1 & 1 \end{bmatrix} \begin{matrix} y & z & x \\ \\ \\ x & y & z \end{matrix} \begin{bmatrix} 1 & -1 & -1 \\ -1 & 1 & -1 \\ -1 & -1 & 1 \\ 1 & 1 & 1 \end{bmatrix} \begin{bmatrix} c \\ a \\ b \\ d \end{bmatrix}.$$

$$\xrightarrow{(acb)}$$

$$\begin{bmatrix} a \\ b \\ c \\ d \end{bmatrix} = \begin{bmatrix} -1 & 1 & -1 \\ -1 & -1 & 1 \\ 1 & -1 & -1 \\ 1 & 1 & 1 \end{bmatrix} \begin{matrix} -y & -z & x \\ \\ \\ x & y & z \end{matrix} \begin{bmatrix} -1 & 1 & -1 \\ 1 & -1 & -1 \\ 1 & 1 & 1 \\ -1 & -1 & 1 \end{bmatrix} \begin{bmatrix} a \\ c \\ d \\ b \end{bmatrix}.$$

$$\xrightarrow{(bcd)}$$

$$\begin{bmatrix} a \\ b \\ c \\ d \end{bmatrix} = \begin{bmatrix} -1 & 1 & -1 \\ -1 & -1 & 1 \\ 1 & -1 & -1 \\ 1 & 1 & 1 \end{bmatrix} \begin{matrix} -y & z & -x \\ \\ \\ x & y & z \end{matrix} \begin{bmatrix} -1 & -1 & 1 \\ 1 & 1 & 1 \\ 1 & -1 & -1 \\ -1 & 1 & -1 \end{bmatrix} \begin{bmatrix} b \\ d \\ c \\ a \end{bmatrix}.$$

$$\xrightarrow{(abd)}$$

$$\begin{bmatrix} a \\ b \\ c \\ d \end{bmatrix} = \begin{bmatrix} -1 & 1 & -1 \\ -1 & -1 & 1 \\ 1 & -1 & -1 \\ 1 & 1 & 1 \end{bmatrix} \begin{matrix} y & -z & -x \\ \\ \\ x & y & z \end{matrix} \begin{bmatrix} 1 & 1 & 1 \\ -1 & -1 & 1 \\ -1 & 1 & -1 \\ 1 & -1 & -1 \end{bmatrix} \begin{bmatrix} d \\ b \\ a \\ c \end{bmatrix}.$$

$$\xrightarrow{(adc)}$$

Matrix Descriptions

The correspondences among the 3-strings-with-(mod 2)-quipu are tabulated in Tables 3.1 and 3.2.

Table 3.1. Correspondences among signed permutation matrices, 3-strings-with-(mod 2)-quipu, and elements in the alternating group A_4, part 1.

Matrix	Quipu	A_4 element
(e_1, e_2, e_3)	⇌ ||| ⇌	$(a)(b)(c)(d)$
$(e_1, -e_2, -e_3)$	⇌ ||| ⇌	$(ab)(cd)$
$(-e_1, -e_2, e_3)$	⇌ ||| ⇌	$(ac)(bd)$
$(-e_1, e_2, -e_3)$	⇌ ||| ⇌	$(ad)(bc)$

Table 3.2. Correspondences among signed permutation matrices, 3-strings- with-(mod 2)-quipu, and elements in the alternating group A_4, part 2.

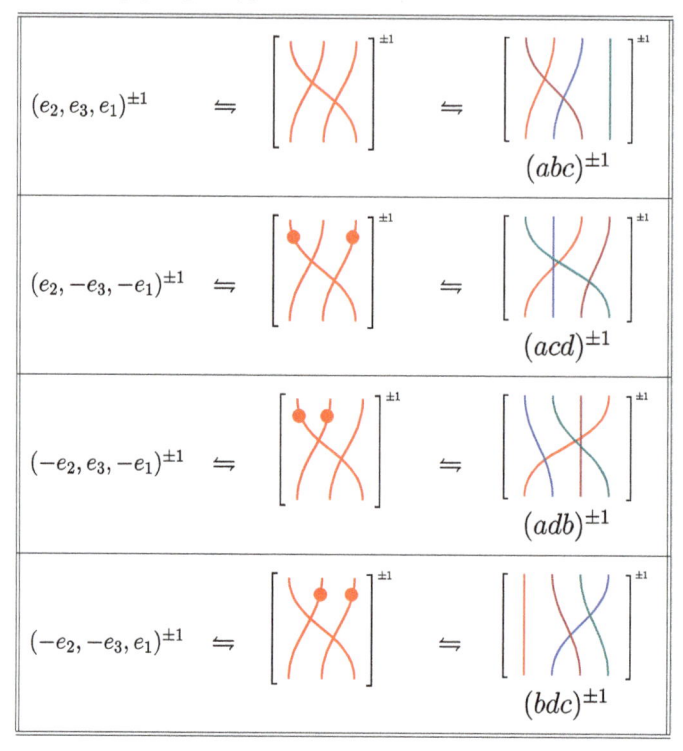

$(e_2, e_3, e_1)^{\pm 1}$	⇆	$\left[\begin{array}{c}\text{\includegraphics}\end{array}\right]^{\pm 1}$	⇆	$\left[\begin{array}{c}\text{\includegraphics}\end{array}\right]^{\pm 1}$ $(abc)^{\pm 1}$
$(e_2, -e_3, -e_1)^{\pm 1}$	⇆	$\left[\begin{array}{c}\text{\includegraphics}\end{array}\right]^{\pm 1}$	⇆	$\left[\begin{array}{c}\text{\includegraphics}\end{array}\right]^{\pm 1}$ $(acd)^{\pm 1}$
$(-e_2, e_3, -e_1)^{\pm 1}$	⇆	$\left[\begin{array}{c}\text{\includegraphics}\end{array}\right]^{\pm 1}$	⇆	$\left[\begin{array}{c}\text{\includegraphics}\end{array}\right]^{\pm 1}$ $(adb)^{\pm 1}$
$(-e_2, -e_3, e_1)^{\pm 1}$	⇆	$\left[\begin{array}{c}\text{\includegraphics}\end{array}\right]^{\pm 1}$	⇆	$\left[\begin{array}{c}\text{\includegraphics}\end{array}\right]^{\pm 1}$ $(bdc)^{\pm 1}$

Theorem 2. *The group G that is the semi-direct product $G = (\mathbb{Z}/2)^3 \rtimes \mathbb{Z}/3$ described in Fig. 3.5 is isomorphic to the cartesian product $\mathbb{Z}/2 \times A_4$ of the cyclic group of order 2 and the 12-element alternating group that is isomorphic to the group of symmetries of the tetrahedron depicted in Fig. 3.9.*

Proof. To begin, we seek a subgroup H of $G = (\mathbb{Z}/2)^3 \rtimes \mathbb{Z}/3$ that is isomorphic to $\mathbb{Z}/2$, for which every element of H commutes with A_4, such that every element in G can be written as a product kh, where $k \in A_4$ and $h \in H$.

Note that $H = \{(e_1, e_2, e_3), (-e_1, -e_2, -e_3)\}$ is isomorphic to $\mathbb{Z}/2$. The quipu representation

$$\text{[quipu diagram]} = \begin{bmatrix} -1 & 0 & 0 \\ 0 & -1 & 0 \\ 0 & 0 & -1 \end{bmatrix}$$

makes it clear that the non-identity element of H commutes with every element of G. The identity commutes with everything.

It is a fairly direct exercise to represent an element of G that has an odd number of quipu as a product $(-e_1, -e_2, -e_3) \cdot k$, where $k \in A_4$. If there are three quipu present, let $k = (e_1, e_2, e_3)$, (e_2, e_3, e_1), or (e_3, e_1, e_2). If there is only one, it is in position 1, 2, or 3 at the top of the string diagram. Then let k have two quipu on the complementary strings.

Suppose that $g \in G$ can be written as $g = h_1 k_1 = h_2 k_2$, where $h_i \in H$ and $k_i \in A_4$. Then $h_2^{-1} h_1 = k_2 k_1^{-1} \in A_4 \cap H = \{1\}$. So, that $h_1 = h_2$ and $k_1 = k_2$.

An isomorphism $\mathbb{Z}/2 \times A_4 \xleftarrow{\phi} (\mathbb{Z}/2)^2 \rtimes \mathbb{Z}/3$ is defined by $\phi(-e_1, -e_2, -e_3) = (-1, 1)$, while $\phi(k) = (1, k)$. Here, the $\mathbb{Z}/2$ factor of the cartesian product is written multiplicatively as $\mathbb{Z}/2 = \{\pm 1\}$. In general, $\phi(hk) = (h, k)$. By the above paragraph, this value is well defined. Moreover, $\phi((h_1 k_1)(h_2 k_2)) = \phi((h_1 h_2)(k_1 k_2)) = (h_1 h_2, k_1 k_2) = \phi(h_1 k_1)\phi(h_2 k_2)$. So, ϕ is a homomorphism. It is surjective and injective by construction. This completes the proof. □

3.4. Signed permutation

In this section, the group of *signed permutations*

$$(\mathbb{Z}/2)^n \rtimes \Sigma_n$$

and its action (that is induced by a matrix representation) upon the n-dimensional cube

$$C_n = \{(x_1, x_2, \ldots, x_n) \in \mathbb{R}^n : |x_\ell| = 1, \text{ for all } \ell = 1, 2, \ldots, n\}$$

will be described for all $n = 1, 2, \ldots$ The main focus will be the 3-dimensional case $(\mathbb{Z}/2)^3 \rtimes \Sigma_3$. In particular, two subgroups that are isomorphic to Σ_3 will be given by means of strings with (and without) quipu. Another important subgroup that is isomorphic to Σ_4 will be presented via a 3-strings-with-quipu description. This representation coincides with the group of rigid symmetries of the cube C_3. Both the group G defined in Section 3.3 and Σ_4 are subgroups of $(\mathbb{Z}/2)^3 \rtimes \Sigma_3$. The intersection between G and this representation of Σ_4 is isomorphic to the alternating group A_4.

Section 3.4.1 begins with some background material on determinants of matrices and signs of permutations as matters of review. These ideas have been used in passing within several discussions above. But a more formal introduction might be helpful. In particular, some signs can easily be read from strings-with-(mod 2)-quipu diagrams.

Some aspects of the 4-dimensional (hyper)-cube C_4 will be described in Section 3.4.4.

The 3-strings-with-(mod 2)-quipu diagrams for Σ_4 will be important for many reasons. One, in particular, is that these diagrams will be directly compared with others for a variety of extensions of Σ_4. There is an analogue (Theorem 3) of Theorem 2 that defines an isomorphism between $(\mathbb{Z}/2)^3 \rtimes \Sigma_3$ and $\mathbb{Z}/2 \times \Sigma_4$.

The current section provides a simple instance of the general construction that will be given in Section 3.5.

3.4.1. Signs of permutations and determinant

The *determinant* of an $(n \times n)$-matrix A whose (i, j)th entry (ith-row, jth-column) is a_j^i is defined to be the quantity

$$\det(A) = \sum_{\sigma \in \Sigma_n} (-1)^{t(\sigma)} \prod_{j=1}^n a_j^{\sigma(j)},$$

where $(-1)^{t(\sigma)}$ is the *sign of the permutation* $\sigma \in \Sigma_n$. The exponent $t(\sigma)$ is defined to be the number of transpositions $t_\ell = (\ell, \ell+1)$ used to express σ in terms of the standard generators. While this number is not well defined, its parity is.

That definition contains a lot of nested adjectival phrases and notational nuances. Let's unpack it in a step-by-step fashion. In Section 2.1.1 of Chapter 2, the group of permutations was defined to be the set of functions

$$\Sigma_n = \{[n] \xleftarrow{\sigma} [n] : \sigma \text{ is bijective}\}$$

from $[n] = \{1, 2, \ldots, n\}$ to itself. Such a permutation can be achieved by successively transposing adjacent elements in $[n]$. The generating (and standard) transpositions t_ℓ are written in cycle notations as $(\ell, \ell+1)$. The transposition t_ℓ is presented as

$$\Big| \; \Big| \ldots \Big\rtimes \ldots \Big| \; .$$
$$1 \quad 2 \qquad \ell \;\; \ell+1 \qquad n$$

The parity (oddness or evenness) of the number of t_ℓs that are needed to write σ does not depend on the representation thereof. For example, if w represents σ, then so too does wt_1^2. So, while $t(\sigma)$ is not well defined, the sign $(-1)^{t(\sigma)}$ is well defined.

To further understand the terms $\prod_{j=1}^n a_j^{\sigma(j)}$ in the sum that describes the determinant, we ask you to think about an $(n \times n)$-chessboard upon which n rooks are configured. The chess piece called a *rook* can move vertically or horizontally from its initial position. So, if the rook r appears in the ith row, jth column of the board, and there are no other pieces in the way, it can move to anywhere in the ith row or to anywhere in the jth column. It is possible to configure n rooks on an $(n \times n)$ board so that they are mutually benevolent. That is, no rook can attack another, or each rook has a full + shape of possible motions.

One such configuration is for the rooks to be positioned at entries $(1,1), (2,2), \ldots, (n,n)$. This will be called the *base configuration*. If $\sigma \in \Sigma_n$ is any permutation, then rooks configured in positions $(\sigma(1), 1), (\sigma(2), 2), \ldots, (\sigma(n), n)$ are also mutually benevolent.

The product $\prod_{j=1}^n a_j^{\sigma(j)}$ of the matrix entries corresponds to the multiplication among the matrix entries that appear in the locations of a mutually benevolent rook configuration. Exactly one entry is chosen from any row or column.

The signed product $(-1)^{t(\sigma)} \prod_{j=1}^{n} a_j^{\sigma(j)}$ associates to the product in a rooks' configuration the sign of the permutation in comparison to the base configuration.

Finally, one adds up all the terms $(-1)^{t(\sigma)} \prod_{j=1}^{n} a_j^{\sigma(j)}$ over all permutations.

The result of computing the determinant of a general (3×3)-matrix is illustrated:

$$\det \begin{bmatrix} a & b & c \\ d & e & f \\ g & h & i \end{bmatrix} = aei - afh + bgf - bdi + cdh - ceg.$$

There are more sophisticated definitions of the determinant. For example, it is the unique alternating multilinear form that takes the value 1 upon the identity matrix [7]. To develop the determinant from that point of view would take us much farther away from the discussion than even this short excursion did.

If the matrix whose determinant is to be computed is a permutation matrix, as defined in Section 2.3.3 of Chapter 2, then there is only one non-zero term in the expression,

$$\sum_{\sigma \in \Sigma_n} (-1)^{t(\sigma)} \prod_{j=1}^{n} a_j^{\sigma(j)},$$

namely, the configuration of 1s in that permutation corresponds to the rooks' position. Any other configuration of rooks will have a factor of 0 present.

The determinants of strings-with-(mod 2)-quipu, can be computed by counting the number of crossings between pairs of strings, adding to this the number of quipu and reducing modulo 2. Let d denote this count, then det $= (-1)^d$. That methodology will be reviewed after the description of signed permutation matrices is given in general.

The *alternating group* A_n consists of those permutations in Σ_n that can be written by using an even number of standard generators. In the representation that uses permutation matrices, the sign of the determinant of the matrix that represents an element in A_n is positive.

3.4.2. The group and a matrix representation

In Fig. 3.5, 24 group elements were listed as strings-with-(mod 2)-quipu. The corresponding matrix representations were defined and enumerated in Tables 3.1 and 3.2 that precede Theorem 2. The strings involved were either the identity permutation or one of the 3-cycles (123) or (132) in Σ_3. The elements of the larger group $(\mathbb{Z}/2)^3 \rtimes \Sigma_3$ also allow for the transpositions (12), (23), and (13).

In matrix form,

$$(\mathbb{Z}/2)^3 \rtimes \Sigma_3 = \{(\epsilon_1 e_{\sigma 1}, \epsilon_2 e_{\sigma 2}, \epsilon_3 e_{\sigma 3}) : \epsilon_\ell = \pm 1 \ \& \ \sigma \in \Sigma_3\}.$$

Figrue 3.10 contains a list of the original elements, and Fig. 3.11 contains a list of the remaining elements. The elements are colored blue if the determinant of the corresponding matrix is negative, and they are colored red when the determinant is positive.

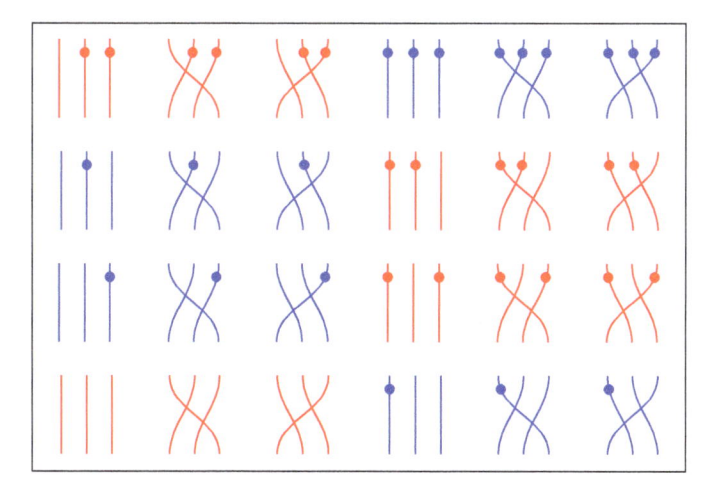

Figure 3.10. Elements of $(\mathbb{Z}/2)^3 \rtimes \mathbb{Z}/3$.

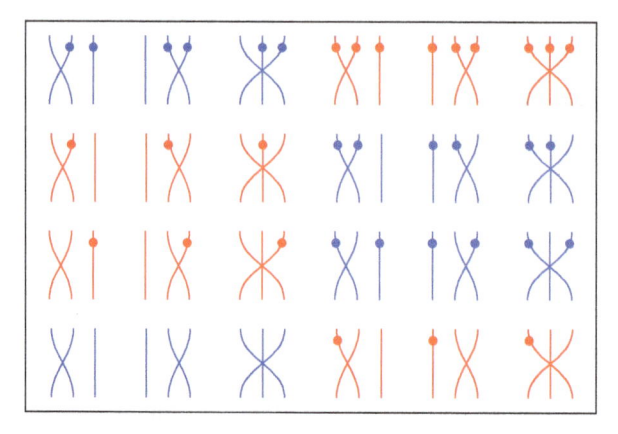

Figure 3.11. The remaining elements in $(\mathbb{Z}/2)^3 \rtimes \Sigma_3$.

The elements in red that are depicted in these figures form a subgroup of $(\mathbb{Z}/2)^3 \rtimes \Sigma_3$ that is isomorphic to Σ_4. The isomorphism will be given by way of studying the action of the group of signed permutations upon the cube, C_3. The tables that precede Theorem 2 will be extended to include the elements in Fig. 3.11.

In order to tabulate these correspondences, the cube from Fig. 3.8 is replicated in Fig. 3.12, and its vertices are labeled $\pm a = \pm(-1, 1, -1)$, $\pm b = \pm(-1, -1, 1)$, $\pm c = \pm(1, -1, -1)$, and $\pm d = \pm(1, 1, 1)$. The diagonals of the cube, $-a \xrightarrow{A} a$, $-b \xrightarrow{B} b$, $-c \xrightarrow{C} c$, and $-d \xrightarrow{D} d$, are transformed among themselves by the matrices that correspond to the group presented in Figs. 3.10 and 3.11.

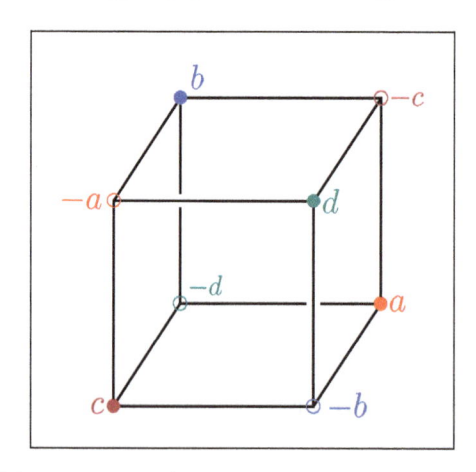

Figure 3.12. Antipodal vertices in the cube.

Matrix Descriptions

In Tables 3.1 and 3.2 that are immediately above Theorem 2, the actions of the red-colored elements in Fig. 3.10 have been compiled. To complete the compilation for the red elements in Fig. 3.11, we first observe that

$$\left[\;\substack{\bullet} \!\!\diagup\!\!\diagdown \; | \; \right]^{-1} = \;\substack{\bullet} \!\!\diagdown\!\!\diagup \; | \;,$$

$$\left[\; | \;\substack{\bullet} \!\!\diagup\!\!\diagdown \; \right]^{-1} = \; | \;\substack{\bullet} \!\!\diagdown\!\!\diagup \;,$$

and finally,

$$\left[\;\substack{\bullet} \!\!\asymp\!\!\; \right]^{-1} = \;\substack{\bullet} \!\!\asymp\!\! \;,$$

We compute

$$\begin{bmatrix} -a \\ -b \\ -c \\ -d \end{bmatrix} = \begin{bmatrix} 1 & -1 & 1 \\ 1 & 1 & -1 \\ -1 & 1 & 1 \\ -1 & -1 & -1 \end{bmatrix} \overset{-y\ \ x\ \ z}{\underset{x\ \ y\ \ z}{\diagup\!\!\!\diagdown}} \begin{bmatrix} 1 & 1 & 1 \\ -1 & 1 & -1 \\ -1 & -1 & 1 \\ 1 & -1 & -1 \end{bmatrix} \begin{bmatrix} d \\ a \\ b \\ c \end{bmatrix}.$$

$$\xrightarrow{\;(ADCB)\;}$$

$$\begin{bmatrix} -a \\ -b \\ -c \\ -d \end{bmatrix} = \begin{bmatrix} 1 & -1 & 1 \\ 1 & 1 & -1 \\ -1 & 1 & 1 \\ -1 & -1 & -1 \end{bmatrix} \overset{y\ \ x\ \ -z}{\underset{x\ \ y\ \ z}{\diagup\!\!\!\diagdown}} \begin{bmatrix} -1 & 1 & -1 \\ 1 & 1 & 1 \\ 1 & -1 & -1 \\ -1 & -1 & 1 \end{bmatrix} \begin{bmatrix} a \\ d \\ c \\ b \end{bmatrix}.$$

$$\xrightarrow{\;(BD)\;}$$

$$\begin{bmatrix} -a \\ -b \\ -c \\ -d \end{bmatrix} = \begin{bmatrix} 1 & -1 & 1 \\ 1 & 1 & -1 \\ -1 & 1 & 1 \\ -1 & -1 & -1 \end{bmatrix} \overset{-y\ -x\ -z}{\underset{x\ \ y\ \ z}{\bigtimes\!\mid}} \begin{bmatrix} 1 & -1 & -1 \\ -1 & -1 & 1 \\ -1 & 1 & -1 \\ 1 & 1 & 1 \end{bmatrix} = \begin{bmatrix} c \\ b \\ a \\ d \end{bmatrix}.$$

$$\xrightarrow{\quad(AC)\quad}$$

$$\begin{bmatrix} -a \\ -b \\ -c \\ -d \end{bmatrix} = \begin{bmatrix} 1 & -1 & 1 \\ 1 & 1 & -1 \\ -1 & 1 & 1 \\ -1 & -1 & -1 \end{bmatrix} \overset{x\ -z\ y}{\underset{x\ \ y\ \ z}{\mid\!\bigtimes}} \begin{bmatrix} 1 & -1 & -1 \\ 1 & 1 & 1 \\ -1 & -1 & 1 \\ -1 & 1 & -1 \end{bmatrix} = \begin{bmatrix} c \\ d \\ b \\ a \end{bmatrix}.$$

$$\xrightarrow{\quad(ACBD)\quad}$$

$$\begin{bmatrix} -a \\ -b \\ -c \\ -d \end{bmatrix} = \begin{bmatrix} 1 & -1 & 1 \\ 1 & 1 & -1 \\ -1 & 1 & 1 \\ -1 & -1 & -1 \end{bmatrix} \overset{-x\ z\ y}{\underset{x\ \ y\ \ z}{\mid\!\bigtimes}} \begin{bmatrix} -1 & 1 & -1 \\ -1 & -1 & 1 \\ 1 & 1 & 1 \\ 1 & -1 & -1 \end{bmatrix} = \begin{bmatrix} a \\ b \\ d \\ c \end{bmatrix}.$$

$$\xrightarrow{\quad(CD)\quad}$$

$$\begin{bmatrix} -a \\ -b \\ -c \\ -d \end{bmatrix} = \begin{bmatrix} 1 & -1 & 1 \\ 1 & 1 & -1 \\ -1 & 1 & 1 \\ -1 & -1 & -1 \end{bmatrix} \overset{-x\ -z\ -y}{\underset{x\ \ y\ \ z}{\mid\!\bigtimes}} \begin{bmatrix} -1 & -1 & 1 \\ -1 & 1 & -1 \\ 1 & -1 & -1 \\ 1 & 1 & 1 \end{bmatrix} = \begin{bmatrix} b \\ a \\ c \\ d \end{bmatrix}.$$

$$\xrightarrow{\quad(AB)\quad}$$

$$\begin{bmatrix} -a \\ -b \\ -c \\ -d \end{bmatrix} = \begin{bmatrix} 1 & -1 & 1 \\ 1 & 1 & -1 \\ -1 & 1 & 1 \\ -1 & -1 & -1 \end{bmatrix} \begin{bmatrix} -1 & -1 & 1 \\ 1 & 1 & 1 \\ -1 & 1 & -1 \\ 1 & -1 & -1 \end{bmatrix} \begin{bmatrix} b \\ d \\ a \\ c \end{bmatrix}.$$

$$\xrightarrow{(ABDC)}$$

$$\begin{bmatrix} -a \\ -b \\ -c \\ -d \end{bmatrix} = \begin{bmatrix} 1 & -1 & 1 \\ 1 & 1 & -1 \\ -1 & 1 & 1 \\ -1 & -1 & -1 \end{bmatrix} \begin{bmatrix} 1 & 1 & 1 \\ -1 & -1 & 1 \\ 1 & -1 & -1 \\ -1 & 1 & -1 \end{bmatrix} \begin{bmatrix} d \\ b \\ c \\ a \end{bmatrix}.$$

$$\xrightarrow{(AD)}$$

$$\begin{bmatrix} -a \\ -b \\ -c \\ -d \end{bmatrix} = \begin{bmatrix} 1 & -1 & 1 \\ 1 & 1 & -1 \\ -1 & 1 & 1 \\ -1 & -1 & -1 \end{bmatrix} \begin{bmatrix} -1 & 1 & -1 \\ 1 & -1 & -1 \\ -1 & -1 & 1 \\ 1 & 1 & 1 \end{bmatrix} \begin{bmatrix} a \\ c \\ b \\ d \end{bmatrix}.$$

$$\xrightarrow{(BC)}$$

The results of the calculation are compiled in Tables 3.3–3.5 that indicate the correspondences among the remaining 3-strings-with-quipu, their matrix representation, and their associated elements in Σ_4.

Exercise 8. Verify in several cases that these correspondences are respected under compositions. In the 3-strings-with-quipu case and the 4-strings case of Σ_4, the compositions are vertical juxtaposition. In the matrix case, matrix multiplication is the composition operation.

Table 3.3. 3-strings-with-(mod 2)-quipu representatives of elements in Σ_4, part 1.

$(e_2, -e_1, e_3)$	⇌	(1432)
$(e_2, e_1, -e_3)$	⇌	(24)
$(-e_2, -e_1, -e_3)$	⇌	(13)
$(-e_2, e_1, e_3)$	⇌	(1234)

Table 3.4. 3-strings-with-(mod 2)-quipu representatives of elements in Σ_4, part 2.

$(e_1, e_3, -e_2)$	$\rightleftharpoons$ ... $\rightleftharpoons$	(1324)
$(-e_1, e_3, e_2)$	$\rightleftharpoons$ ... $\rightleftharpoons$	(34)
$(-e_1, -e_3, -e_2)$	$\rightleftharpoons$ ... $\rightleftharpoons$	(12)
$(e_1, -e_3, e_2)$	$\rightleftharpoons$ ... $\rightleftharpoons$	(1423)

Table 3.5. 3-strings-with-(mod 2)-quipu representatives of elements in Σ_4, part 3.

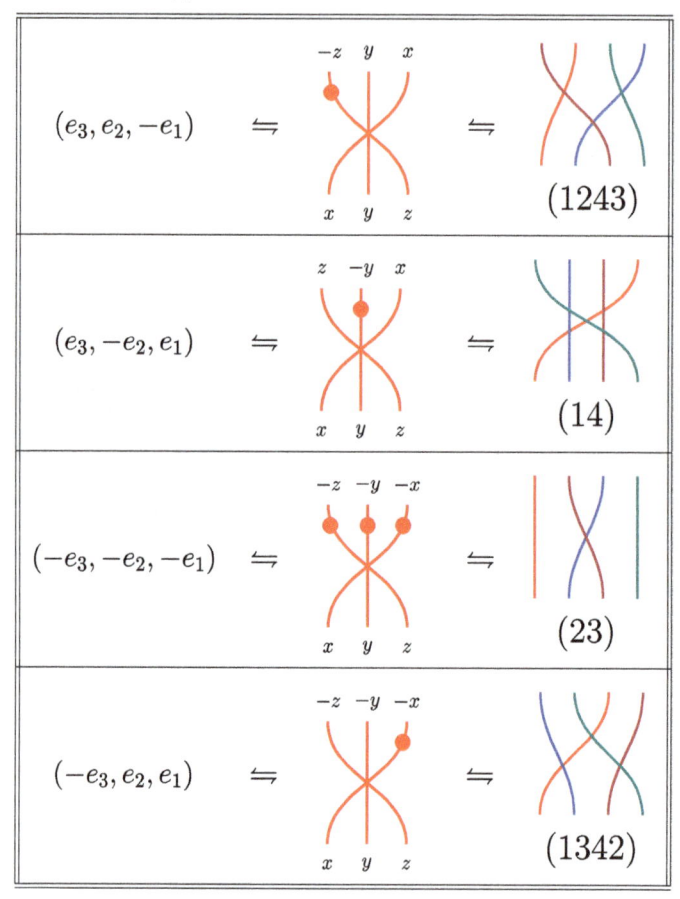

$(e_3, e_2, -e_1)$	⇋	[diagram: $-z$ y x / x y z]	⇋ [diagram] (1243)
$(e_3, -e_2, e_1)$	⇋	[diagram: z $-y$ x / x y z]	⇋ [diagram] (14)
$(-e_3, -e_2, -e_1)$	⇋	[diagram: $-z$ $-y$ $-x$ / x y z]	⇋ [diagram] (23)
$(-e_3, e_2, e_1)$	⇋	[diagram: $-z$ $-y$ $-x$ / x y z]	⇋ [diagram] (1342)

3.4.3. Subgroups

The permutation group $\Sigma_3(\hat{4}) \subset \Sigma_4$ is defined to be those permutations that fix the element 4 among $\{1, 2, 3, 4\}$. The 3-strings-with-(mod 2)-quipu representatives for these elements are compiled and depicted in Fig. 3.13.

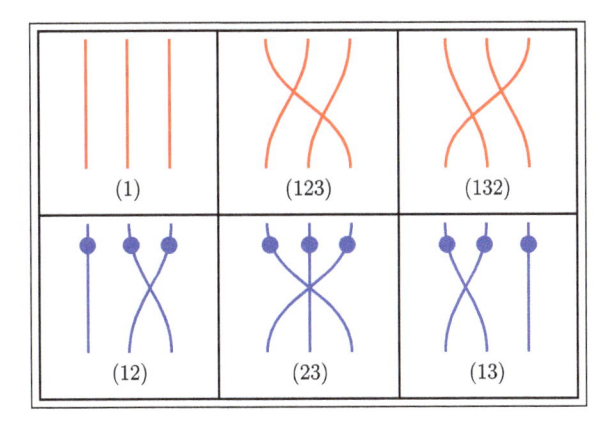

Figure 3.13. The group Σ_3 revisited.

More generally, $\Sigma_3(\hat{\ell})$ for $\ell = 2, 3, 4$ consists of those elements that fix the element ℓ. All of these subgroups can be represented with elements that are 3-strings-with-(mod 2)-quipu. For example, the 3-strings-with-(mod 2)-quipu that are compiled in Fig. 3.14 correspond to the elements in $\Sigma_3(\hat{3})$. However, the labels on the elements are as if they are in $\Sigma_3 = \Sigma_3\{1, 2, 3\}$.

We are about to develop an alternate point of view to obtain the diagrams depicted in Fig. 3.14, colored purple, and their labelings. Let us explain.

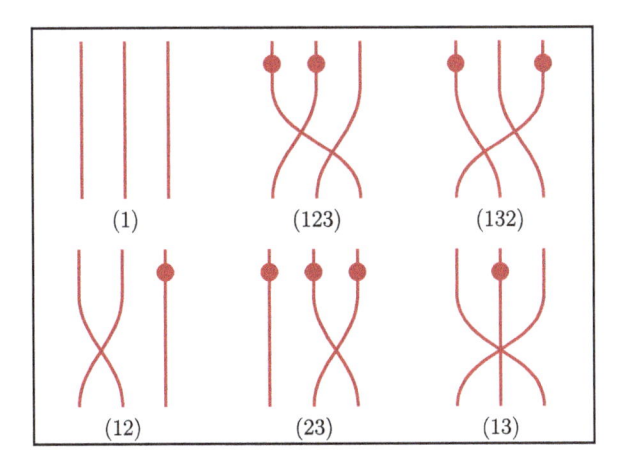

Figure 3.14. Another subgroup isomorphic to Σ_3.

Let $H = \{1, (23)\}$ denote a subgroup of Σ_3 that is isomorphic to $\mathbb{Z}/2$. We order $H = (1, (23))$ and consider H and its ordered cosets, $(12)H = ((12), (123))$ and $(13)H = ((13), (132))$. Label these as $\mathbf{1} = H$, $\mathbf{2} = (12)H$, and $\mathbf{3} = (13)H$. Act upon the cosets by the elements of Σ_3. The equation

$$(12)\mathbf{1} = (12)((1), (23)) = ((12), (123)) = [\mathbf{2}, 0]$$

means that the ordered coset $\mathbf{1}$ is moved to the ordered coset $\mathbf{2}$ and that the order is preserved. Similarly,

$$(12)\mathbf{2} = (12)((12), (123)) = ((1), (23)) = [\mathbf{1}, 0].$$

Meanwhile,

$$(12)\mathbf{3} = (12)((13), (132)) = ((132), (13)) = [\mathbf{3}, 1]$$

indicates that even though the coset $\mathbf{3}$ is mapped to itself, its order has been reversed.

In a similar manner, we compute that

$$(23)\mathbf{1} = [\mathbf{1}, 1]; \quad (23)\mathbf{2} = [\mathbf{3}, 1]; \quad (23)\mathbf{3} = [\mathbf{2}, 1].$$

So, (23) interchanges the second and third cosets, maps the first to itself, but reverses the order of all three of these.

Since Σ_3 is generated by these two elements, $(13) = (12)(23)$ $(12) = (23)(12)(23)$, $(123) = (12)(23)$, and $(132) = (23)(12)$ diagrams can be vertically juxtaposed to determine the remaining elements. Figure 3.15 contains the calculations.

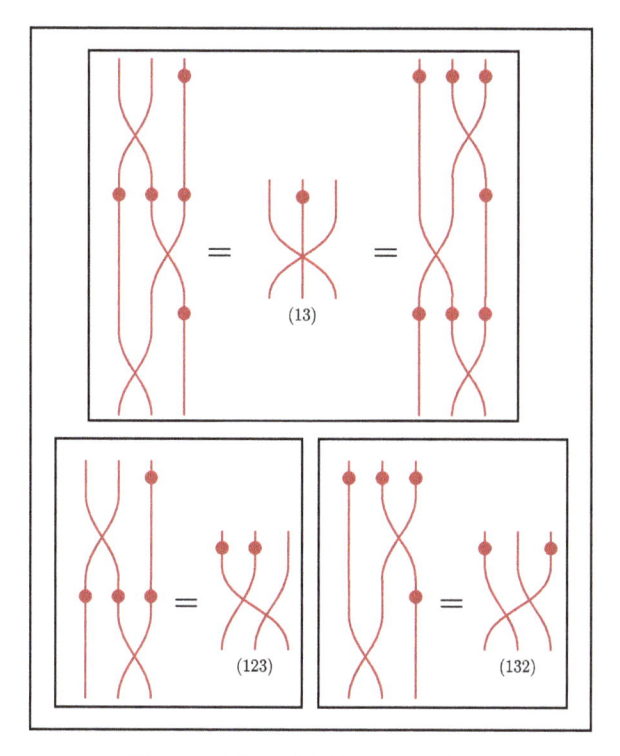

Figure 3.15. Verifying relations.

On the other hand, we compute directly

$$(13)\mathbf{1} = (13)((1),(23)) = ((13)(132)) = [\mathbf{3},0],$$
$$(13)\mathbf{2} = (13)((12),(123)) = ((123)(12)) = [\mathbf{2},1],$$
$$(13)\mathbf{3} = (13)((13),(132)) = ((1)(23)) = [\mathbf{1},0].$$

Furthermore,

$$(123)\mathbf{1} = (123)((1),(23)) = ((123)(12)) = [\mathbf{2},1],$$
$$(123)\mathbf{2} = (123)((12),(123)) = ((13)(132)) = [\mathbf{3},0],$$
$$(123)\mathbf{3} = (123)((13),(132)) = ((23)(1)) = [\mathbf{1},1].$$

Finally,

$$(132)\mathbf{1} = (132)((1),(23)) = ((132)(13))[\mathbf{3},1],$$
$$(132)\mathbf{2} = (132)((12),(123)) = ((23)(1)) = [\mathbf{1},1],$$
$$(132)\mathbf{3} = (132)((13),(132)) = ((12)(123)) = [\mathbf{2},0].$$

The details that were provided here might seem extreme. The same calculation could be performed in fewer steps, or we could have just accepted the results of the diagrammatic calculations. On the other hand, we are novices, and we expect that you are as well. So for all of our benefits, we don't mind providing more details — at least this early in the discussion.

Let us observe from Table 3.3 that the diagram that we chose to represent the transposition (12) corresponds to the permutation $(24) \in \Sigma_4$. Moreover, from Table 3.4, the diagram for (23) represents $(12) \in \Sigma_4$. Figure 3.16 demonstrates that the compositions of permutations in either group Σ_3 or Σ_4 agree.

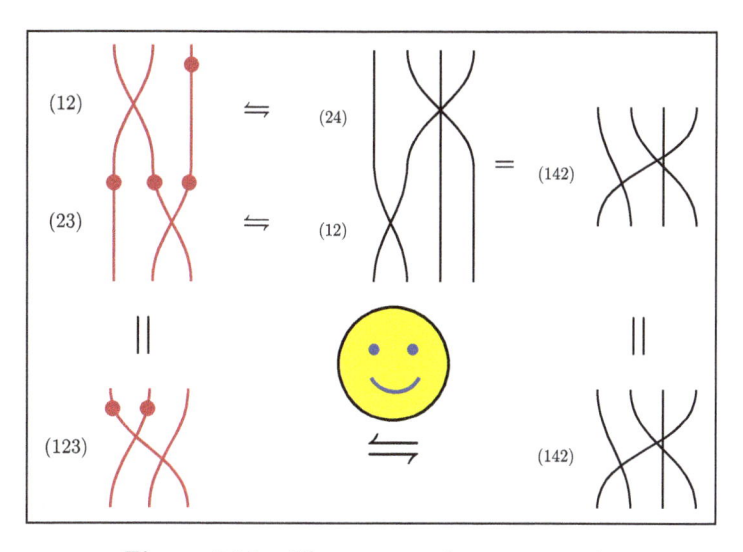

Figure 3.16. The correspondences commute.

In similar fashions, $(23)(12) = (132) \in \Sigma_3$ and the corresponding elements of the diagrams are (12), (24), and $(124) \in \Sigma_4$. We noted above that the permutations Σ_4 which correspond to the diagrams depicted in Fig. 3.14 are the elements of the subgroup $\Sigma_3(\hat{3})$ — those permutations in Σ_4 which fix 3. Figure 3.16 is one of a few calculations that demonstrates an isomorphism between the groups $\Sigma_4(\hat{\ell})$ and Σ_3.

110   Quipu: Decorated Permutation Representations of Finite Groups

Looking back at a prior topic, consider the following computation:

$$\begin{bmatrix} -a \\ -b \\ -c \\ -d \end{bmatrix} = \begin{bmatrix} 1 & -1 & 1 \\ 1 & 1 & -1 \\ -1 & 1 & 1 \\ -1 & -1 & -1 \end{bmatrix} \overset{-z \; y \; x}{\underset{x \; y \; z}{\diagup\!\!\!\diagdown}} \begin{bmatrix} -1 & -1 & 1 \\ 1 & 1 & 1 \\ -1 & 1 & -1 \\ 1 & -1 & -1 \end{bmatrix} = \begin{bmatrix} b \\ d \\ a \\ c \end{bmatrix}.$$

$$\xrightarrow{\quad\quad\quad\quad} $$
$$(ABDC)$$

It is a strange corruption of the typical uses of rows and columns in a matrix product. The vector $-a = [1, -1, 1]^t$ should be written as a column vector. In standard notation, the calculation indicates

$$\begin{bmatrix} 0 & 0 & -1 \\ 0 & 1 & 0 \\ 1 & 0 & 0 \end{bmatrix} \cdot \begin{bmatrix} 1 & 1 & -1 & -1 \\ -1 & 1 & 1 & -1 \\ 1 & -1 & 1 & -1 \end{bmatrix} = \begin{bmatrix} -1 & 1 & -1 & 1 \\ -1 & 1 & 1 & -1 \\ 1 & 1 & -1 & -1 \end{bmatrix}.$$

The columns of the (3×4) source matrix (the factor on the right) correspond in order to the vectors $-a, -b, -c, -d$, and the columns of the result of the calculation correspond, respectively, to b, d, a, c. In the standard notation, the (3×3) signed permutation, that is the factor on the left, transposes the rows and changes the signs of some (well only one in this case) of the rows.

In the diagrammatic setting, the vectors $-a$ through $-d$ are arranged as rows. These are fed from the bottom through the strings-with-quipu as if they were inserted like a can of tennis balls fed through a system of pneumatic tubes. In this way, the columns are permuted and signed.

A lot of mental transformations among rows, columns, and diagrams are needed. Some are downright confusing.

Theorem 3. *The group* $(\mathbb{Z}/2)^3 \rtimes \Sigma_3$ *that is described by the collection of its elements in Figs. 3.10 and 3.11 is isomorphic to* $\mathbb{Z}/2 \times \Sigma_4$.

Proof. Let H denote the subgroup

$$H = \begin{bmatrix} \begin{array}{c}\text{[three red vertical strings]}\end{array} , \begin{array}{c}\text{[three blue strings with quipu on top]}\end{array} \end{bmatrix}$$

$$= [(e_1, e_2, e_3), (-e_1, -e_2, -e_3)] \ .$$

Recall that the elements that are depicted in red among the elements in Figs. 3.10 and 3.11 form a subgroup that is isomorphic to Σ_4. It is not difficult to determine that $h\sigma = \sigma h$ for all $h \in H$ and for all $\sigma \in \Sigma_4$. The elements in blue in those figures each can be written uniquely as a product $(-1)\sigma$, where $-1 = (-e_1, -e_2, -e_3)$ and $\sigma \in \Sigma_4$ represents a permutation. Look at the location of the quipu at the top of the strings for the items in blue, and choose σ to have quipu at the complementary locations.

As in Theorem 2, let $x = h\sigma$ denote an element in $(\mathbb{Z}/2)^3 \rtimes \Sigma_3$. By a slight abuse of notation, we can write $x = (\pm)\sigma$. As before, this expression is unique. Define the isomorphism

$$\mathbb{Z}/2 \times \Sigma_4 \xleftarrow{\phi} (\mathbb{Z}/2)^3 \rtimes \Sigma_3$$

by $\phi((\pm)\sigma) = (\pm 1, \sigma)$. There is a sufficient amount of information to determine that ϕ is surjective, injective, and a homomorphism. This completes the proof. $\square$

Let us briefly contemplate $(\mathbb{Z}/2)^n \rtimes \Sigma_n$ for the smaller values of n. These are $n = 1, 2$. At $n = 1$, there is only the identity function that is a bijection from $[1]$ to itself. So, $(\mathbb{Z}/2)^1 \rtimes \Sigma_1$ is $\mathbb{Z}/2$ and represented in strings-with-quipu diagrams as

$$T = \{\ |\ ,\ \dot{|}\ \}.$$

In case $n = 2$, $(\mathbb{Z}/2)^2 \rtimes \Sigma_2$ is represented as the 8-element dihedral group

$$D = \{\,|\;|\,,\bowtie,\,|\;\bowtie,\,|\bowtie\,|,\,|\bowtie\,|,\,|\bowtie\,|,\,|\bowtie\,\},$$

as described in Section 3.2. In both cases, the groups can be understood as the set of symmetries of the n-cube

$$C_n = \{(x_1, x_2, \ldots, x_n) \in \mathbb{R}^n : |x_\ell| = 1, \text{ for all } \ell = 1, 2, \ldots, n\}.$$

When $n = 1$, C_1 is an interval of length 2. When $n = 2$, C_2 is a square of area 4.

For larger n, the word "cube" is misleading; C_n is the nth power of the interval $[-1, 1]$. It has hypervolume 2^n. Often, it is called a "hypercube," but sometimes that word is used to denote the 4-dimensional cube, or tesseract, which we describe in Section 3.4.4. In English, a cube is a solid object like a brick or a block of ice. We don't have an adequate word for the n-dimensional cube C_n.

The group of signed permutations $\mathcal{H}_n = (\mathbb{Z}/2)^n \rtimes \Sigma_n$ is sometimes called the *hyperoctahedral group*. Where the geometric figure that is called the "hyperoctahedron" is the convex hull of the union of coordinate segments

$$\cup_{\ell=1}^{n}[-e_\ell, e_\ell].$$

That figure has 2^n $(n-1)$-dimensional facets that each resembles Δ_{n-1}. It is dual to the n-cube in the same way that the octahedron is dual to the 3-dimensional cube. The signed permutation group $\mathcal{H}_n$, then, is the group of symmetries of the n-dimensional cube C_n.

For n large, the signed permutation group $\mathcal{H}_n$ is huge. It has $2^n \cdot n!$ elements. Even in the case $n = 4$, we might want to cut the group down by a bit. First, we can consider the alternating signed permutations $(\mathbb{Z}/2)^4 \rtimes A_4$. Since A_4 only has 12 elements, that doesn't seem so bad. Next, we can restrict to signed alternating permutations where the matrix representations have positive determinant. The resulting group has $8 \times 12 = 96$ elements.

That group is not as large as one other that we will consider, but for now it seems daunting to explore it. On the other hand, consider the alternating elements that are depicted in Fig. 3.7. A full table 4-strings-with-quipu that represent the alternating, determinant 1, signed permutations would consist of eight copies of this table in which there are an even number of (mod 2)-quipu upon the strings. There may be a clever arrangement of these 96 elements in a geometric configuration, or there may be an alternative set of quipu that can be used to describe the group. A group of size less than 100 can be explored manually and with a great deal of patience!

3.4.4. The 4-dimensional cube

This section is a short digression from the algebraic realm to the geometric one. The 4-dimensional cube is the set

$$C_4 = \{(x_1, x_2, x_3, x_4) \in \mathbb{R}^4 : |x_\ell| \leq 1\}.$$

That set will also be shifted and shrunk so that its coordinates are all 0s or 1s. That shift occurs so that the labels of the vertices are labeled by juxtaposed expressions such as 0101 with commas and parenthesis omitted so that they fit nicely on the page.

Later, in Section 4.2.2, similar point sets that live upon the unit 3-dimensional sphere $S^3 \subset \mathbb{R}^4$ will be studied and discussed.

There are amusing animations that are available upon various parts of the internet of the hypercube C_4 as it rotates in 4-space. But this was not always the case. Among the first computer graphics of a hypercube was one from the middle to late 1970s at Tom Banchoff's lab at Brown University. The machine that rendered an oscilloscopic view of this object cost close to $200,000 at the time. In the mid-18th century, d'Alembert first spoke of higher dimensions [17], and in an 1884 essay, Charles Hinton [6] began the popularization of the subject.

Many ideas about 4-dimensional space were used and abused by spiritualists who were contemporaries of Hinton or some who followed him. Some of the more metaphysical aspects attributed to higher-dimensional objects persist in more modern writings. For example, in Lewis's *Perelandra* [16], the author creates a mythos in which the gods/planets Mars and Venus are multi-dimensional creatures and he ascribes similar characteristics to angels. In his system (which seems to be borrowed from Hinton), spiritual entities are not etherial so much as they are higher dimensional. Compare these ideas with those in Abbott's *Flatland*. See also Dalí's *Corpus Hypercubus*.

The first view of the hypercube that we present (Fig. 3.17) is derived from the projection of 4-space into the plane as given by the matrix

$$P = \begin{bmatrix} 0 & 2 & 3 & 2 \\ 3 & 2 & 0 & -2 \end{bmatrix}.$$

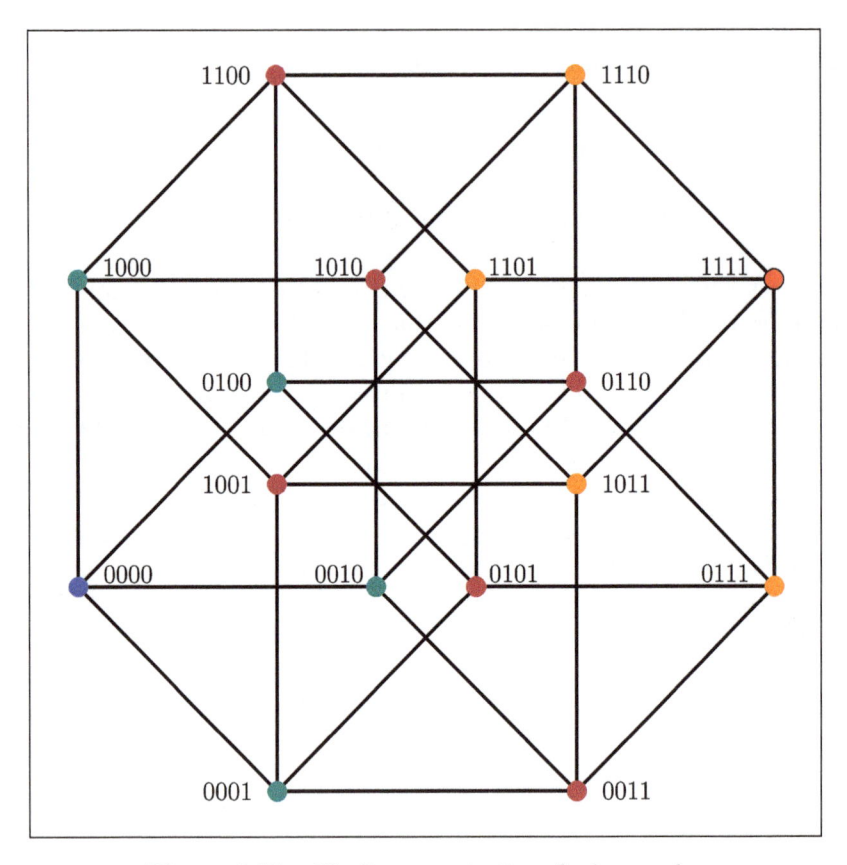

Figure 3.17. The linear projection of a hypercube.

Since $2\sqrt{2} \approx 2.8$ is close to 3, the outline of the figure appears to be a regular hexagon, but it is not. In general, when one is given a piece of graph paper in which there are equally spaced vertical and horizontal rulings, it is not too difficult to sketch an analogous figure.

The illustration in Fig. 3.18 is sometimes called a perspective projection. It is analogous to looking straight and downward at an open box. Yet, a perspective projection should be a 3-dimensional figure, and the pages of this book are best modeled as 2-dimensional rectangles.

In both illustrations, the vertices have been colored. In Fig. 3.17, the vertices that have the same number of 0s are colored similarly. The blue vertex is all 0s. The teal vertices have three 0s in their coordinates. The purple vertices have two 0s. The oranges have one 0, and the red vertex has none.

Were the hypercube to be dipped through space, vertex first, the cross-sections that contain these vertices would be vertex, tetrahedron, octahedron, tetrahedron, vertex. The pattern is $1, 4, 6, 4, 1$ which, we hope, is known to the reader as a row of combinatorial coefficients from what is known as Pascal's triangle.

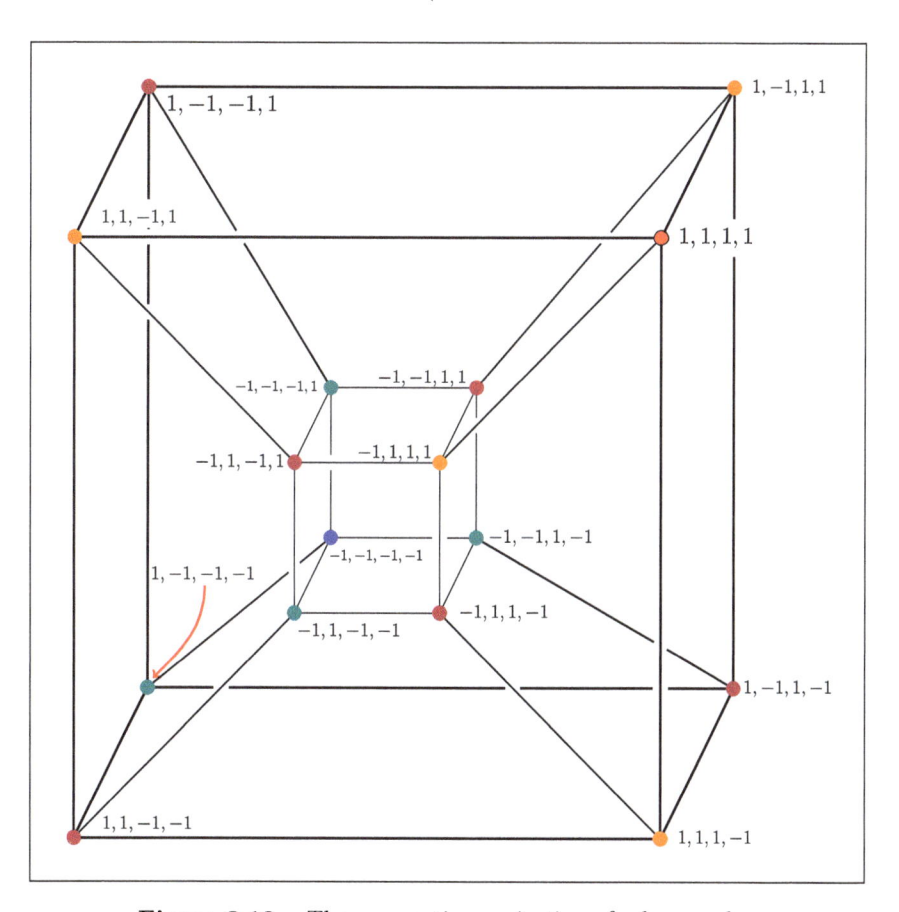

Figure 3.18. The perspective projection of a hypercube.

The group of symmetries of the hypercube is the group $(\mathbb{Z}/2)^4 \rtimes \Sigma_4$. It has $16 \times 24 = 384$ elements.

Figure 3.18 will be modified later (Section 4.2.2) so that its vertices correspond to points that are of length 1 as defined in the following paragraph, and it is a schematic of a stereographic projection of the 3-dimensional sphere. The vertices of Fig. 3.18 vertices are colored using a scheme similar to that above, but the number of (-1)s, rather than the number of 0s, controls the colors.

The length of a point $p = (w, x, y, z)$ in 4-dimensional space is

$$|p| = \sqrt{w^2 + x^2 + y^2 + z^2}.$$

So, any of the 16 points $(\pm 1, \pm 1, \pm 1, \pm 1)/2$ have length

$$\sqrt{1/4 + 1/4 + 1/4 + 1/4} = 1.$$

The points

$$a = (1, 1, 1, 1)/2, \quad b = (1, 1, 1, -1)/2,$$
$$c = (1, 1, -1, -1)/2, \quad d = (1, 1, -1, 1)/2$$

form the vertices of the large square face that is most visible in the analogous picture. They will be important for studies of some of the subgroups of the 3-dimensional sphere.[5]

These figures are introduced here in case a diligent reader wants to explore the symmetries by means of permuting the coordinates and changing their signs.

---

[5]In the text above this paragraph, the labels $a, b, c,$ and d were vertices of a 3-dimensional figure. Those labels have no relationship to the current labels. Henceforth, we will reserve the letters a–d to denote these points in the 3-dimensional sphere.

3.5. Semi-direct products

In this section, we recall the definition of a semi-direct product of groups and demonstrate that strings-with-quipu can be used to represent such groups.

Let G and H denote finite groups where both are given as subgroups of permutation groups, say $G \subset \Sigma_m$ and $H \subset \Sigma_n$. For example, if $h \in H$, then the map $f_h : k \mapsto hk$ represents H as a subgroup of the set of permutations of the underlying set of H. Consider the *semi-direct product* $W = G^n \rtimes H$ in which H acts on the factors of G^n by permuting the coordinates. The group W fits into a split seseq

$$1 \longleftarrow H \xleftarrow{p} W \xleftarrow{i} G^n \longleftarrow 1.$$

To say the sequence is split is to assert that there is a group homomorphism $H \xrightarrow{s} W$ such that $ps = \mathrm{Id}_H$. The group W can be written as $G^n H$, or equivalently, for every element $w \in W$, there are unique elements $g_1, \ldots, g_n \in G$ and an element $h \in H$ such that $w = (g_1, \ldots, g_n)h$. The subgroup G^n is normal and $G^n \cap H = \{0\}$.

For convenience of computation, we express the permutation group Σ_n of which H is a subgroup as the set of permutation matrices. As before, let

$$e_j = [0, \ldots, 0, \underbrace{1}_{j\text{th}}, 0 \ldots, 0]^t$$

denote the jth standard unit vector in $\mathbb{R}^n$. Then a permutation $\sigma \in \Sigma_n$ is given as the matrix $[e_{\sigma 1}, e_{\sigma 2}, \ldots, e_{\sigma n}]$. In our experience, there is often some miscommunication in this formula even when we have presented it earlier. So, for example, the 3-cycle $(123) \in \Sigma_3$ corresponds to the matrix

$$(123) \leftrightarrow \begin{bmatrix} 0 & 0 & 1 \\ 1 & 0 & 0 \\ 0 & 1 & 0 \end{bmatrix} = (e_2, e_3, e_1) = (e_{\sigma 1}, e_{\sigma 2}, e_{\sigma 3}),$$

where $\sigma = (123)$. Let $g_1, g_2, \ldots, g_n \in G$ denote elements in G. Let 1 denote the identity element of G. Then an element of $(\vec{g}, \sigma) \in W$ is expressed so that the entry g_i appears in the ith row, for example,

$$\begin{bmatrix} g_1 & 0 & 0 \\ 0 & g_2 & 0 \\ 0 & 0 & g_3 \end{bmatrix} \cdot \begin{bmatrix} 0 & 0 & 1 \\ 1 & 0 & 0 \\ 0 & 1 & 0 \end{bmatrix} = \begin{bmatrix} 0 & 0 & g_1 \\ g_2 & 0 & 0 \\ 0 & g_3 & 0 \end{bmatrix}.$$

Thus, if $\Delta(\vec{g})$ denotes the matrix in which g_1-g_n appear along the diagonal and 0s appear elsewhere, then the element $(\vec{g}, \sigma) \in W$ is the matrix product $\Delta(\vec{g}) \cdot (e_{\sigma 1}, e_{\sigma 2}, \ldots, e_{\sigma n})$. When two such matrix representatives are multiplied, only one non-zero entry appears in any row and column as a product of pairs of elements in G. In this way, computations in the *semi-direct product of G^n and H* are computed as formal matrix products.

We have been told that this representation of semi-direct products is "well known." We also recall when we were told this. At that time, we didn't know about such a representation. A well-known mathematical fact is a bit oxymoronic, isn't it? Even if there are 50,000 mathematicians in the world, a well-known mathematical fact is not likely to be known by a much more broad audience. Even if the readership of this book is on the order of $10,000$ (which would be huge for a mathematics book), this well-known fact will be obscure. Well, now you, at least, know it.

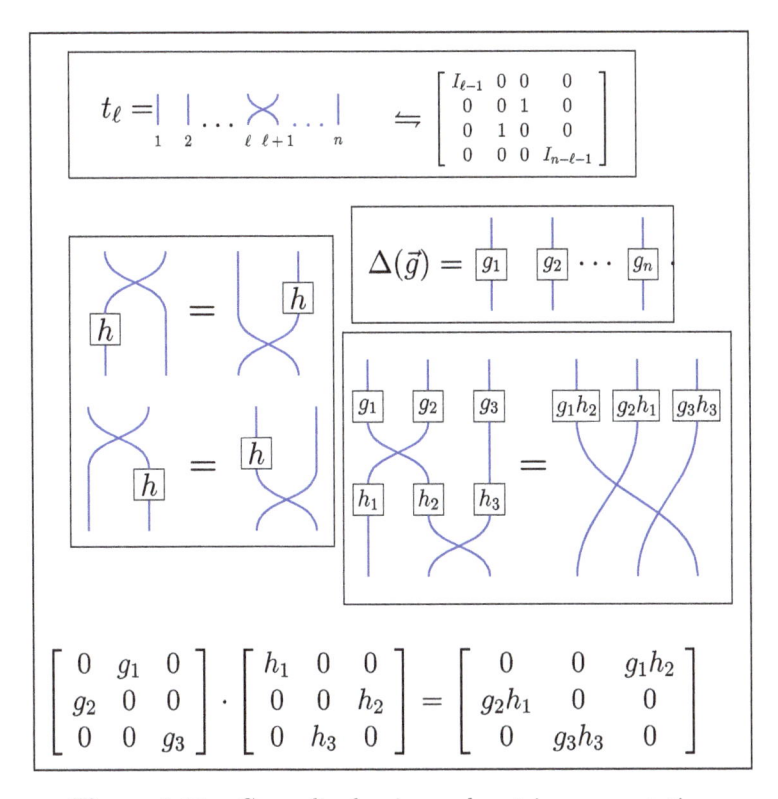

Figure 3.19. Generalized quipu and matrix representation.

The semi-direct product construction of $G^n \rtimes H$, where $H \subset \Sigma_n$ is thought of as a permutation, can be realized by n-strings-diagrams-with-G-valued quipu. The standard generators $t_\ell = (\ell, \ell+1)$ of the symmetric group Σ_n is drawn as a string diagram as in Fig. 3.19.

In this illustration, the element $\Delta(\vec{g})$ is also depicted, and a sample product is shown as a string product in matrix form.

Elements in the semi-direct product are drawn as generalized quipu that are labeled by elements in G. If G also has a permutation representation (as will often be the case here), then the permutation representation of the semi-direct product is obtained by bundling strings together and implementing the representatives of the elements $g_1, \ldots, g_n$ at the top of the diagrams. This string replication is exactly what happened when the quipu were (mod n)-quipu.

More generally, we will construct the elements of G iteratively. The group G is likely to be fairly small with a large cyclic subgroup. For example, in Section 2.3.3 of Chapter 2, elements in the dihedral group D_{2n} were represented by 2-strings pictures with (mod n)-quipu. If the dihedral group is a subgroup of a larger group that interests us, we can use these 2-strings pictures to provide dihedral quipu.

In the human endeavor of spinning fibers to make thread, long fibers are combined and overlapped until isolated long filaments appear. Of course, each filament is manufactured as a conglomerate, and the conception of this conglomerate as a single entity is only metaphoric. Imagine, for example, a sturdy nautical rope. It is thought of as filaments wound around a central core to form a bundle of filaments that are twisted further. In much the same way, we can see the groups that we present here as a bundle of permutations. Unfortunately, words such as bundle, grouping, etc. have technical meanings in mathematics, and within this paragraph, we mean none of these things.

Precisely, we partition groups into sets of cosets and study the group actions on its coset space. In the finite group case, subgroups can be thought of as permutations. These are spun together to make stronger ropes.

3.6. Proof of Theorem 1

To begin, we recall.

Theorem 1. *Let G denote a finite group of order nk. Let H denote a subgroup of order k. Then there is an inclusion $G \subset H^n \rtimes \Sigma_n$, where the second factor permutes the coordinates of H^n.*

Let us also recall the notation and exemplify it with one of our previous examples. The group G is of order $|G| = nk$. For example, the order of the dihedral group D_8 is $2 \cdot 4$. The subgroup H has order $|H| = k$. In the example, let $H = \langle x \rangle$ denote the cyclic subgroup of order 4 that is generated by the 1/4-rotation x. The conclusion of the theorem in this case is that D_8 is isomorphic to a subgroup of the semi-direct product $(\mathbb{Z}/4)^2 \rtimes \Sigma_2$. The illustration in Fig. 2.35 indicates this representation.

The general proof will be a consequence of studying the permutation action of G upon itself by left multiplication: for $g' \in G$, let $g : g' \mapsto gg'$. Denote the elements of H by $h_1, h_2, \ldots, h_k$. In fact, let us observe that this is an arbitrary ordering of H. We write $(H) = (h_1, h_2, \ldots, h_k)$ to indicate H as an ordered set. Choose cosets $a_1 H, \ldots, a_n H$, and observe that this, too, indicates an ordering so that as an ordered set $(G) = (a_1 H, \ldots, a_n H)$. It is safe to assume that $a_1 = e$ is the identity of G, and therefore, $a_1 H = H$.

The action of G upon itself leads to a permutation action of g upon the set of cosets.

In the example of $G = D_8$, order H as $(1, x, x^2, x^3)$. There is only one non-trivial coset rH that is ordered as $rH = (r, rx, rx^2, rx^3)$. The group G is written as $(G) = ((1, x, x^2, x^3), (r, rx, rx^2, rx^3))$.

So, for $g \in G$, there is a permutation $\eta_g = \eta \in \Sigma_n$ such that $g(a_1 H, \ldots, a_n H) = (a_{\eta 1} H, a_{\eta 2} H, \ldots, a_{\eta n} H)$. Let us write $ga_i H = a_j H$ to indicate that $\eta i = j$. In the notation of this paragraph alone, the equation $ga_i H = a_j H$ indicates an equality of cosets. So, we really only know that $ga_i \in a_j H$, but we don't know which element

it may be, for example,
$$x(1, x, x^2, x^3) = (x, x^2, x^3, 1),$$
while
$$xr(1, x, x^2, x^3) = rx^3(1, x, x^2, x^3) = r(x^3, 1, x, x^2).$$
Therefore, there is a permutation $\sigma_{i,j} = \sigma \in \Sigma_k$ such that $(ga_ih_1, \ldots, ga_ih_k) = (a_jh_{\sigma 1}, \ldots, a_jh_{\sigma k})$. In the example, the two permutations are both 4-cycles.

Compare the current situation to that of Section 3.5, in particular Fig. 3.19. The element $g \in G$ under consideration permutes the cosets among each other and meanwhile acts upon each coset in a quipu-like fashion.

That is to say that there is an injective homomorphism[6] $G \subset (\Sigma_k)^n \rtimes \Sigma_n$. In the example, $D_8 \subset (\Sigma_4)^2 \rtimes \Sigma_2$.

To complete the proof, we establish that the permutation representation of g acting upon any coset corresponds to the action of H upon itself.

Since $ga_iH = a_jH$, we have that there is an $h_{i,j} \in H$ with $h_{i,j} = a_j^{-1}ga_i$. So, $h_{i,j}h_\ell = a_j^{-1}ga_ih_\ell = a_j^{-1}a_jh_{\sigma\ell} = h_{\sigma_{i,j}\ell}$. In other words, the permutation action between ga_iH and a_jH coincides with multiplication by some element $h_{i,j} \in H$. Therefore, $G \subset (H)^n \rtimes \Sigma_n$ as desired. This completes the proof. $\square$

3.6.1. An alternative view of the dihedral group D_8

In this section, Theorem 1 is exemplified and explicated by means of a representation of the dihedral group D_8 that uses elements of the Klein 4-group as quipu. In this way, a (2×2)-matrix representation of $\left[(\mathbb{Z}/2)^2\right]^2 \rtimes \mathbb{Z}/2$ will be given that encompasses the generalized quipu.

Recall that the 2-strings-with-(mod 2)-quipu illustration of the Klein 4-group is indicated as in Fig. 3.20.

---

[6]The text has merely shown that there is an embedding. The reader is asked to show that it is a group homomorphism.

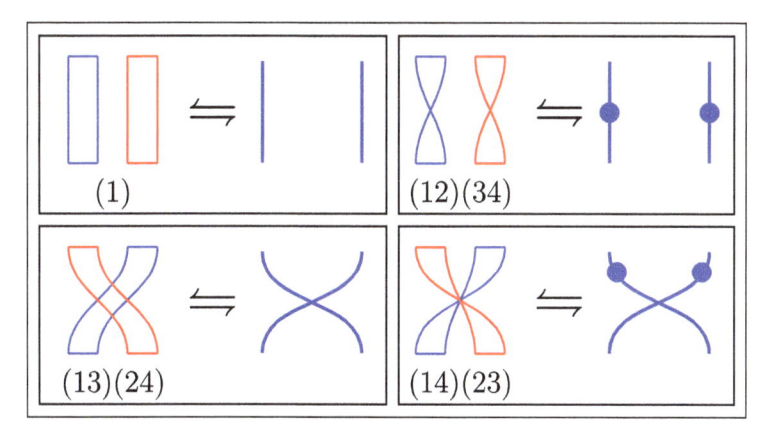

Figure 3.20. The Klein 4-group, K_4.

These strings-with-quipu diagrams can also be obtained by considering the group K_4 in terms of the ordered cosets, $T = ((1), (12)(34))$ and $(13)(24)T = ((13)(24), (14)(23))$, and following the proof of Theorem 1. We quickly compute the actions of $(12)(34)$, $(13)(24)$, and $(14)(23)$ upon this ordered pair of cosets. First, write $\mathbf{1} = T$, and $\mathbf{2} = (13)(24)T$.

Then

$$(12)(34)\mathbf{1} = ((12)(34), (1)) = [\mathbf{1}, 1],$$
$$(12)(34)\mathbf{2} = ((14)(23), (13)(24)) = [\mathbf{2}, 1],$$

$$(13)(24)\mathbf{1} = ((13)(24), (14)(23)) = [\mathbf{2}, 0],$$
$$(13)(24)\mathbf{2} = ((1), (12)(34)) = [\mathbf{1}, 0],$$

$$(14)(23)\mathbf{1} = ((14)(23), (12)(24)) = [\mathbf{2}, 1],$$
$$(14)(23)\mathbf{2} = ((12)(34), (12)(34)) = [\mathbf{1}, 1].$$

Thus, the element $(12)(34)$ transposes the elements in the cosets, $(13)(24)$ transposes the cosets but leaves each coset in order, and $(14)(23)$ transposes the cosets and the orders of the elements.

Recall further that the dihedral group D_8 is given by the presentation

$$D_8 = \langle x, r : x^4 = r^2 = 1, x^{-1}r = rx \rangle.$$

There is a subgroup that is isomorphic to K_4. This will be ordered and represented as

$$K = ((1, x^2), (r, rx^2)).$$

By also using the coset

$$xK = ((x, x^3), (rx^3, rx)),$$

the dihedral group is written as $D_8 = K \cup xK$. Throughout this section, the relationships $x^3 = x^{-1}$, $xr = rx^3$, and $x^3r = rx$ will be used without any further fanfare. Please observe that

$$x^2K = ((x^2, 1), (rx^2, r)) = \boxed{}K.$$

We also have

$$rK = ((r, rx^2), (1, x^2)) = \boxtimes K,$$

and

$$rxK = ((rx, rx^3), (x^3, x)) = \boxtimes xK.$$

In this way, we obtain the 2-strings-with-K_4-quipu representations of x and r that are depicted in Fig. 3.21. By the representations of these elements or by directly computing the actions $x^\ell K$ and $rx^\ell K$ upon the ordered and partitioned cosets, the remaining elements in the tables are determined.

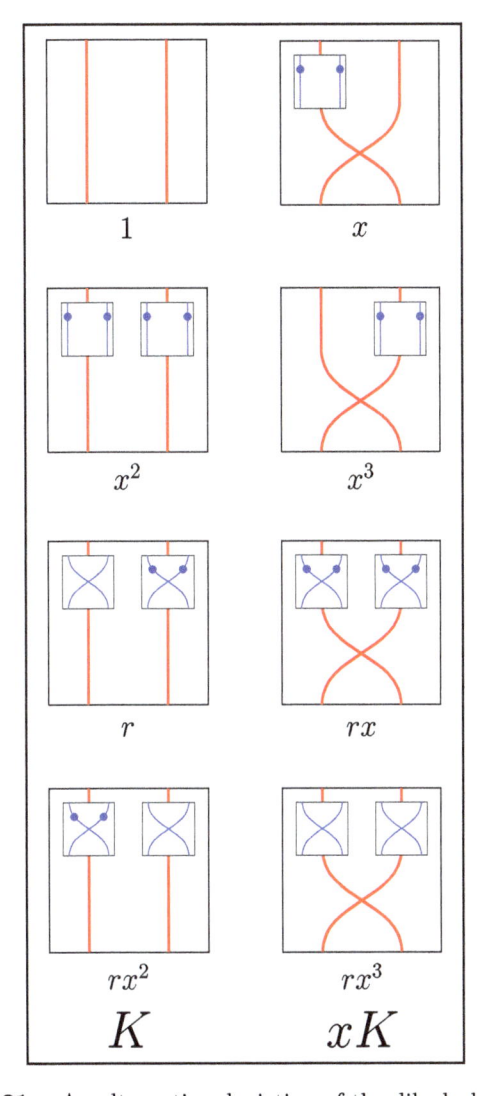

Figure 3.21. An alternative depiction of the dihedral groups D_8.

These string diagrams can be written in matrix form where the non-zero entries are elements of the Klein 4-group K_4. Since K_4 is isomorphic to $\mathbb{Z}/2 \times \mathbb{Z}/2$ and since the operation upon matrices is multiplicative, rather than additive, a correspondence between diagrams and vectors in $K = \{(\pm 1, \pm 1)\}$ will be established.

$$\boxed{\cdot\cdot} \leftrightharpoons (1,-1); \quad \boxtimes \leftrightharpoons (-1,1); \quad \boxtimes \leftrightharpoons (-1,-1).$$

These are multiplied coordinate-wise:

$$(a,b) \cdot (c,d) = (ac, bd).$$

Then

$$x \;\leftrightharpoons\; \begin{bmatrix} 0 & (1,-1) \\ 1 & 0 \end{bmatrix},$$

$$r \;\leftrightharpoons\; \begin{bmatrix} (-1,1) & 0 \\ 0 & (-1,-1) \end{bmatrix}.$$

The remaining six matrices that represent elements in the group D_8 can be obtained by performing the matrix multiplications, for example,

$$rx = \begin{bmatrix} (-1,1) & 0 \\ 0 & (-1,-1) \end{bmatrix} \begin{bmatrix} 0 & (1,-1) \\ 1 & 0 \end{bmatrix} = \begin{bmatrix} 0 & (-1,-1) \\ (-1,-1) & 0 \end{bmatrix}.$$

It is worthwhile to compare these matrix representations to those that were obtained in Section 3.2. In particular, there is a fun method to map the matrices with $(\mathbb{Z}/2 \times \mathbb{Z}/2)$ entries to the signed permutation matrices, namely, send (ϵ, δ) to $\epsilon\delta$, where $\epsilon, \delta = \pm 1$. In this way, the quipu $\boxtimes$ is sent to the identity. See Fig. 3.22.

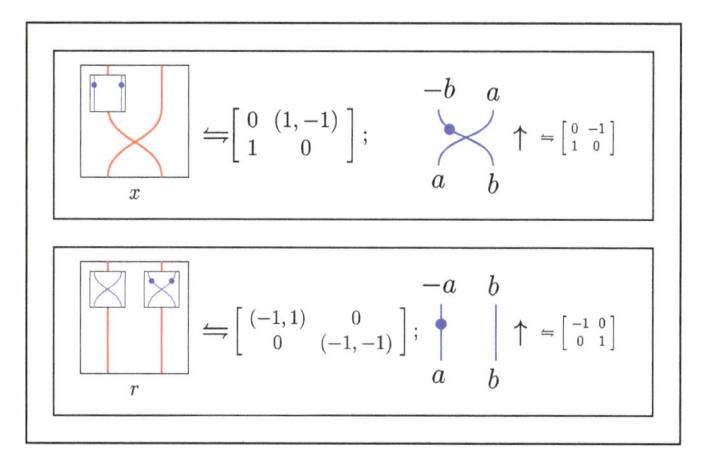

Figure 3.22. Different matrix representations of the elements of D_8.

Similar matrix representatives can be found for the original 2-strings-with-(mod 4)-quipu pictures that were given in Section 2.3.3 of Chapter 2. Since the non-zero entries should be elements of order 4, it makes sense to use complex numbers to represent them. Recall that the so-called imaginary number i has the properties, $i^2 = -1$, $i^3 = -i$, and $i^4 = 1$. Then we have

$$x \leftrightharpoons \;\leftrightharpoons \begin{bmatrix} i & 0 \\ 0 & -i \end{bmatrix},$$

$$r \leftrightharpoons \;\leftrightharpoons \begin{bmatrix} 0 & 1 \\ 1 & 0 \end{bmatrix}.$$

The reader is urged to form the matrix products that correspond to x, x^2, x^3, rx, rx^2, and rx^3 in order to verify that these matrices correspond to the diagrams depicted in Fig. 2.35.

3.6.2. An alternative view of the alternating group A_4

In this section, Theorem 1 is further exemplified and explicated by means of a representation of the alternating group A_4 that uses elements of the Klein 4-group as quipu. In this way, (3×3)-matrix representation of $\left[(\mathbb{Z}/2)^2\right]^3 \rtimes \mathbb{Z}/3$ that is substantially different from that given in Section 3.3.1 is found. Despite its differences, it encompasses the generalized quipu in that the non-zero entries are elements of $K = \{(\pm 1, \pm 1)\}$.

We consider

$$K_4 = ((1), (12)(34), (13)(24), (14)(23))$$

as an ordered subgroup of the alternating group A_4 and then subsequently consider the actions of the elements in K_4 upon these ordered cosets. The cosets in A_4 of K_4 are listed using a naming convention dependent upon the initial representatives:

$$\left.\begin{array}{l} [\mathbf{1}] = [(1), (12)(34), (13)(24), (14)(23)] \\ [\mathbf{123}] = [(123), (134), (243), (142)] \\ [\mathbf{132}] = [(132), (234), (124), (143)] \end{array}\right\} = A_4.$$

Since the inverses of the elements of the coset $[\mathbf{123}]$ are in $[\mathbf{132}]$, any element in $[\mathbf{123}]$ takes $[\mathbf{1}]$ to $[\mathbf{123}]$, $[\mathbf{123}]$ to $[\mathbf{132}]$, and $[\mathbf{132}]$ to $[\mathbf{1}]$. Similarly, the elements in $[\mathbf{1}]$ fix the cosets set-wise because K_4 is normal. Any element in $[\mathbf{132}]$ takes $[\mathbf{1}]$ to $[\mathbf{132}]$, $[\mathbf{123}]$ to $[\mathbf{1}]$, and $[\mathbf{132}]$ to $[\mathbf{123}]$. That is, the elements in $[\mathbf{123}]$ act as the 3-cycle (123) on this ordered set of cosets. The elements in $[\mathbf{132}]$ act as the 3-cycle (132), and the elements in $[\mathbf{1}]$ act as the identity. Yet, within any coset, the action is governed by K_4. This paragraph is easier to understand, as indicated in Fig. 3.23, which also adorns these elements in $\mathbb{Z}/3$ with K_4-quipu. That illustration is the result of a thorough calculation, but some shortcuts may be found.

It is nice to observe some patterns within Fig. 3.23. The columns in the illustration correspond to the cosets. The sequence of 3-quipu from left to right in each element is a cyclic permutation of the representatives for $p = (12)(34)$, $q = (13)(24)$, and $r = (14)(23)$ in that cyclic order — so, (pqr), (qrp), or (rpq). The elements that correspond to the standard representative [**1**], [**123**], and [**132**] have identity quipu. Compare with Theorem 6.

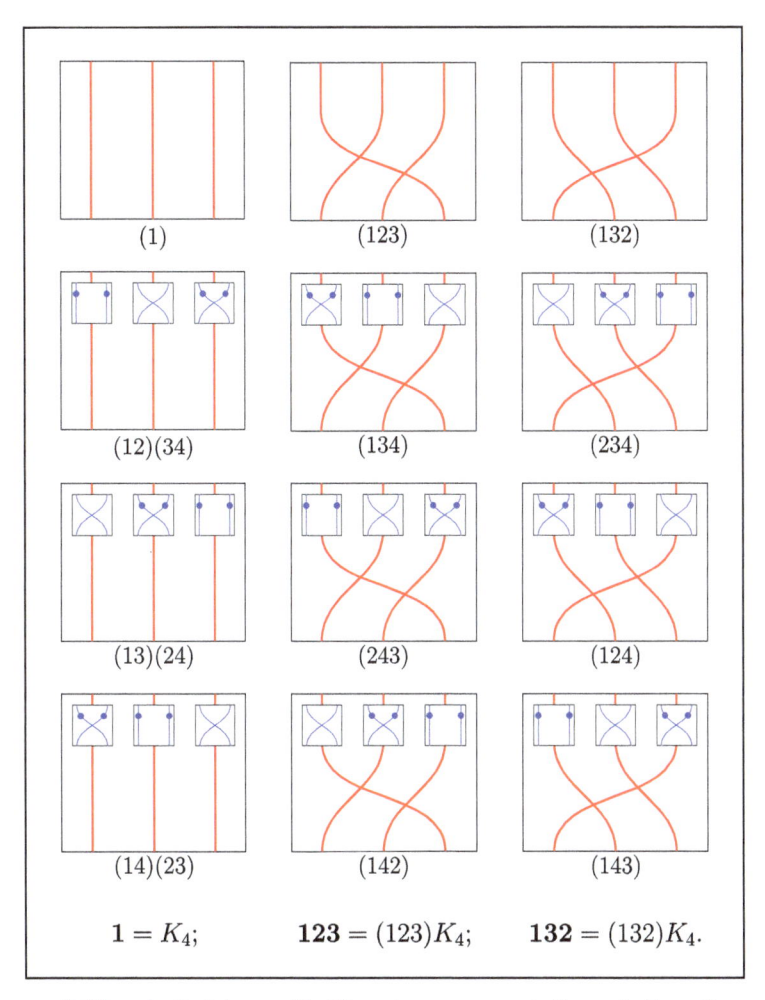

Figure 3.23. A 3-strings-with-K_4-quipu representation of the alternating group A_4.

For the purposes of providing some visual emphasis, the 3-strings-with-K_4-quipu have been shrunk and the 3-strings-with-(mod 2)-quipu have been placed to the right of these in Fig. 3.24. Please note that the crossings in the diagrams for K_4 correspond to the modulo 2 quipu in the previous depictions. In this case, we only need to keep track of the modulo 2 quipu to specify the elements in A_4.

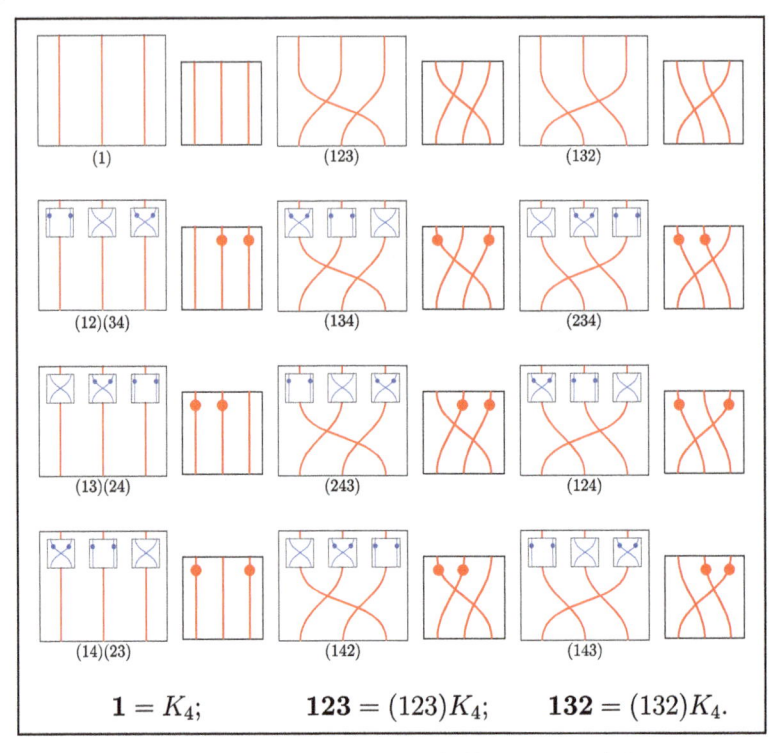

$$1 = K_4; \quad 123 = (123)K_4; \quad 132 = (132)K_4.$$

Figure 3.24. Comparing the 3-strings-with-K_4-quipu and the 3-strings-with-(mod 2)-quipu for the alternating group.

To complete this section, a litany of matrix representatives for the elements in the cosets, [**1**] and [**123**] will be tabulated. There are

Matrix Descriptions

many ways of homomorphically mapping the Klein 4-group K_4 to $\mathbb{Z}/2 \times \mathbb{Z}/2 = \{(\pm 1, \pm 1)\}$. Since the Fig. 3.24 mapped $(14)(23)$ and $(13)(24)$ to a non-trivial (mod 2)-quipu, we will map as indicated: $(12)(34) \mapsto (-1,-1)$, $(13)(24) \mapsto (-1,1)$, and $(14)(24) \mapsto (1,-1)$. In this way, we compare the matrix representations in which the non-zero entries are $(\pm 1, \pm 1)$ to the signed permutation matrices of Tables 3.6 and 3.7.

Table 3.6. Comparing the matrix representations of elements in the alternating group A_4, part 1.

$\begin{bmatrix} 1 & 0 & 0 \\ 0 & 1 & 0 \\ 0 & 0 & 1 \end{bmatrix}$	$\begin{bmatrix} (1,1) & 0 & 0 \\ 0 & (1,1) & 0 \\ 0 & 0 & (1,1) \end{bmatrix}$	(1)
$\begin{bmatrix} 1 & 0 & 0 \\ 0 & -1 & 0 \\ 0 & 0 & -1 \end{bmatrix}$	$\begin{bmatrix} (-1,-1) & 0 & 0 \\ 0 & (-1,1) & 0 \\ 0 & 0 & (1,-1) \end{bmatrix}$	(12)(34)
$\begin{bmatrix} -1 & 0 & 0 \\ 0 & -1 & 0 \\ 0 & 0 & 1 \end{bmatrix}$	$\begin{bmatrix} (-1,1) & 0 & 0 \\ 0 & (1,-1) & 0 \\ 0 & 0 & (-1,-1) \end{bmatrix}$	(13)(24)
$\begin{bmatrix} -1 & 0 & 0 \\ 0 & 1 & 0 \\ 0 & 0 & -1 \end{bmatrix}$	$\begin{bmatrix} (1,-1) & 0 & 0 \\ 0 & (-1,-1) & 0 \\ 0 & 0 & (-1,1) \end{bmatrix}$	(14)(23)

Table 3.7. Comparing the matrix representations of elements in the alternating group A_4, part 2.

$\begin{bmatrix} 0 & 0 & 1 \\ 1 & 0 & 0 \\ 0 & 1 & 0 \end{bmatrix}^{\pm 1}$	$\begin{bmatrix} 0 & 0 & (1,1) \\ (1,1) & 0 & 0 \\ 0 & (1,1) & 0 \end{bmatrix}^{\pm 1}$	± 1 $(123)^{\pm 1}$
$\begin{bmatrix} 0 & 0 & -1 \\ 1 & 0 & 0 \\ 0 & -1 & 0 \end{bmatrix}^{\pm 1}$	$\begin{bmatrix} 0 & 0 & (1,-1) \\ (-1,-1) & 0 & 0 \\ 0 & (-1,1) & 0 \end{bmatrix}^{\pm 1}$	± 1 $(134)^{\pm 1}$
$\begin{bmatrix} 0 & 0 & 1 \\ -1 & 0 & 0 \\ 0 & -1 & 0 \end{bmatrix}^{\pm 1}$	$\begin{bmatrix} 0 & 0 & (-1,-1) \\ (-1,1) & 0 & 0 \\ 0 & (1,-1) & 0 \end{bmatrix}^{\pm 1}$	± 1 $(243)^{\pm 1}$
$\begin{bmatrix} 0 & 0 & -1 \\ -1 & 0 & 0 \\ 0 & 1 & 0 \end{bmatrix}^{\pm 1}$	$\begin{bmatrix} 0 & 0 & (-1,1) \\ (1,-1) & 0 & 0 \\ 0 & (-1,-1) & 0 \end{bmatrix}^{\pm 1}$	± 1 $(142)^{\pm 1}$

There are two small things that are worth mentioning about these matrix representations.

First, instead of using the vectors $\{(\pm 1, \pm 1)\}$ to represent the elements of K_4, it might make more sense to represent these as matrices:

$$\begin{array}{c} \boxed{\vert\vert} \leftrightharpoons \begin{bmatrix} -1 & 0 \\ 0 & -1 \end{bmatrix}, \\[1em] \boxed{\asymp} \leftrightharpoons \begin{bmatrix} 0 & 1 \\ 1 & 0 \end{bmatrix}, \\[1em] \boxed{\asymp} \leftrightharpoons \begin{bmatrix} 0 & -1 \\ -1 & 0 \end{bmatrix}. \end{array}$$

In this way, the resulting (3×3)-matrices become (6×6)-block matrices in which all but three blocks are of the form $\begin{bmatrix} 0 & 0 \\ 0 & 0 \end{bmatrix}$. With these choices, the map to the signed permutation matrices is to take the determinant of the (2×2)-blocks.

The second remark is more subtle. Recall from Chapter 2 that each blue string in the pictures represents a pair of, perhaps thinner, and perhaps black, strings-with (mod 2)-quipu representing adjacent transpositions between them. The crossings between the blue strings represents a rotated hash mark, ✻. In the same way, each red string in the pictures (in this section) for the elements of A_4 represents four of the smaller black strings. So, the 3-red-strings picture with K_4-quipu summarizes a 12-black-strings picture.

Okay, you may not think that is subtle yet. But the diagram for an element is, in fact, a summary of the action of that element upon the other elements of the group. Label the bottoms and tops of the diagrams with the elements in the congregated order

$$((1, 12.34, 13.24, 14.23), (123, 134, 243, 142), (132, 234, 124, 143)).$$

Parentheses are omitted to fit the expression upon one line, for example, $12.34 = (12)(34)$. Then the strings, read from bottom to top, tell you to which element g'' an element g' goes under the action of g

that is represented by the 3-strings-with-K_4-quipu diagram of g. So, $g'' = g \cdot g'$.

There were a lot of words in that paragraph, so Fig. 3.25 indicates the meaning for one example.

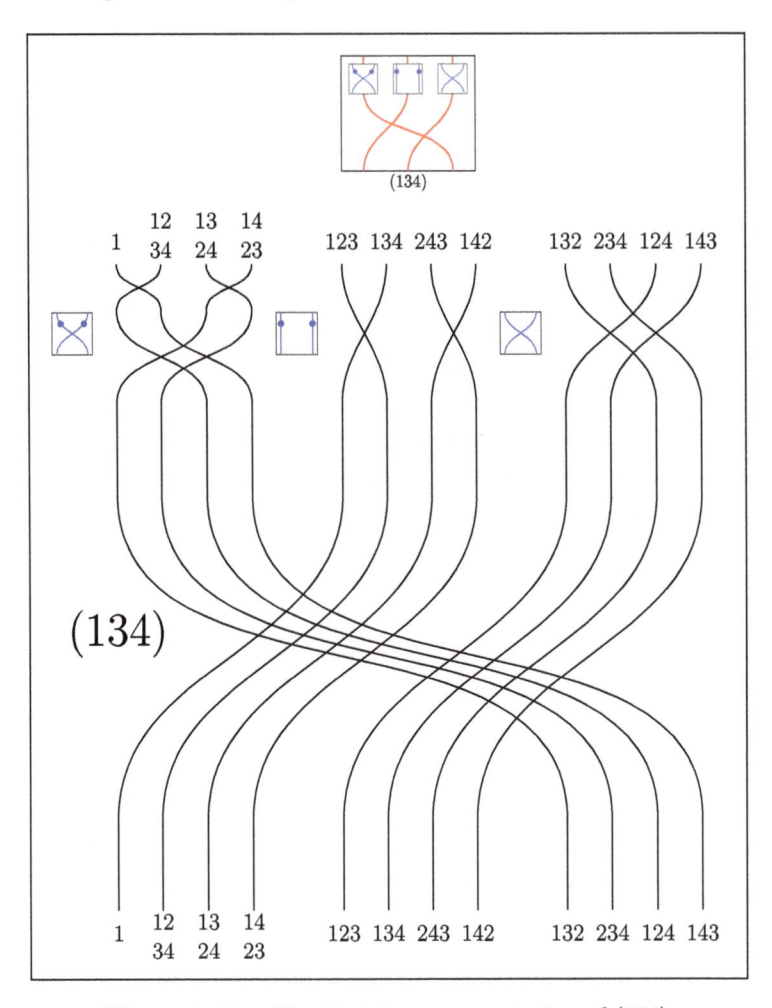

Figure 3.25. The 12-strings representation of (134).

To put this diagram into a more algebraic context, observe that $(134)(123) = (124)$, and the string that starts at the bottom location (123) ends at the top location (124). To repeat, these 3-strings-with-K_4-quipu diagrams each represent a row in the multiplication table for the corresponding element.

Chapter 4

The 3-Dimensional Sphere is a Group

As the title suggests, the chapter describes a group structure of the 3-dimensional sphere. In addition to its group structure, we also would like to explore it from a topological point of view. Moreover, the group structure and the space will be identified with the set of (2×2)-unitary matrices over the complex numbers. Fear not! That matrix group will be explicitly described.

The adjective "topological" has two meanings that we also will look into. The first literal meaning of topological has to do with defining a topology upon a set. A *topology* is a collection of subsets that are called *open sets* and are closed under arbitrary union and finite intersection. By default, the entire set is said to be open as is the empty set. The idea of topology is to give a notion of proximity without having an explicit distance function defined. Still all of the topological spaces we consider have distance functions defined upon them. In these cases, open sets are unions of open balls,[1] whose centers and radii vary.

The second use of the word is not so well defined. When a topologist says that a space is to be considered topologically, the meaning is "up to a homeomorphism."[2] So, a topologist considers a circle and a square to be the same. But in a more nuanced way, a topological

---

[1] An "open ball" is one for which its boundary points are not included, and it will be described in a rigorous manner subsequently.

[2] The word *homeomorphism* will be given a rigorous definition soon.

property is one in which the explicit geometric description does not matter so much as other more space-like properties. A topological property is more like a quality than a quantity.

For example, it is quite difficult to define the dimension of a space. Dimension is a quality that some spaces may have, and the dimension of such spaces is one of its intrinsic topological properties. Twentieth century topologists found ways in which dimension could be measured. These will not be explored herein. However, we have used, and we will use, the term *dimension* in a more colloquial sense.

Some topological techniques are really combinatorial. At an operational level, we construct spaces by sewing or gluing together standard pieces. Sewing and gluing are synonyms to a topologist. They are colloquialisms for the more technical notion of quotient topology. When boundaries of cells are computed in the following, you may note that the calculations are more combinatorial in nature than they are analytical.

The 3-*dimensional sphere* is defined to be the point set

$$S^3 = \{(w, x, y, z) \in \mathbb{R}^4 : w^2 + x^2 + y^2 + z^2 = 1\}.$$

The definition of its group structure will be postponed until Section 4.2. But for a short while, alternative and topological descriptions will be given. Furthermore, despite it being an intrinsically 3-dimensional object, the 3-sphere is not such a familiar space. Even more so, its group structure is curious, and it invites exploration.

In Section 4.1, the n-dimensional simplex, Δ_n, and the n-dimensional cube, C_n, will be reviewed and identified with an n-dimensional (round) ball. The simplex and cube have some combinatorial quantities at their substructure. Their boundaries will be discussed (Section 4.1.2) as will aspects of their topology. In particular, we will assert (without proof) that these spaces are the same in a topological sense. In Section 4.2, the multiplication on the 3-sphere is defined that turns it into a group. Then two families of strings-with-quipu representations for the *quaternions*

$$Q_8 = \{\pm 1, \pm i, \pm j, \pm k\}$$

will be given and proven to give representations that are isomorphic to Q_8. One of these is generalized to the dicyclic groups. The dicyclic groups map two-to-one onto the dihedral groups.

In Section 4.2.4, the strings-with-(mod 4)-quipu are used to demonstrate that the 3-sphere is isomorphic to the group SU(2) of (2×2)-special unitary matrices. These are matrices with entries in the complex numbers, $\mathbb{C}$, whose determinant is 1 and that are *unitary*. The inverse of a unitary matrix A is its conjugate transpose. If $A = \begin{bmatrix} \alpha & \beta \\ \gamma & \delta \end{bmatrix}$ is the matrix, its *conjugate transpose* is $A^* = \begin{bmatrix} \bar{\alpha} & \bar{\gamma} \\ \bar{\beta} & \bar{\delta} \end{bmatrix}$ where, for example, $\bar{\alpha} = x - iy$ when $\alpha = x + iy$.

Section 4.2.2, describes the set of equatorial 2-dimensional spheres by means of stereographic projection. The quaternions Q_8 sit upon the intersections among these spheres. A stylized illustration of the elements in the binary tetrahedral group is presented in this section even though the group will not be discussed in detail until Section 5.2 of Chapter 5 and Chapter 6.

4.1. Balls and their boundaries

This section begins with a slight review. The review has only a little to do with the current chapter, but it has a lot to do with some topological and algebraic aspects that may be unfamiliar to you. Be careful. Not everything here is review, and some ideas will be used later in different contexts.

Please recall that the n-dimensional simplex is the convex hull of the standard unit vectors $e_1, e_2, \ldots, e_{n+1}$, i.e.,

$$\Delta_n = \left\{ (x_1, x_2, \ldots, x_{n+1}) \in \mathbb{R}^{n+1} : \sum_{\ell=1}^{n+1} x_\ell = 1, \ \& \ 0 \le x_\ell \right\}.$$

Recall further that the n-dimensional cube of volume 2^n is the set

$$C_n = \{(x_1, x_2, \ldots, x_n) \in \mathbb{R}^n : -1 \le x_\ell \le 1\}.$$

Even though these sets have different symmetries, they both are considered to be n-dimensional balls. To continue the parade of point sets, the *n-dimensional ball* is the set

$$B^n = \left\{ (x_1, x_2, \ldots, x_n) \in \mathbb{R}^n : \sum_{\ell=1}^{n} x_\ell^2 \le 1 \right\}.$$

More generally, an *open ball* of radius r that is centered at a point $(y_1, y_2, \ldots, y_n) = \vec{y} \in \mathbb{R}^n$ is given as the point set,

$$o\left(B_r^n(\vec{y})\right) = \left\{ (x_1, x_2, \ldots, x_n) \in \mathbb{R}^n : \left(\sum_{\ell=1}^{n} (x_\ell - y_\ell)^2\right)^{1/2} < r \right\}.$$

When we write that Δ_n and C_n are "considered to be n-dimensional balls," we are saying that each is homeomorphic to B^n, and a *homeomorphism* is a continuous bijection that has a continuous inverse.

The *standard* or *metric topology* of n-space, $\mathbb{R}^n$, is defined by specifying a family of sets that is open. The whole space, $\mathbb{R}^n$, is open and so is the empty set. The arbitrary union of open sets is open, and the finite intersection of open sets is open. An *open subset* of $\mathbb{R}^n$ is one that can be written as an arbitrary union of open balls. A *continuous function* is one for which the inverse image of any open set is open. Thus, for every $n = 1, 2, 3, \ldots$, the space $\mathbb{R}^n$ is a topological space. The topology is a *metric topology* because the open sets are constructed by means of open balls, and these are constructed by means of the (Euclidean) distance function

$$d(\vec{x}, \vec{y}) = \left(\sum_{\ell=1}^{n} (x_\ell - y_\ell)^2\right)^{1/2}.$$

There are other notions of distance that give the same family of open sets, but these will not be discussed here.

In topology, a set is *closed* if and only if its complement is open. There are two operations that are defined and associated to a subset Y of a topological space. The *interior of Y* is the union of all open sets that are contained within it. The *closure of Y* is the intersection of all the closed sets that contain Y. The *frontier* or *boundary* of a set is the set-theoretic difference between its closure and its interior.

These point-set notions are not very intuitive, but their implementations are. In particular, there is quite a lot of subtlety in the point-set notions of open and closed sets that is a bit less onerous at the level of $\mathbb{R}^n$ and in metric spaces in general. We beg the forgiveness of the reader for the text not delving into these aspects of point-set topology in detail. We hope that the introductory remarks are sufficient. They are included here for the sake of "completeness," but with the understanding of the authors that much more investigation may be needed on the part of the reader. Therefore, the somewhat technical details that are needed to demonstrate that certain sets are indeed boundaries will be delegated to either your imagination or to an alternative text of your choosing.

The boundary of the n-dimensional ball B^n is denoted by S^{n-1}. It is called the $(n-1)$-dimensional sphere, and it is traditional to write the boundary operator as ∂. For example, $\partial B^n = S^{n-1}$. It is not so difficult to write the boundary in the round ball case:

$$\partial B^n = S^{n-1} = \left\{ (x_1, x_2, \ldots, x_n) \in \mathbb{R}^n : \sum_{\ell=1}^{n} x_\ell^2 = 1 \right\}.$$

In a few paragraphs, the boundaries of both the simplex, Δ_n, and the cube, C_n, will be presented.

That the sphere is $(n-1)$-dimensional is an observation about its intrinsic nature. Sure, you are thinking of it as a subset of n-dimensional space. But think a little bit more profoundly.

The circle, S^1, is 1-dimensional in that the location of points on a circle can be determined by means of a directed displacement angle from a fixed reference point. We think of the angle $\pi/3$ as the point on the unit circle that is 60° counter-clockwise from the 3 o'clock point on the right. Thus, in terms of a clock face, $\pi/3$ is at 1 o'clock. The hours themselves are expressions of one degree of freedom.

The 2-dimensional sphere, S^2, is a mathematical abstraction of our home planet. Longitude and latitude are two directions that describe your current location unless, of course, you are reading the text while traveling above the surface of the earth. For example, the authors met at roughly 35° north and 128° east. The location

($0°E, 0°N$) is off the coast of Africa, west of Gabon and south of Ghana.

Thinking more mathematically about the unit circle S^1, the angle θ subtends the point whose cartesian coordinates are $(\cos(\theta), \sin(\theta))$. In general, the circle S^1 has a commutative group structure that is determined by means of adding angles. Add the 0 angle to any angle θ, and θ remains fixed. The Eulerian expression $e^{i\theta} = \cos(\theta) + i\sin(\theta)$ can be used to define an associative multiplication:

$$e^{i\theta} \cdot e^{i\phi}$$
$$= (\cos(\theta) + i\sin(\theta)) \cdot (\cos(\phi) + i\sin(\phi))$$
$$= [\cos(\theta)\cos(\phi) - \sin(\theta)\sin(\phi)]$$
$$+ i[\sin(\theta)\cos(\phi) - \cos(\theta)\sin(\phi)]$$
$$= (\cos(\theta + \phi) + i\sin(\theta + \phi)) = e^{i(\theta+\phi)}.$$

Inverses are determined by complex conjugation:

$$\left(e^{i\theta}\right)^{-1} = e^{-i\theta} = \cos(\theta) - i\sin(\theta).$$

This last equation is informed by the parity identities $\cos(-\theta) = \cos(\theta)$ and $\sin(-\theta) = -\sin(\theta)$.

It is worth mentioning the 0-dimensional sphere, $S^0 = \{-1, 1\}$, while we are on this digression. It is the boundary of the 1-dimensional ball or interval $[-1, 1]$. That interval is the description of both C_1 and B^1. We haven't emphasized this fact in the past text, but the 0-dimensional sphere has the structure of a group since $(-1)(1) = (1)(-1) = -1$ and $(1)(1) = (-1)(-1) = 1$. It is isomorphic to $\mathbb{Z}/2$, and even though we didn't emphasize the spherical nature, we did use this group structure repeatedly. It is represented by (mod 2)-quipu.

That group structure on S^0 and the group structure on S^1 that is given by means of adding rotations are aspects that will have analogues once we consider the unit 3-dimensional sphere S^3. But before we go further with the subject of the chapter, let us do a little work to express the boundaries of the n-simplex, Δ_n, and the n-cube, C_n.

The boundaries of polyhedral sets consist of polyhedra that are of one smaller dimension, and they are endowed with an orientation sense. Orientations are geometric analogues of the determinants of matrices. In general, getting the signs of things right is a notorious personal problem through which we all suffer. Rest assured. The authors spent much time that is unseen in order to get many signs correct. If some are wrong, we regret the typographic errors.

4.1.1. Boundaries of cubes

We begin with the n-cube by building up from the interval $[-1, 1] = \{x \in \mathbb{R} : -1 \le x \le 1\}$ through the square and the 3-dimensional cube before embarking upon the general case.

As a point set, the boundary of $[-1, 1]$ is the set $\{-1, 1\}$. But let us examine the orientations. The closed line segment $[-1, 1]$ should be considered to be an oriented arrow $\{-1\} \longrightarrow \{1\}$. So, its boundary is a *formal difference*

$$\partial[-1, 1] = \{1\} - \{-1\}.$$

Here, the integer enclosed in braces (either $\{1\}$ or $\{-1\}$) should be thought of us a symbol or a name of the corresponding point and not as a number.

Next up, let us consider the square:

$$[-1, 1] \times [-1, 1] = \{(x, y) : -1 \le x, y \le 1\}.$$

In this case, the use of the character ∂ that denotes boundary is motivated, in part, because the boundary acts according to the product rule, but with a sign that is associated.

$$\begin{aligned}
&\partial\left([-1, 1] \times [-1, 1]\right) \\
&= ((\partial[-1, 1]) \times [-1, 1]) - ([-1, 1] \times \partial ([-1, 1])) \\
&= ((\{1\} - \{-1\}) \times [-1, 1]) - ([-1, 1] \times (\{1\} - \{-1\})) \\
&= (\{1\} \times [-1, 1] - \{-1\} \times [-1, 1]) \\
&\quad - ([-1, 1] \times \{1\} - [-1, 1] \times \{-1\}) \\
&= (\{1\} \times [-1, 1] - \{-1\} \times [-1, 1]) \\
&\quad - [-1, 1] \times \{1\} + [-1, 1] \times \{-1\}.
\end{aligned}$$

There is a *slang notation* that will help us understand the four pieces of the boundary as well as the signs of these pieces. Abbreviate
$$[-1,1] \times [-1,1] = \{(x,y) \in \mathbb{R}^2 : -1 \le x, y \le 1\}$$
to the expression (x,y). Then
$$\partial(x,y) = \partial(x)y - x\partial(y)$$
$$= \underbrace{(1,y)}_{A} - \underbrace{(-1,y)}_{B} - \underbrace{(x,1)}_{C} + \underbrace{(x,-1)}_{D}.$$

In a more rigorous notational scheme, the positive interval $+[-1,1]$ will be denoted either as
$$\{-1\} \longrightarrow \{1\} \quad \text{or as} \quad \begin{array}{c} \{1\} \\ \uparrow \\ \{-1\} \end{array},$$
and the negative interval $-[-1,1]$ will be denoted as
$$\{-1\} \longleftarrow \{1\} \quad \text{or as} \quad \begin{array}{c} \{1\} \\ \downarrow \\ \{-1\} \end{array}.$$

There are four segments to the boundary of the square:
$$A = \{(1,y)\} : \{1\} \times [-1,1] = \{1\} \boxed{\begin{array}{c} \{1\} \\ \uparrow \\ \{-1\} \end{array}},$$

$$B = \{(-1,y)\} : -\{-1\} \times [-1,1] = \{-1\} \boxed{\begin{array}{c} \{1\} \\ \downarrow \\ \{-1\} \end{array}},$$

$$C = \{(x,1)\} : -[-1,1] \times \{1\} = \boxed{\{-1\} \longleftarrow \{1\}} \{1\},$$

and

$$D = \{(x,-1)\} : [-1,1] \times \{-1\} = \boxed{\{-1\} \longrightarrow \{1\}} \{-1\}.$$

These notationally intense formalisms are meant to convey the fairly simple geometric idea that is expressed in Fig. 4.1. The boundary of a square consist of its four edges and these are oriented anti-clockwise as they wrap around the center of the square.

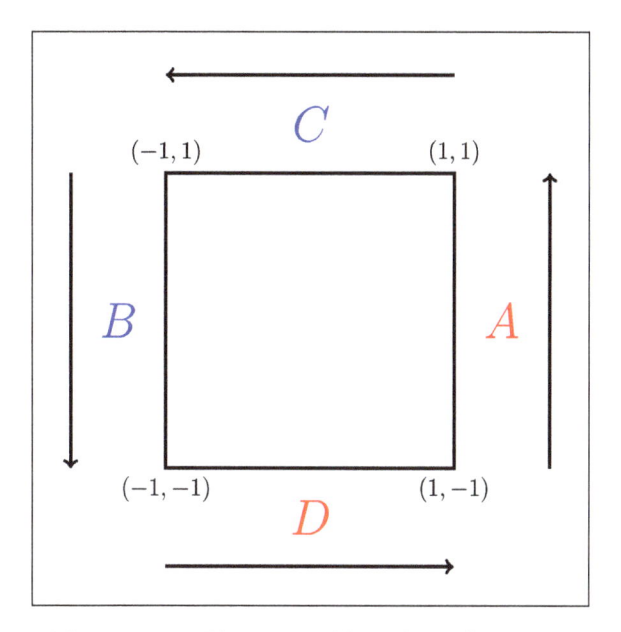

Figure 4.1. The oriented boundary of a square.

In the case of the cube, let us compute the boundary in a manner similar to the previous calculation, namely,

$$\partial[-1,1]^3 = \partial\left([-1,1] \times [-1,1]^2\right)$$
$$= \Big((\partial[-1,1]) \times [-1,1]^2\Big) - \Big([-1,1] \times (\partial[-1,1]^2)\Big)$$
$$= \Big(\{1\} \times [-1,1]^2 - \{-1\} \times [-1,1]^2\Big)$$
$$\quad - \Big([-1,1] \times (\partial[-1,1]^2)\Big)$$
$$= \Big(\{1\} \times [-1,1]^2 - \{-1\} \times [-1,1]^2\Big)$$
$$\quad - \Big([-1,1] \times (\{1\} \times [-1,1] - \{-1\} \times [-1,1])$$
$$\quad - ([-1,1]^2 \times \{1\} - [-1,1]^2 \times \{-1\})\Big).$$

There are six square faces of the cube. In order to describe them and to keep track of the sign that is associated to the face, we will write $(\pm 1, y, z)$, $(x, \pm 1, z)$, and $(x, y, \pm 1)$ for these faces with the understanding that $-1 \leq x, y, z \leq 1$. In this slang, we obtain

$$\partial(x, y, z) = (1, y, z) - (-1, y, z)$$
$$- (x, 1, z) + (x, -1, z)$$
$$+ (x, y, 1) - (x, y, -1).$$

Let us name the (x, y) squares top $(+)$ and bottom (-1). Name the (x, z) squares left $(-)$ and right $(+)$, and name the (y, z) squares front $(+)$ and back $(-)$. Then the top, left, and front faces are positive. These are indicated in the illustration of the cube (Fig. 4.2) by means of red arrows that are perpendicular to the respective faces and pointing outwardly.

In these conventions, the oriented perpendiculars to the faces of the cube all are outward pointing. It may be a little disturbing that the (x, z) face has a perpendicular that points left, but that is a convention determined by means of a right-hand rule.

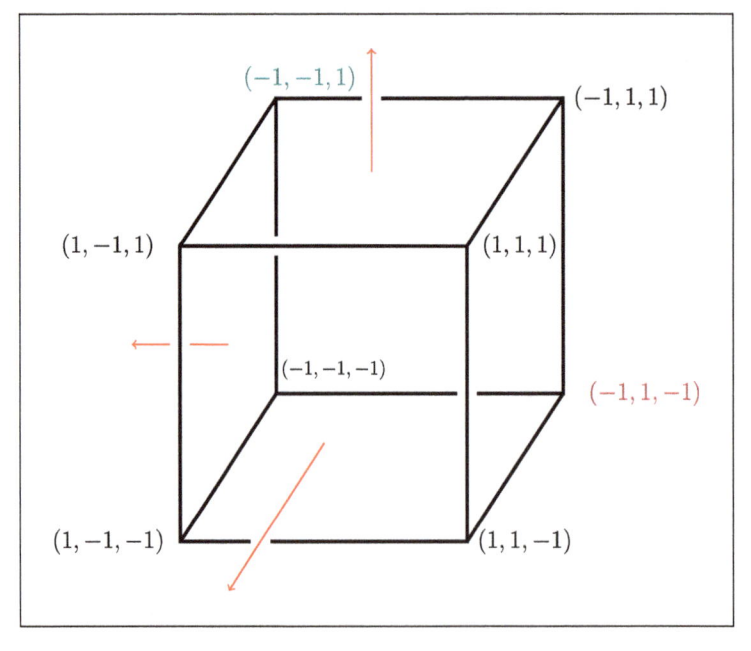

Figure 4.2. The oriented faces of the cube.

In general, for an n-cube, opposite $(n-1)$-dimensional faces are oriented oppositely. One is positive and the other is negative.

Sequentially, adjacent pairs of faces will have their signs differ from the priors. Be patient and let us explain.

An inductive method is used to compute the boundary of an n-cube, namely,

$$\partial[-1,1]^n = \partial\left([-1,1] \times [-1,1]^{n-1}\right)$$
$$= \left((\partial[-1,1]) \times [-1,1]^{n-1}\right) - \left([-1,1] \times (\partial[-1,1]^{n-1})\right).$$

There is a small trick with things that act like boundary operators satisfying a signed product rule. The *signed product rule* is given by a formula:

$$\partial(PQ) = (\partial P)Q + (-1)^{d(P)} P(\partial Q),$$

where $d(P)$ is a *degree* or dimension of the object P. In the case above, the interval $[-1,1]$ has dimension 1, and so, the sign that is associated to $\left([-1,1] \times (\partial[-1,1]^{n-1})\right)$ is negative.

In the slang notation,

$$\partial(x_1, x_2, \ldots, x_n) = \sum_{\ell=1}^{n} (-1)^{\ell+1} \big[(x_1, \ldots, x_{\ell-1}, 1, x_{\ell+1}, \ldots, x_n)$$
$$- (x_1, \ldots, x_{\ell-1}, -1, x_{\ell+1}, \ldots, x_n)\big]$$
$$= \big[(1, x_2, \ldots, x_n) - (-1, x_2, \ldots, x_n)\big]$$
$$+ \big[(x_1, -1, x_3, \ldots, x_n) - (x_1, 1, x_3, \ldots, x_n)\big]$$
$$+ \cdots$$
$$+ (-1)^{n+1} \big[(x_1, x_2 \ldots, 1) - (x_1, x_2, \ldots, -1)\big].$$

It may be a more painful exercise than you care to do right now, but it can be shown that if $n = a + b$, then the formula

$$\partial[-1,1]^n = \partial\left([-1,1]^a \times [-1,1]^b\right)$$
$$= \left((\partial[-1,1]^a) \times [-1,1]^b\right) + (-1)^a \left([-1,1]^a \times (\partial[-1,1]^b)\right)$$

yields the same boundary with the same signs once all of the subsequent boundaries are taken.

We are rarely interested in boundaries of cube larger than dimension 4. So, to finalize this section, let's compute the boundary of the 4-cube:

$$\partial(w,x,y,z) = (1,x,y,z) - (-1,x,y,z)$$
$$+ (w,-1,y,z) - (w,1,y,z)$$
$$+ (w,x,1,z) - (w,x,-1,z)$$
$$+ (w,x,y,-1) + (w,x,y,1).$$

We refer the reader to Figure 3.18. In descriptive terms, the boundary reads, respectively, as (big cube minus small cube) plus (back frustum minus front frustum) plus (right frustum minus left frustum) plus (bottom frustum minus top frustum).

4.1.2. Boundaries of simplices

The 1-simplex, Δ_1, is the interval in the plane

$$\Delta_1 = \{(x,y) \in \mathbb{R}^2 : x+y = 1; 0 < x,y\}.$$

With the understanding that $x + y = 1$, we adopt and adapt[3] the slang notation from the previous section and write

$$\partial(x,y) = (0,1) - (1,0).$$

The 2-simplex or triangle, Δ_1, is the set

$$\Delta_2 = \{(x,y,z) \in \mathbb{R}^2 : x+y+z = 1; 0 < x,y,z\}.$$

The boundary consists of the three edges $(0,x,y)$, where $x+y = 1$ and $0 < x,y$, $(x,0,z)$, where $x+z = 1$ and $0 < x,z$, and $(0,y,z)$, where $y+z = 1$ and $0 < y,z$. We use the technique of alternating the signs to orient these. Again, in slang notation,

$$\partial(x,y,z) = +(0,y,z) - (x,0,z) + (x,y,0).$$

The oriented edges are indicated in Fig. 4.3.

---

[3]In the slang of the current section, (x,y) no longer denotes a square, but instead denotes the diagonal interval in the northeastern region of the plane. Similarly, (x,y,z) denotes an equilateral triangular face in the positive region of 3-space (Fig. 4.3).

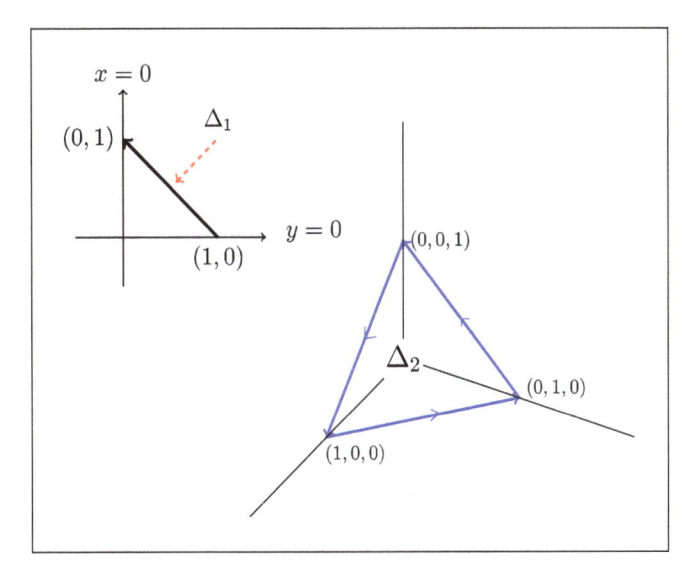

Figure 4.3. The boundaries of Δ_1 and Δ_2.

The 3-dimensional simplex Δ_3 is a tetrahedron that we consider as

$$\Delta_3 = \{(w, x, y, z) \in \mathbb{R}^4 : w + x + y + z = 1; 0 < w, x, y, z\}.$$

The four oriented triangular faces in the boundary are given in slang notation as

$$\partial(w, x, y, z) = (0, x, y, z) - (w, 0, y, z) + (w, x, 0, z) - (w, x, y, 0).$$

In Fig. 4.4, the orientations of the faces are indicated by means of red arrows that point outward along each face. The "right-hand rule" for an orientation of a face works like this. Consider the face with $W = (1, 0, 0, 0)$, $Y = (0, 0, 1, 0)$, and $Z = (0, 0, 0, 1)$ as its vertices. This is the larger triangle on the the back of the figure. The cyclic ordering of the vertices is (W, Y, Z). Curl your right hand from W to Y to Z and your thumb points inward. But these are the vertices of the triangle $(w, 0, y, z)$. So, that face is negatively oriented. The resulting perpendicular vector is outward pointing.

The general boundary of Δ_n is written in slang as

$$\partial(x_1, \ldots, x_{n+1}) = \sum_{\ell=1}^{n+1} (-1)^{\ell+1} (x_1, \ldots, x_{\ell-1}, 0, x_{\ell+1}, \ldots, x_{n+1}).$$

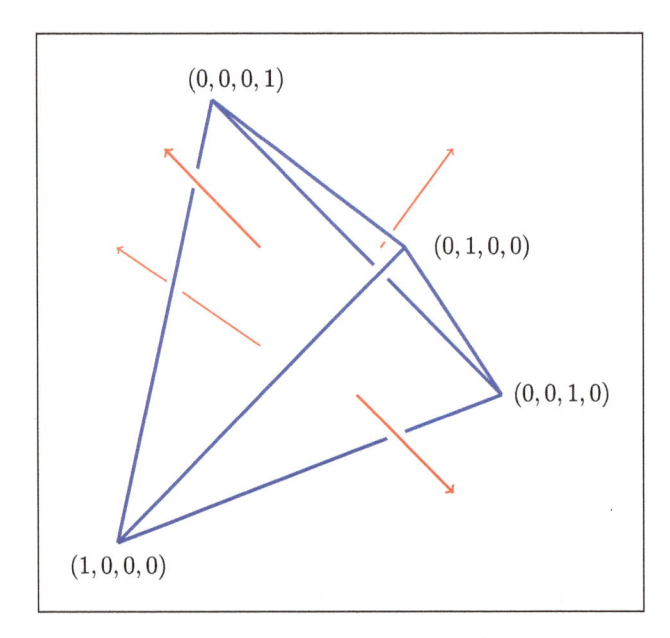

Figure 4.4. The boundary of Δ_3.

4.1.3. Taking the boundary twice

An appealing aspect of the boundary operation is that when it is applied twice, the result is 0. We write $\partial \circ \partial = 0$ to indicate that the boundaries of the boundaries are algebraically 0. We will start by considering the simplices. A general slang term in the expression of the boundary is

$$(-1)^{\ell+1}(x_1, \ldots, x_{\ell-1}, 0, x_{\ell+1}, \ldots, x_{n+1}).$$

This indicates the oriented $(n-1)$-dimensional simplex, $\Delta_{n-1}(\ell)$, that lies on the n-dimensional wall $x_\ell = 0$ of $\mathbb{R}^{n+1}$. Let's pretend that the idea of an orientation makes sense and that it generalizes the tangent and perpendicular arrows that appear in the illustrations. It is true that the orientations do both of these things, and they can be related to the determinants of matrices that are constructed by way of the vertices of this simplex $\Delta_{n-1}(\ell)$. But for now, we really only need to work with the sign $(-1)^{\ell+1}$ in a formal way.

Before we embark upon computing $\partial\left(\partial(x_1, x_2, \ldots, x_n)\right)$ in the general case, we work first with the triangle and then with the tetrahedron:

$$\partial(x,y,z) = \begin{vmatrix} +(0,y,z) & -(x,0,z) & +(x,y,0) \\ \downarrow \partial & \downarrow \partial & \downarrow \partial \\ +(0,0,z) & -(0,0,z) & +(0,y,0) \\ -(0,y,0) & +(x,0,0) & -(x,0,0) \end{vmatrix}.$$

The six slang terms cancel.

In the case of the tetrahedron,

$$\partial(w,x,y,z) =$$

$$\begin{vmatrix} +(0,x,y,z) & -(w,0,y,z) & +(w,x,0,z) & -(w,x,y,0) \\ \downarrow \partial & \downarrow \partial & \downarrow \partial & \downarrow \partial \\ +(0,0,y,z) & -(0,0,y,z) & +(0,x,0,z) & -(0,x,y,0) \\ -(0,x,0,z) & +(w,0,0,z) & -(w,0,0,z) & +(w,0,y,0) \\ +(0,x,y,0) & -(w,0,y,0) & +(w,x,0,0) & -(w,x,0,0) \end{vmatrix}.$$

The cancelation between terms follow a nice pattern. Please observe it. So, in the general case,

$$\partial(x_1, x_2, \ldots, x_{n+1}) =$$

$$\begin{array}{cccc}
+(0, x_2, \ldots, x_{n+1}) & -(x_1, 0, \ldots, x_{n+1}) & \cdots & +(-1)^n(x_1, \ldots, x_n, 0) \\
\downarrow \partial & \downarrow \partial & \vdots & \downarrow \partial \\
(0, 0, x_3, \ldots, x_{n+1}) & -(0, 0, x_3, \ldots, x_{n+1}) & \cdots & +(-1)^n(0, x_2, \ldots, x_n, 0) \\
-(0, x_2, 0, \ldots, x_{n+1}) & +(x_1, 0, 0, \ldots, x_{n+1}) & \cdots & -(-1)^n(x_1, 0, \ldots, x_n, 0) \\
\cdots & \cdots & & \cdots \\
+(0, x_2, \ldots, x_n, 0) & -(x_1, 0, \ldots, x_n, 0) & \cdots & -(x_1, \ldots, x_{n-1}, 0, 0).
\end{array}$$

The first two terms in the first row cancel, then the third term in the first row cancels with the first term in the second row, and so forth. The last term in the first row cancels with the last term in the first column. Inductively, the pattern continues until the last two terms in the last row cancel.

———————————————————

The computations above will be mimicked in the cases of C_2 and C_3. In the case of C_2, here is the computation:

$$\partial(x,y) = \begin{array}{|c|c|c|c|} \hline +(1,y) & -(-1,y) & -(x,1) & +(x,-1) \\ \downarrow \partial & \downarrow \partial & \downarrow \partial & \downarrow \partial \\ \hline +(1,1) & -(-1,1) & -(1,1) & +(1,-1) \\ -(1,-1) & +(-1,-1) & +(-1,1) & -(-1,-1) \\ \hline \end{array}.$$

Please observe that the terms cancel.

In the case of C_3, we have

$$\partial(x,y,z)$$

$$= \begin{array}{|c|c|} \hline +(1,y,z)-(-1,y,z) & -(x,1,z)+(x,-1,z) \\ \downarrow \partial & \downarrow \partial \\ \hline (1,1,z)-(-1,1,z) & -(1,1,z)+(1,-1,z) \\ -(1,-1,z)+(-1,-1,z) & +(-1,1,z)-(-1,-1,z) \\ \hline -(1,y,1)+(-1,y,1) & +(x,1,1)-(x,-1,1) \\ +(1,y,-1)+(-1,y,-1) & -(x,1,-1)-(x,-1,1) \\ \hline \end{array}$$

$$+ \begin{array}{|c|} \hline +(x,y,1)-(x,y,-1) \\ \downarrow \partial \\ \hline +(1,y,1)-(1,y,-1) \\ -(-1,y,1)+(-1,y,-1) \\ \hline -(x,1,1)+(x,1,-1) \\ +(x,-1,1)-(x,-1,-1) \\ \hline \end{array}.$$

Each slang term in this expansion represents an oriented edge. A term that represents such an edge, for example, $(-1,y,1)$ appears once positively and once negatively. In each block of the computation, there are four terms. There are three columns each of which contains two blocks that are indexed as rows. The four terms in the first row, first column cancel with the four terms in the first row, second column. The four terms

$$+(1,y,1)-(1,y,-1)-(-1,y,1)+(-1,y,-1)$$

in the first row, third column cancel with the four terms in the second row, first column. The four terms in the second row, second column cancel with the four terms in the second row, third column. The cancelation patterns are $+(A-B-C+D)$ canceling with $-(A-C-B+D)$.

Exercise 9.

(1) Mimic the style of this computation. Compute $\partial(w, x, y, z)$ and arrange eight terms in a long horizontal row. Then compute the boundary of each of these eight terms and arrange the terms vertically. For each term such as $(1, x, y, z)$, there will be six terms. The term $(1, x, y, z)$ represents a cubical face of the 4-cube, and the resulting six terms represent the six faces of that cube. Your computation should have 48 terms before implementing the cancelations. Then observe that terms cancel in similar blocks: first row, first column cancels with first row, second column. As you move along the first row, the canceling terms appear in the first column.

(2) Work out the inductive step to demonstrate that

$$\partial \circ \partial (x_1, x_2, \ldots, x_n) = 0.$$

When you encounter homology (either from an algebraic or a topological point of view), expect that similar computations will recur. A good organizational mental template often facilitates the computations. Or so our experience tells us.

The 1 simplex Δ_1 is the edge $x + y = 1$ with $0 < x, y$. Its oriented boundary is the formal difference

$$\partial \Delta_1 = (0, 1) - (0, 1).$$

By convention, the boundary of a single point is 0. So, in this case, the boundary of the boundary of the edge Δ_1 is also 0. In the same way,

$$\partial \circ \partial [-1, 1] = \partial(\{1\}) - \partial(\{-1\}) = 0 - 0 = 0.$$

There are other more subtle ways of considering the boundaries of 0-dimensional objects. In these unreduced theories, the boundary of a point is a formal symbol $*$. But the boundary of any other point is the same symbol $*$. So, the computation looks similar. For example, in the case of C_1,

$$\partial \circ \partial [-1, 1] = \partial(\{1\}) - \partial(\{-1\}) = * - * = 0.$$

4.2. The unit vectors i, j, and k

The 3-*dimensional sphere* S^3 is given to be

$$S^3 = \{(w, x, y, z) : w^2 + x^2 + y^2 + z^2 = 1\}.$$

The standard basis elements $e_1 = [1, 0, 0, 0]^t$, $e_2 = [0, 1, 0, 0]^t$, $e_3 = [0, 0, 1, 0]^t$, and $e_4 = [0, 0, 0, 1]^t$ are relabeled as 1, i, j, and k, respectively. The following multiplication rules are defined upon these elements:

$$i \cdot j = k = -j \cdot i,$$
$$j \cdot k = i = -k \cdot j,$$
$$k \cdot i = j = -i \cdot k,$$

and

$$i \cdot i = j \cdot j = k \cdot k = -1.$$

By convention, the element (-1) commutes with any of the other elements.

All of our own calculations among i, j, and k are facilitated by means of using the standard diagram:

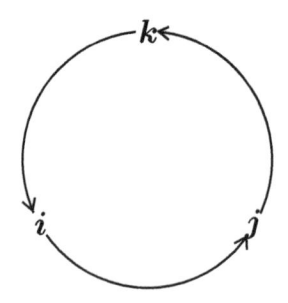

It indicates that if you multiply two elements in cyclic alphabetical order, the product is the next element. See also Fig. 4.9. If you multiply out of order, the product is the negative of the other element. Each of i, j, and k squares to -1.

Moreover, at several junctures, Table 4.1 (and variations upon it that depend upon signs) is helpful.

The 3-Dimensional Sphere is a Group

Table 4.1. A table of products among i, j, k.

$\cdot$	1	i	j	k
1	1	i	j	k
i	i	-1	k	$-j$
j	j	$-k$	-1	i
k	k	j	$-i$	-1

In particular suppose that $a_\ell, b_\ell, c_\ell, d_\ell \in \mathbb{R}$ and $a_\ell^2 + b_\ell^2 + c_\ell^2 + d_\ell^2 = 1$ for $\ell = 1, 2$. Table 4.2 contains all of the terms in the product between these two elements.

Table 4.2. Products between a pair of elements in S^3.

$\cdot$	a_2	$b_2 i$	$c_2 j$	$d_2 k$
a_1	$a_1 a_2$	$a_1 b_2 i$	$a_1 c_2 j$	$a_1 d_2 k$
$b_1 i$	$a_2 b_1 i$	$-b_1 b_2$	$b_1 c_2 k$	$-b_1 d_2 j$
$c_1 j$	$a_2 c_1 j$	$-b_2 c_1 k$	$-c_1 c_2$	$c_1 d_2 i$
$d_1 k$	$a_2 d_1 k$	$b_2 d_1 j$	$-c_2 d_1 i$	$-d_1 d_2$

It is possible to tabulate the squares of the coefficients of 1, i, j, and k in the product $\vec{X}_1 \cdot \vec{X}_2$, where $\vec{X}_1 = a_1 + b_1 i + c_1 j + d_1 k$ and $\vec{X}_2 = a_2 + b_2 i + c_2 j + d_2 k$ that appears in Table 4.2 in about two hours time.[4] The mixed terms cancel and the result is

$$(a_1 a_2 - b_1 b_2 - c_1 c_2 - d_1 d_2)^2 + (a_1 b_2 + a_2 b_1 + c_1 d_2 - c_2 d_1)^2$$
$$+ (a_1 c_2 - b_1 d_2 + a_2 c_1 + b_2 d_1)^2 + (a_1 d_2 + b_1 c_2 - b_2 c_1 + a_2 d_1)^2$$
$$= a_1^2 (a_2^2 + b_2^2 + c_2^2 + d_2^2) + b_1^2 (a_2^2 + b_2^2 + c_2^2 + d_2^2)$$
$$+ c_1^2 (a_2^2 + b_2^2 + c_2^2 + d_2^2) + d_1^2 (a_2^2 + b_2^2 + c_2^2 + d_2^2)$$
$$= [a_1^2 + b_1^2 + c_1^2 + d_1^1](1) = 1.$$

---

[4] We listened to both sides of a favorite LP while executing the calculation and nearly completed it before having to find another record. But such calculations are fraught with errors, so double the alotted time.

As a guide to the calculation, we work the much more simple example which shows that the multiplication in the circle S^1 is closed:

$\cdot$	c	$d\boldsymbol{i}$
a	ac	$ad\boldsymbol{i}$
$b\boldsymbol{i}$	$bc\boldsymbol{i}$	$-bd$

Then

$$(ac - bc)^2 + (ad + bd)^2 = a^2c^2 - 2acbd + b^2c^2 + a^2d^2 + 2abcd + b^2d^2$$
$$= a^2(c^2 + d^2) + b^2(c^2 + d^2)$$
$$= [a^2 + b^2](1) = 1.$$

In the 3-sphere case, we compiled the squares in four (4×4)-tables, and looked for cancelations among the terms of the form $\pm a_* b_* c_* d_*$. These are the terms that we described as mixed above.

In this discussion, we are not trying to be malicious when we assign the intricacies of the calculation to you. We just expect that you'll find them more interesting if you work them instead of reading about them. As a small consolation, we demonstrate that

$$(w + x\boldsymbol{i} + y\boldsymbol{j} + z\boldsymbol{k})^{-1} = w - x\boldsymbol{i} - y\boldsymbol{j} - z\boldsymbol{k}:$$

$\cdot$	w	$-x\boldsymbol{i}$	$-y\boldsymbol{j}$	$-z\boldsymbol{k}$
w	w^2	$-wx\boldsymbol{i}$	$-wy\boldsymbol{j}$	$-wz\boldsymbol{k}$
$x\boldsymbol{i}$	$wx\boldsymbol{i}$	x^2	$-xy\boldsymbol{k}$	$xz\boldsymbol{j}$
$y\boldsymbol{j}$	$wy\boldsymbol{j}$	$xy\boldsymbol{k}$	y^2	$-yz\boldsymbol{i}$
$z\boldsymbol{k}$	$wz\boldsymbol{k}$	$-xz\boldsymbol{j}$	$yz\boldsymbol{i}$	z^2

The off-diagonal elements cancel, and the sum of the squares along the diagonal is 1:

$$w^2 + x^2 + y^2 + z^2 = 1.$$

The associativity of the product follows from the associativity of multiplication of real numbers and the associativity that occurs in the quaternions, $Q_8 = \{\pm 1, \pm i, \pm j, \pm k\}$. That latter associativity is also a routine calculation.

This completes the description of the group structure upon the 3-sphere, S^3.

4.2.1. The quaternions: Part 1

Two diagrammatic representations of the quaternions are developed herein. One or the other will be used in the subsequent descriptions of the dicyclic groups and the binary polyhedral groups. Recall that the quaternions is the 8-element group $Q_8 = \{\pm 1, \pm i, \pm j, \pm k\}$, where the products of (half of) these elements are presented in Table 4.1.

The subgroup $H = (-1, 1)$ is ordered as indicated.[5] The ordered cosets are $iH = (-i, i)$, $jH = (-j, j)$, and $kH = (-k, k)$. They are put in the order (H, iH, jH, kH). Recall that (-1) commutes with every element in Q_8 and that its action upon these cosets is to reverse the order in a given coset. There is a 4-strings-with-(mod 2)-quipu representation of (-1) that is presented as follows:

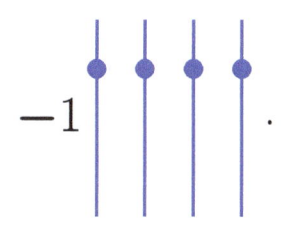

That $iH = iH$ is tautological. Meanwhile, $iiH = -H = [H, 1]$, and $i(jH) = kH$, while $ikH = -jH = [jH, 1]$. The notation $[jH, 1]$ indicates the 2-element coset jH with its orientation reversed. Similar calculations give the actions of j and k upon the ordered cosets. In Fig. 4.5, the 4-strings-with-(mod 2)-quipu representations for Q_8 are compiled.

---

[5]The subgroup is the set $H = \{-1, 1\}$; it is written as an ordered set, but shouldn't be confused with an open interval, nor with the northwest corner of a square.

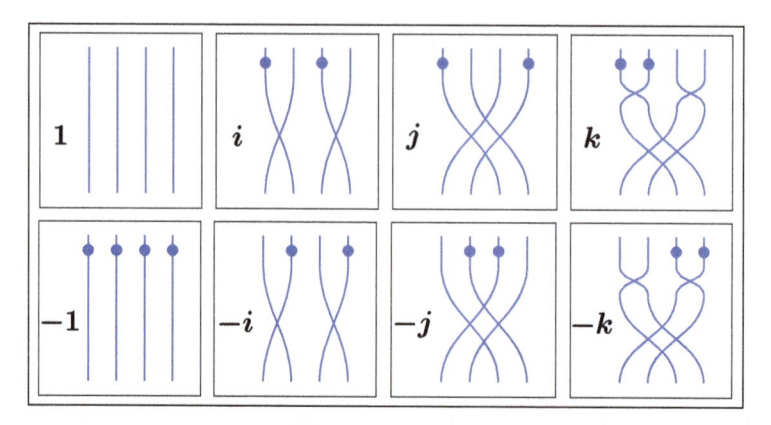

Figure 4.5. The 4-strings-with-(mod 2)-quipu illustration of Q_8.

Let us label the cosets as $[1]$, $[i]$, $[j]$, and $[k]$, and then consider the quipu diagram of the element k, for example. The (mod 2)-quipu can be read as if they are negatives:

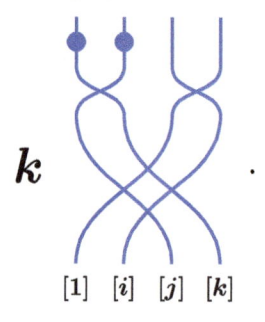

The diagram then reads that k acts as $k(1) = k$, $k(i) = j$, $k(j) = -i$, and, of course, $k(k) = -1$. We pronounce the equations as "k upon 1 is k, k upon i is j," and so forth. The diagrams for the other elements of Q_8 can be read similarly. This idea also appears in a paper by Kauffman [10].

The (mod 2)-quipu diagrams for the quaternions Q_8 that are depicted in Fig. 4.5 suggest a surjective projection $K_4 \xleftarrow{p} Q_8$ onto the Klein 4-group K_4. The values of this homomorphism are $p(\pm 1) = 1$, $p(\pm i) = (12)(34)$, $p(\pm j) = (13)(24)$, and $p(\pm k) = (14)(23)$. The projection is obtained by ignoring the quipu upon the strings. So, there is a seseq

$$0 \longleftarrow K_4 \xleftarrow{p} Q_8 \xleftarrow{i} \mathbb{Z}/2 \longleftarrow 0,$$

The 3-Dimensional Sphere is a Group 159

in which the image of the inclusion i is the subgroup $H = \{\pm 1\}$. On the other hand, this seseq does not split, so Q_8 is not a semi-direct product. See Section 8.4.2 and Table 8.5 of Chapter 8.

Please recall the identities that are expressed in Figs. 2.7 and 2.13 as well as the notion that the thick blue strings satisfy the ordinary rules $t_i^2 = 1$, $t_i t_{i+1} t_i = t_{i+1} t_i t_{i+1}$ and $t_i t_j = t_j t_i$ when $1 < |i - j|$ that are defining relations of permutations. Then the relations of the quaternions can be visualized using the quipu diagrams as in Fig. 4.6.

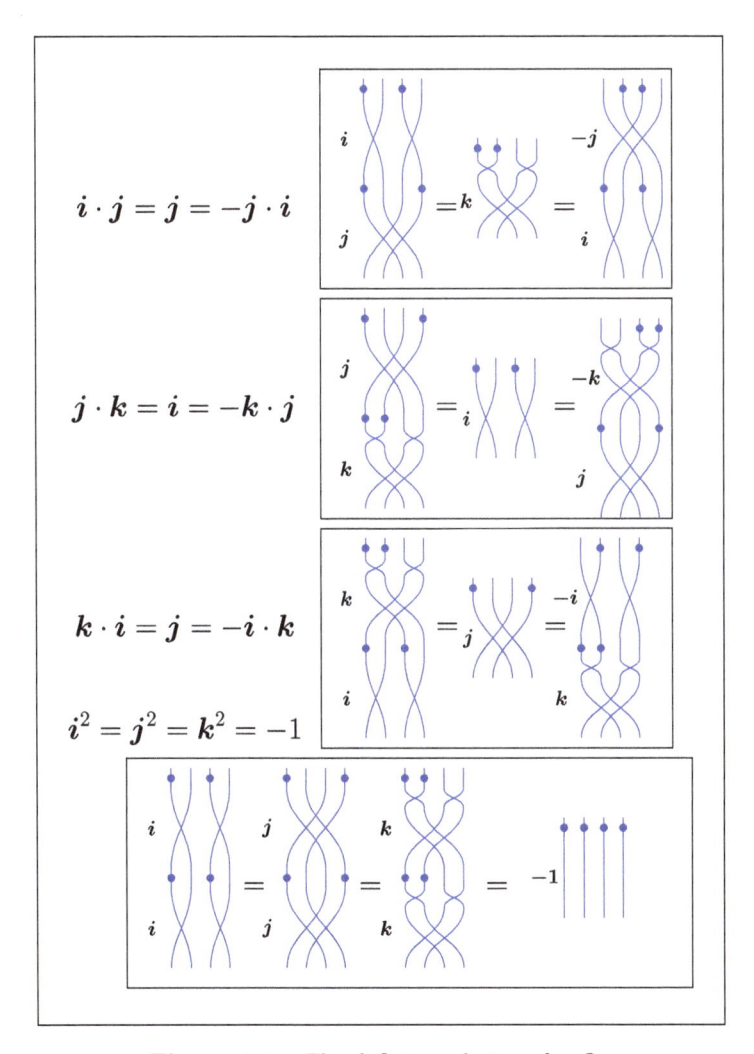

$i \cdot j = j = -j \cdot i$

$j \cdot k = i = -k \cdot j$

$k \cdot i = j = -i \cdot k$

$i^2 = j^2 = k^2 = -1$

Figure 4.6. The defining relations for Q_8.

4.2.2. Stereograph projection

In Section 4.2.3, another description of the quaternions will be given, in which the ordered subgroup $I = (-1, -i, 1, i)$ and its coset $jI = (-j, k, -j, -k)$ are used to give a 2-strings-with-(mod 4)-quipu description of Q_8. In Section 4.3, a similar method will be used to describe the dicyclic groups. Therein, quipu of higher moduli are also used. In all these cases, the subgroup and its coset lie upon a pair of linked circles in the 3-sphere S^3.

In general, it would be nice to have a working image of the 3-sphere that is not populated by angels and devils, as in the case in Dante's description of the universe [13]. Just as the 2-sphere does not embed into the plane, the 3-sphere does not embed into $\mathbb{R}^3$. But we can see almost all of both in a single Euclidean image by means of *stereographic projection*.

To illustrate the idea, we first consider the circle, and then the 2-sphere, and finally the 3-sphere. By that time, it should be easy to write down an explicit formula that gives a homeomorphism

$$\mathbb{R}^n \longleftarrow S^n \backslash \{(\underbrace{0, 0, \ldots, 0, 1}_{(n+1)\text{-coord.s}})\}$$

from the sphere minus a polar point to n-dimensional Euclidean space. One way in which we think of S^n as being n-dimensional is via such a homeomorphism.

A clever reader will wonder: "How do we actually know that $\mathbb{R}^n$ is n-dimensional?" The answer to this is one of those miraculous results of 20th century mathematics that is known as invariance of domain. But the proof of invariance of domain is not relevant to the rest of this book.

In Fig. 4.7, a diagram indicates the projection from the unit circle $S^1 = \{(x, y) : x^2 + y^2 = 1\}$ to the line $\mathbb{R}$. The line segment

$$\vec{L}_x(t) = (0, 1) + t(x, -1) = (tx, 1 - t) \quad \text{for } t \in [0, 1]$$

is indicated as a dotted arc. It intersects the circle and terminates at $x \in \mathbb{R}$. The family $\{\vec{L}_x(t) : x \in \mathbb{R}\}$ gives the projection from $S^1 \backslash (0, 1)$ to $\mathbb{R}$.

The 3-Dimensional Sphere is a Group

Or if you prefer, define $\mathbb{R} \xleftarrow{p} (S^1 \backslash (0,1))$ by $p(x,y) = x/(1-y)$. Since $y \neq 1$, the function is defined and is continuous on $S^1 \backslash (0,1)$. This expression was found by considering the time the line $\vec{N}_{(x,y)}(t) = (tx, 1 + t(y-1))$ intersects the line $\mathbb{R} \times \{0\}$ in the plane.

For reasons to be seen, it is amusing to determine where the line $\vec{L}_x(t)$ intersects the unit circle. Instead of determining "where," first we find "when." That is, we solve the equation

$$(tx)^2 + (1-t)^2 = 1$$

for t. Then $t = 0$, or $t = 2/(x^2+1)$. The coordinates of the point on the unit circle S^1 that correspond to $t = 2(x^2+1)^{-1}$ are

$$\left(\frac{2x}{x^2+1}, \frac{x^2-1}{x^2+1} \right).$$

Suppose that $x \in \mathbb{R}$ is a rational number, say $x = p/q$. Then

$$\left(\frac{2x}{x^2+1}, \frac{x^2-1}{x^2+1} \right) = \left(\frac{2pq}{p^2+q^2}, \frac{p^2-q^2}{p^2+q^2} \right),$$

where $(2pq, p^2-q^2, p^2+q^2)$ is a Pythagorean triple. For example, if $x = 7/5$, then $2 \cdot 7 \cdot 5 = 70$, $49 - 25 = 24$, and $49 + 25 = 74$. Meanwhile, $576 + 4900 = 5476 = 74^2$. Or we can divide everything by 2 to obtain the Pythagorean triple $(12, 35, 37)$. Since $144 + 1225 = 1369$, this is also a Pythagorean triple. This method of generating such triples is probably as old as civilization. We still find it amusing.

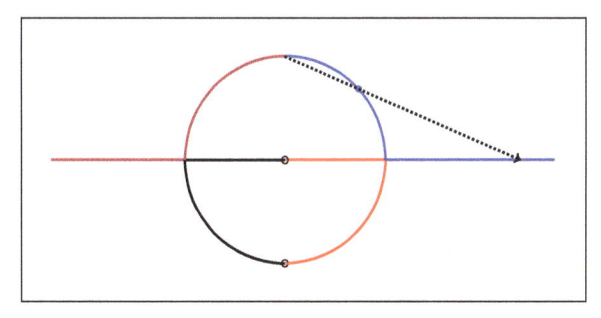

Figure 4.7. Stereographic projection from the circle S^1 to the line $\mathbb{R}$.

In the case of the 2-sphere, a line segment
$$\vec{L}_x(t) = (tx, ty, (1-t)) \quad \text{for } t \in [0,1]$$
is illustrated in Fig. 4.8. The times at which it intersects the sphere are $t = 0$ and $t = \frac{2}{x^2+y^2+1}$. The coordinates of the point on the sphere that correspond to $(x, y) \in \mathbb{R}^2$ are
$$\left(\frac{2x}{x^2+y^2+1}, \frac{2y}{x^2+y^2+1}, \frac{x^2+y^2-1}{x^2+y^2+1} \right).$$

The stereographic projection function $\mathbb{R}^2 \xleftarrow{p} S^2 \setminus \{(0,0,1)\}$ is defined by $p(x, y, z) = (x/(z-1), y/(z-1), 0)$ when $x^2 + y^2 + z^2 = 1$ and $z \neq 1$. This formula can be derived by taking the line $\vec{L}_{(x,y,z)}(t) = (tx, ty, 1+t(z-1))$ that starts at $(0,0,1)$ and passes through the point (x, y, z) and then computing the time at which $1 + t(z-1) = 0$. Since it is rational, and $z \neq 1$, it is continuous. Its inverse is given by
$$(x, y) \mapsto \left(\frac{2x}{x^2+y^2+1}, \frac{2y}{x^2+y^2+1}, \frac{x^2+y^2-1}{x^2+y^2+1} \right).$$

Thus, stereographic projection upon the punctured 2-sphere is a homeomorphism.

The *equatorial great circles* $x^2 + z^2 = 1$ and $y^2 + z^2 = 1$ go to the coordinate arcs in the plane, as indicated in the figure, and the equator $x^2 + y^2 = 1$ is fixed by the projection. Please note a notational slang, that is analogous to some used above, has been adopted.

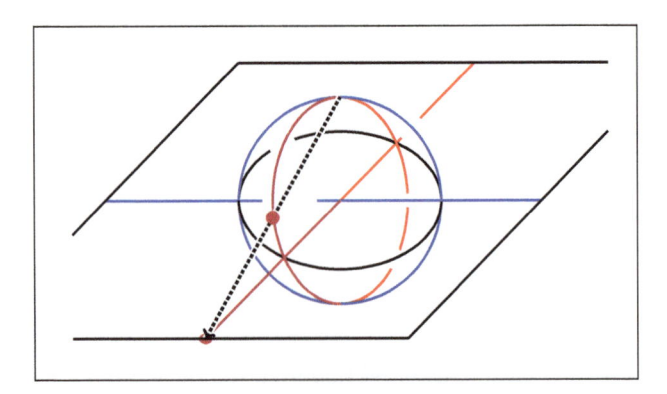

Figure 4.8. Stereographic projection from the 2-sphere S^2 to the plane $\mathbb{R}^2$.

In the case of the 3-sphere, the stereographic projection that we consider uses the point $(1, 0, 0, 0)$ from which to project. Define

$$\mathbb{R}^3 \xleftarrow{p} S^3 \setminus \{(1, 0, 0, 0)\}$$

by

$$p((w, x, y, z)) = \left(\frac{x}{(1 - w)}, \frac{y}{(1 - w)}, \frac{z}{(1 - w)} \right).$$

In a manner similar to that above, the image of (w, x, y, z) under p is determined by computing the time t at which $\vec{L}_{(w,x,y,z)}(t) = (1 + t(w - 1), tx, ty, tz))$ intersects the horizontal (x, y, z)-space: $1 + t(w - 1) = 0$.

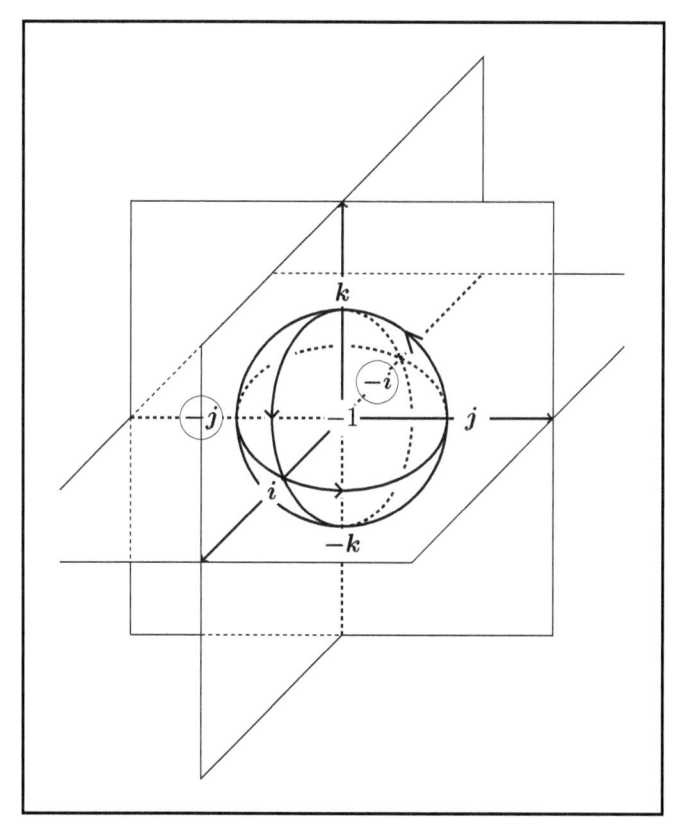

Figure 4.9. Stereographic projection from the 3-sphere S^3 to 3-space $\mathbb{R}^3$.

It doesn't seem possible, in this case, to draw a figure that is analogous to Fig. 4.7 or 4.8. Indeed, the 3-dimensional sphere S^3 cannot be embedded into $\mathbb{R}^3$. Instead, the image of several pieces of the 3-sphere are indicated in Fig. 4.9. The equatorial 2-spheres $w^2 + x^2 + y^2 = 1$, $w^2 + x^2 + z^2 = 1$, and $w^2 + y^2 + z^2 = 1$ map to the coordinate planes in space. The sphere $x^2 + y^2 + z^2 = 1$ maps to the round sphere that surrounds the image of $(-1, 0, 0, 0)$ that is labeled -1 in the figure. The identity element $(1, 0, 0, 0)$ is the "point at infinity" that is missing in the illustration. The set of eight intersection points among any three of these spheres corresponds to the points of Q_8.

In the positive region of 3-space as illustrated in Fig. 4.9, there is a spherical triangle $T = \{(x, y, z) : x^2 + y^2 + z^2 = 1;\ 0 < x, y, z\}$, whose vertices in the figure are labeled $\boldsymbol{i}$, $\boldsymbol{j}$, and $\boldsymbol{k}$. The quarter-circular arcs of this triangle are oriented as $\boldsymbol{i} \to \boldsymbol{j} \to \boldsymbol{k} \to \boldsymbol{i}$. These orientations are depicted to further emphasis the alphabetical signed multiplication among these elements of Q_8.

In Section 4.2.3, we will use the subgroup $I = (-1, -\boldsymbol{i}, 1, \boldsymbol{i})$ and its coset $\boldsymbol{j}I = (-\boldsymbol{j}, \boldsymbol{k}, -\boldsymbol{j}, -\boldsymbol{k})$ to give an alternate quipu description of Q_8. At this time, we observe that these two subgroups appear on the red circle and the blue axis in Fig. 4.10. But remember that that axis is the image of the great circle $\{(w, x, 0, 0) : w^2 + x^2 = 1\}$ in the 3-sphere. Intrinsically, within S^3, these two circles form what is commonly called the *Hopf link*. The orientations on these two curves in Fig. 4.10 demonstrate ordering on I and $\boldsymbol{j}I$.

The 3-Dimensional Sphere is a Group 165

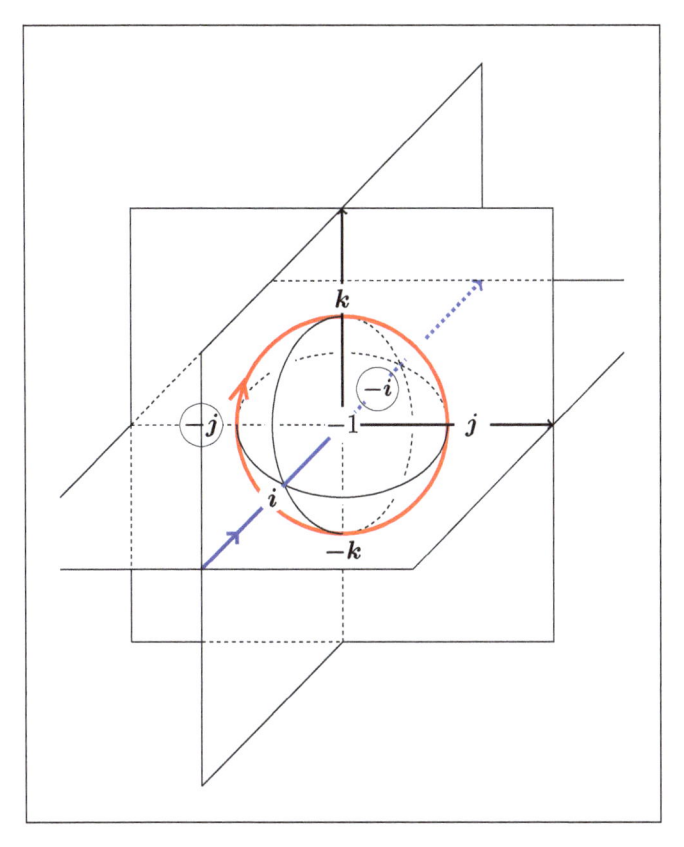

Figure 4.10. The cosets $I = (-1, -i, 1, i)$ and jI upon a Hopf link in S^3.

It is an interesting fact that there is a continuous surjection $S^2 \xleftarrow{h} S^3$ that cannot be contracted to a point. The function is called the *Hopf fibration*. The preimage of any point in the 2-sphere is a circle. The preimage of any pair of points is a pair of linked circles, as indicated in Fig. 4.11. This Hopf link is also the preimage under stereographic projection of the red circle and blue axis in the 3-sphere.

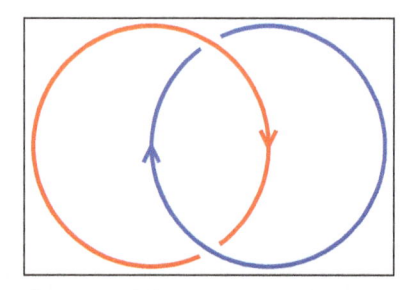

Figure 4.11. A Hopf link in S^3.

The dicyclic groups that will be described in Section 4.3 also lie on this pairs of linked circles.

In general, stereographic projection is a useful tool for studying the 3-sphere S^3. Another finite subgroup that will interest us is called the *binary tetrahedral group*. This group is denoted as $\widetilde{A_4}$, and there is a 2-to-1 group homomorphism to the alternating group A_4. That is a surjection whose kernel is $\mathbb{Z}/2$. While we are drawing pictures, an illustration of 20 of the elements in $\widetilde{A_4}$ is given in Fig. 4.12. The identity is at infinity, and the locations of the three points $-\boldsymbol{i}$, $-\boldsymbol{j}$, and $-\boldsymbol{k}$ should be easy to see by following the major axes backward.

The figure is stylized or merely topological. The edges of the hypercube in the illustration should be curved under stereographic projection.

$\frac{(1,-1,-1,1)}{2} = -b^2$

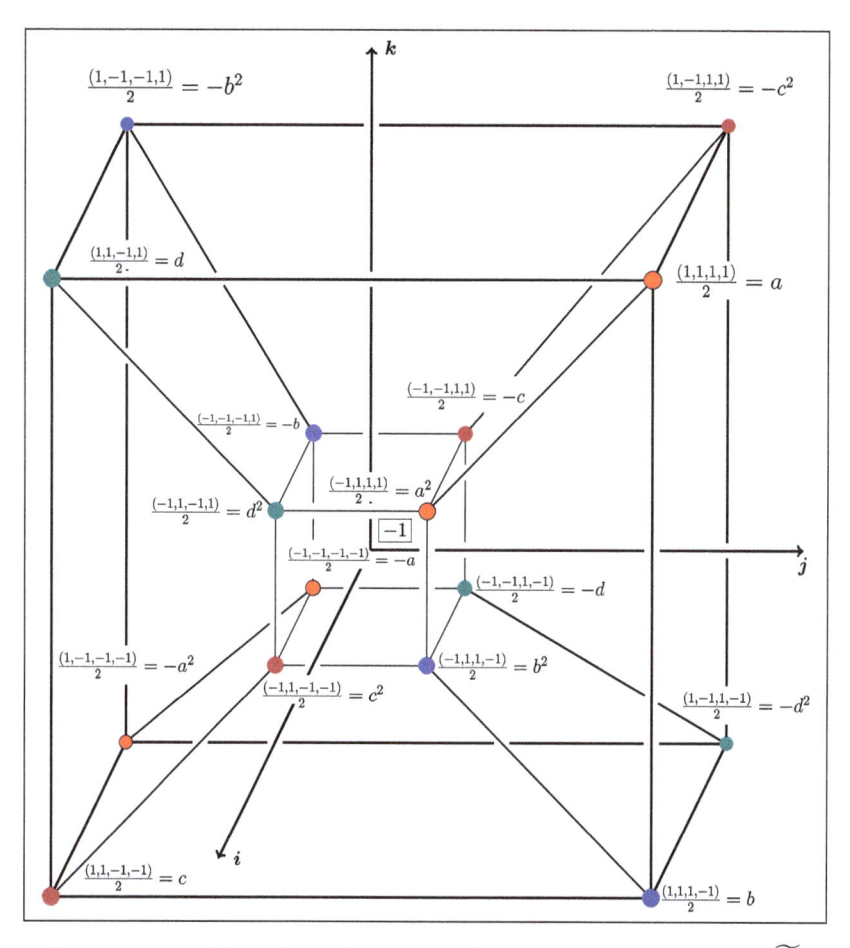

$\frac{(1,-1,1,1)}{2} = -c^2$

$\frac{(1,1,-1,1)}{2.} = d$

$\frac{(1,1,1,1)}{2} = a$

$\frac{(-1,-1,1,1)}{2} = -c$

$\frac{(-1,-1,-1,1)}{2} = -b$

$\frac{(-1,1,1,1)}{2.} = a^2$

$\frac{(-1,1,-1,1)}{2} = d^2$

$\boxed{-1}$

$\frac{(-1,-1,-1,-1)}{2} = -a$

$\frac{(-1,1,1,-1)}{2} = -d$

$\frac{(1,-1,-1,-1)}{2} = -a^2$

$\frac{(-1,1,1,-1)}{2} = b^2$

$\frac{(-1,1,-1,-1)}{2} = c^2$

$\frac{(1,-1,1,-1)}{2} = -d^2$

$\frac{(1,1,-1,-1)}{2} = c$

$\frac{(1,1,1,-1)}{2} = b$

Figure 4.12. Most of the elements in the binary tetrahedral group $\widetilde{A_4}$.

The elements in the binary tetrahedral group $\widetilde{A_4}$ are

$$\pm 1, \pm i, \pm j, \pm k$$

and the following 16 elements:

$$[(\pm 1) + (\pm i) + (\pm j) + (\pm k)]/2.$$

Exercise 10. Use Table 4.1 and variations to compute the powers of the elements

$$a = [1 + i + j + k]/2, \quad b = [1 + i + j - k]/2,$$
$$c = [1 + i - j - k]/2, \quad \text{and} \quad d = [1 + i + j - k]/2.$$

Demonstrate that they are as indicated in Fig. 4.12.

You may find it helpful to copy Table 4.1 upon an electronic note pad several times and edit the entries by attaching signs to the appropriate multiplicands. Correspondingly, edit the signs of the entries of the products. Alternatively, substitute signs into Table 4.2. Also, observe that the factors of $1/2$ among the entries seem to manage themselves. This exercise is inserted here so that you can take a break from the narrative and foreshadow other intricate calculations in $\widetilde{A_4}$ and other finite subgroups of S^3.

4.2.3. The quaternions: Part 2

Consider the ordered subgroup $I = (-1, -i, 1, i)$ and its coset $jI = (-j, k, -j, -k)$. Compute the actions of i and j upon each. We obtain $iI = (-i, 1, i, -1) = [I, 1]$, and $i(jI) = (ij)I = kI = (-k, -j, k, j) = [jI, 3]$. In other words, multiplication by i increments the $(1, i)$ circle positively by $90°$, and multiplication by i increments the (j, k) circle by $270°$. Similarly, multiplication by j moves the $(1, i)$ circle to the (j, k) circle: $jI = jI = [jI, 0]$, while $j(jI) = -I = (1, i, -1, -i) = [I, 2]$. So, that multiplication by j moves the (j, k) circle to the $(1, i)$ circle and twists it by $180°$. The representations of i and j are depicted in Fig. 4.13.

You may compute the actions of k upon the ordered cosets I and jI or multiply the representatives of i and j by vertically juxtaposing as in Fig. 4.15. In either case, the catalogue of elements in Q_8 as represented by 2-strings-with-(mod-4)-quipu appears in Fig. 4.14.

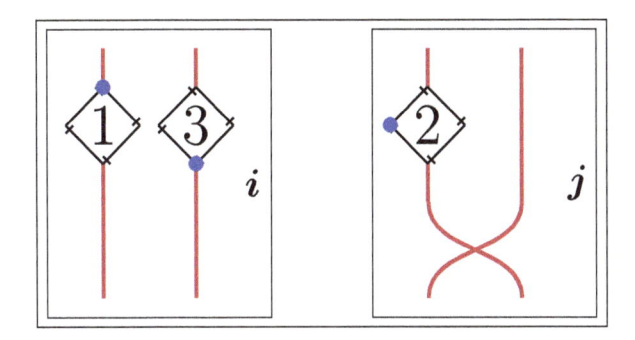

Figure 4.13. The (mod 4)-quipu representations of i and j.

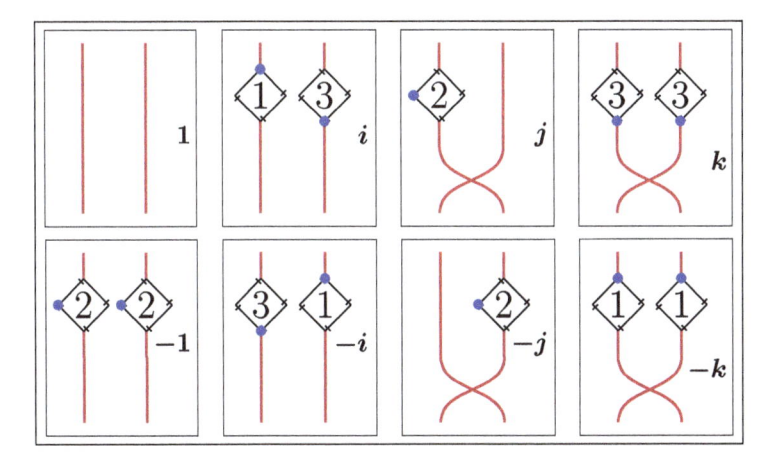

Figure 4.14. The (mod 4)-quipu representations of the quaternions Q_8.

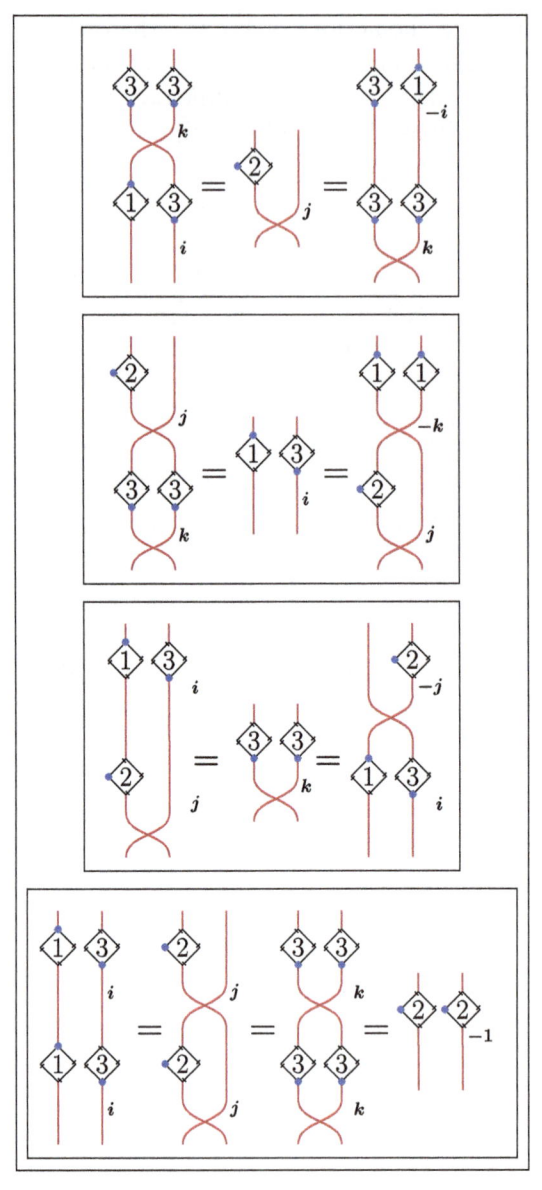

Figure 4.15. The defining relations for Q_8 among 2-strings-with-(mod 4)-quipu.

4.2.4. Matrix representations

In Fig. 3.19, we observed that elements in a semi-direct product with a permutation group could be represented as "permutation matrices with group entries." In the representation of Fig. 4.14, the quipu are elements of a cyclic group of order 4. For example, in the complex numbers $\mathbb{C}$, the powers of i can serve as such a group. Recall that the row entries of the matrix representatives from top to bottom correspond to the (generalized) quipu that appear left to right in the string diagrams. So, the 2-strings-with-(mod 4)-quipu depicted in Fig. 4.15 correspond to the so-called *Pauli matrices*, as indicated.

$$\leftrightharpoons \begin{bmatrix} i & 0 \\ 0 & -i \end{bmatrix},$$

$$\leftrightharpoons \begin{bmatrix} 0 & -1 \\ 1 & 0 \end{bmatrix},$$

and

$$\leftrightharpoons \begin{bmatrix} 0 & -i \\ -i & 0 \end{bmatrix}.$$

These matrices all have trace[6] $= 0$, and their determinants are 1. Each corresponds to a 90° rotation of a complex plane $\mathbb{C}^2$ (which has four real dimensions).

---

[6]$\operatorname{Tr} \begin{bmatrix} a & b \\ c & d \end{bmatrix} = a + d$. $\operatorname{Det} \begin{bmatrix} a & b \\ c & d \end{bmatrix} = ad - bc$.

By dividing the entries by 2, one can obtain the standard basis of the lie algebra $su(2)$, and the Lie bracket $[A, B] = [AB - BA]$ induces the cross-product multiplication that is, perhaps, familiar from vector calculus.

Exercise 11. Recall that the *conjugate transpose* of a matrix $A = \begin{bmatrix} \alpha & \beta \\ \gamma & \delta \end{bmatrix}$ is the matrix $A^* = \begin{bmatrix} \overline{\alpha} & \overline{\gamma} \\ \overline{\beta} & \overline{\delta} \end{bmatrix}$, where, for example, $\overline{\alpha} = x - iy$ when $\alpha = x + iy$.

Determine the conjugate transposes of the matrices $\begin{bmatrix} i & 0 \\ 0 & i \end{bmatrix}$, $\begin{bmatrix} 0 & -1 \\ 1 & 0 \end{bmatrix}$, and $\begin{bmatrix} 0 & -i \\ -i & 0 \end{bmatrix}$, and observe that these are the matrices that correspond, respectively, to $-i$, $-j$, and $-k$. Conclude that these three matrices are unitary.

Compute directly that
$$w\begin{bmatrix} 1 & 0 \\ 0 & 1 \end{bmatrix} + x\begin{bmatrix} i & 0 \\ 0 & -i \end{bmatrix} + y\begin{bmatrix} 0 & -1 \\ 1 & 0 \end{bmatrix} + z\begin{bmatrix} 0 & -i \\ -i & 0 \end{bmatrix}$$
$$= \begin{bmatrix} w + xi & -y - zi \\ y - zi & w - xi \end{bmatrix} = A(w, x, y, z)$$
is a unitary matrix when $w^2 + x^2 + y^2 + z^2 = 1$.

Theorem 4. *The function* $\mathrm{SU}(2) \xleftarrow{A} S^3$ *that is defined by*
$$w + xi + yxj + zxk \mapsto A(w, x, y, z)$$
defines a group isomorphism between the group S^3 and the group of 2×2 unitary matrices.

Proof. For $\ell = 1, 2$, write
$$A_\ell = A(a_\ell, b_\ell, c_\ell, d_\ell) = \begin{bmatrix} a_\ell + b_\ell i & -c_\ell - d_\ell i \\ c_\ell - d_\ell i & a_\ell - b_\ell i \end{bmatrix} = \begin{bmatrix} \alpha_\ell & -\beta_\ell \\ \overline{\beta_\ell} & \overline{\alpha_\ell} \end{bmatrix},$$
where $\alpha_\ell = a_\ell + b_\ell i$ and $\beta_\ell = c_\ell + d_\ell i$. Then the (1, 1)-entry of $A_1 A_2$ is
$$\alpha_1 \alpha_2 - \beta_1 \overline{\beta_2} = (a_1 a_2 - b_1 b_2 + c_1 c_2 + d_1 d_2)$$
$$+ (b_1 a_2 + a_1 b_2 + d_1 c_2 - c_1 d_2)i.$$

Compare these entries with the sum of the coefficients of 1 and i in Table 4.2. Similarly, the $(1,2)$-entry of $A_1 A_2$ is

$$-[\alpha_1 \beta_2 + \beta_1 \bar{\alpha}_2] = -[(a_1 c_2 - b_1 d_2 + c_1 a_2 + d_1 b_2))$$
$$+ (b_1 c_2 + a_1 d_2 + d_1 a_2 - c_1 b_2)i].$$

These terms (with the outer sign neglected) should be compared with the sums of the coefficients of j and k in Table 4.2. Thus, the products are identical in this case.

Now, suppose that $B = \begin{bmatrix} \alpha & \beta \\ \gamma & \delta \end{bmatrix}$ is unitary and of determinant $\alpha\delta - \beta\gamma = 1$. The inverse of a (2×2) determinant 1 matrix is given by the recipe: "Switch the diagonals and change the signs of the off diagonals." Since B is unitary, $B^{-1} = B^*$ — the conjugate transpose of B. We obtain that

$$\begin{bmatrix} \delta & -\beta \\ -\gamma & \alpha \end{bmatrix} = \begin{bmatrix} \bar{\alpha} & \bar{\gamma} \\ \bar{\beta} & \bar{\delta} \end{bmatrix}.$$

Therefore, B is of the form

$$B = \begin{bmatrix} \alpha & \beta \\ -\bar{\beta} & \bar{\alpha} \end{bmatrix}.$$

This shows that the function $\text{SU}(2) \xleftarrow{A} S^3$ is surjective. Injectivity follows by equating the real and imaginary parts of the entries of $A(a_1, b_1, c_1, d_1) = A(a_2, b_2, c_2, d_2)$. This completes the proof. □

From a stylistic point view, it might make sense to end the chapter with the proof of Theorem 4 since its proof seems to be a climactic end to the study of 2-strings-with-(mod 4)-quipu diagrams. However, the quaternions Q_8 are the smallest example of a dicyclic group, and a dicyclic group is related to a dihedral group via a surjective homomorphism that has a $\mathbb{Z}/2$ as its kernel.

Moreover, we are in a good place to describe the 2-to-1 homomorphism $SO(3) \xleftarrow{p} SU(2)$ from the set of (2×2)-unitary matrices to the group of orientation preserving rotations of 3-dimensional space that also can be described as the set of (3×3)-matrices of determinant 1, whose columns form an orthogonal[7] basis for $\mathbb{R}^3$. The following section gives the homomorphism p and presents quipu diagrams for the dicyclic groups.

Returning to stylistic considerations, the following section, then, provides a (literary) resolution so that you can better understand the characters that have been introduced thus far.

4.3. The dicyclic groups

The background material here has been compiled from a number of wikipedia sources, specifically Refs. [14, 15].

The second representation of the group Q_8 is part and parcel of the following diagrammatic descriptions of the dicyclic groups. The dicyclic group of order $4n$ is given by the presentation

$$\mathrm{Dic}_n = \langle \rho, x : \rho^{2n} = 1,\ x^2 = \rho^n,\ \rho x = x\rho^{-1} \rangle.$$

It maps 2-to-1 to the dihedral group D_{2n} of order $2n$.

4.3.1. Revisiting the dihedral group

Before discussing the groups Dic_n, a little more material about the dihedral groups will be included. When

$$\zeta_n = e^{2\pi i/n} = \cos\left(\frac{2\pi}{n}\right) + \sin\left(\frac{2\pi}{n}\right)i,$$

the powers of ζ_n, that appear in *the vertex set* $V_n = \{\zeta_n^\ell : \ell = 0, 1, \ldots, n-1\}$, are equally spaced points upon the unit circle S^1. Therefore, they can serve as the vertices of a regular n-gon for $n = 3, 4, \ldots$ The degenerate case $n = 2$ is represented in Fig. 2.36, and we will postpone further discussions about it.

---

[7]Each column has length 1 in the standard distance, and any pair of columns are perpendicular.

The 3-Dimensional Sphere is a Group

Let P_n denote the regular polygon that has as its set of vertices V_n. The edges of P_n are line segments $\zeta_n^\ell \to \zeta_n^{\ell+1}$, where the exponents are read modulo n. The group of symmetries of P_n is the *dihedral group* which is given by the presentation

$$D_{2n} = \langle s, t : s^n = t^2 = 1, st = ts^{-1} \rangle.$$

We let s denote the transformation $\zeta^\ell \mapsto \zeta^{\ell+1}$ that is defined on P_n. The exponents are read modulo n so that $\zeta^{n-1} \mapsto \zeta^0 = 1$. The action on the vertex set V_n induces an action on the set of edges as well. Let $\overline{\zeta^\ell} = \cos(2\pi\ell/n) - \sin(2\pi\ell/n)i = \zeta^{-\ell}$ denote the complex conjugate of ζ^ℓ for $\ell = 0, 1, \ldots, n-1$. Then t is the transformation $t : \zeta^\ell \mapsto \zeta^{-\ell}$. (The letters that denote these transformations have changed from those that were introduced in Section 2.3.3.)

Label the vertices of P_n: $0 \leftrightarrow \zeta^0$, $1 \leftrightarrow \zeta$, ..., $\ell \leftrightarrow \zeta^\ell$, ..., $(n-1) \leftrightarrow \zeta^{(n-1)}$. Then as elements of the permutation group $\Sigma_n = \Sigma(\{0, 1, \ldots, n-1\})$, the transformations that correspond to s and t are depicted in Fig. 4.16.

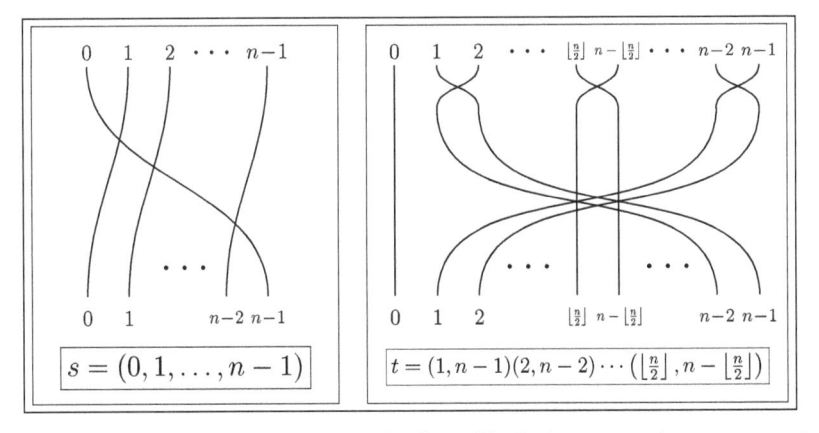

Figure 4.16. The generators of the dihedral group in permutation representation.

The figure indicates the correspondences

$$s \leftrightarrow (0, 1, \ldots, n-1)$$

and

$$t \leftrightarrow (1, n-1)(2, n-2) \cdots \left(\left\lfloor \frac{n}{2} \right\rfloor, n - \left\lfloor \frac{n}{2} \right\rfloor\right),$$

where $\lfloor x \rfloor$ denotes the greatest integer less than or equal to x. In case $n = 2\ell$, then the last cycle $(\lfloor \frac{n}{2} \rfloor, n - \lfloor \frac{n}{2} \rfloor) = (\ell, 2\ell - \ell) = (\ell, \ell)$ indicates that the ℓth vertex is fixed by t. As a matter of fact, so is the zeroth vertex, and t corresponds to a reflection about the horizontal axis of the polygon P_n. In the even case, this axis contains the pair of vertices $\{0, \ell\}$ in P_n. In the odd case, only 0 is fixed. Meanwhile, s corresponds to an anti-clockwise rotation.

It may be difficult to parse the expression for t when $n = 3$. In this case, $t \leftrightarrow (1, n-1)$ since the remaining transpositions are vacuous. Figure 4.17, indicates the generators of D_6 which is also described as the symmetric group $\Sigma_3 = \Sigma_3\{0, 1, 2\}$.

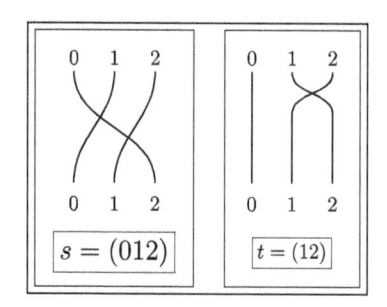

Figure 4.17. The generators of the dihedral group D_6.

For reasons that will be apparent shortly, let $P_n = P_n(x, y)$ denote the polygon P_n in the (x, y)-plane with its "initial" vertex at $(1, 0)$. More specifically, we should say that the vertex labeled 0 is at $[1, 0, 0]^t \in \mathbb{R}^3$. There is a copy of P_n in the (y, z)-plane that will be denoted $P_n(y, z)$. The corresponding vertex 0 of $P_n(y, z)$ is at the point $[0, 1, 0]^t \in \mathbb{R}^3$. Place the vertex ℓ of $P(y, z)$ at the point $[0, \cos\left(\frac{2\pi\ell}{n}\right), \sin\left(\frac{2\pi\ell}{n}\right)]^t$.

4.3.2. Projection from S^3 to the 3-dimensional rotation group

It is a good time to introduce a formula for the projection from $SU(2) = S^3$ to the group, $SO(3)$, of (3×3) orthogonal matrices that are of determinant 1. The "equality" $SU(2) = S^3$ is a consequence of Theorem 4, and it is more convenient for us to write elements therein in the form $w + x\boldsymbol{i} + y\boldsymbol{j} + z\boldsymbol{k}$.

The continuous 2-to-1 covering[8] $SO(3) \xleftarrow{p} S^3$ is given by the formula

$$p(w + xi + yj + zk)$$
$$= \begin{bmatrix} 1 - 2(y^2 + z^2) & 2(xy - zw) & 2(xz + yw) \\ 2(xy + zw) & 1 - 2(x^2 + z^2) & 2(yz - xw) \\ 2(xz - yw) & 2(yz + xw) & 1 - 2(x^2 + y^2) \end{bmatrix}.$$

It is a straightforward, yet tedious, calculation to show that the columns of the matrix form an orthogonal basis for $\mathbb{R}^3$. If, for example, a column of this matrix is of the form $[a, b, c]^t$, then one can show that $a^2 + b^2 + c^2 = 1$ provided, of course, that $w^2 + x^2 + y^2 + z^2 = 1$, and if $[a_\ell, b_\ell, c_\ell]^t$ and $[a_m, b_m, c_m]^t$ are a pair of columns, then one shows that $a_\ell a_m + b_\ell b_m + c_\ell c_m = 0$. The calculation is tedious since it involves quartic functions.

It is easy to see that antipodal points have the same image, $p(-w - xi - yj - zk) = p(w + xi + yj + zk)$, because each entry is a homogeneous degree 2 polynomial.

The geometric representation of the dihedral group D_{2n} assigns (3×3)-orthogonal, determinant 1, matrices to the rotation s and the reflection t. The symmetries of the regular polygon $P_n(y, z)$ are given by the corresponding matrices:

$$s \leftrightarrow S = \begin{bmatrix} 1 & 0 & 0 \\ 0 & \cos\left(\frac{2\pi}{n}\right) & -\sin\left(\frac{2\pi}{n}\right) \\ 0 & \sin\left(\frac{2\pi}{n}\right) & \cos\left(\frac{2\pi}{n}\right) \end{bmatrix}$$

and

$$t \leftrightarrow T = \begin{bmatrix} -1 & 0 & 0 \\ 0 & 1 & 0 \\ 0 & 0 & -1 \end{bmatrix}.$$

From the point of view of the (y, z)-plane, the matrix T reflects the plane about the horizontal y-axis. It does so in 3-space by simultaneously reflecting the x-axis. To imagine this, hold up your right

---

[8]It is a *covering* in the sense that it is a surjective group homomorphism and each point in the image has a small neighborhood, U, for which the preimage is a pair of neighborhoods U_1 and U_2, each of which maps homeomorphically onto U.

hand with the index finger pointing up, the thumb pointing behind your right shoulder, and the middle finger pointing toward your nose. In this (non-standard) reference frame, the middle finger represents the x-axis, the thumb the y-axis, and the index finger represents the z-axis.

Then rotate your hand by pointing your elbow upward, until your index finger points downward. While the position may be uncomfortable, it did not involve a reflection, and the final hand position represents the image of T upon the finger basis.

4.3.3. The dicyclic group as a subset of the 3-sphere

We turn now to include the dicyclic groups into S^3 in such a way that the projection $SO(3) \xleftarrow{p} S^3$ induces the covering p' in the seseq

$$1 \longleftarrow D_{2n} \xleftarrow{p'} \text{Dic}_n \xleftarrow{i} \mathbb{Z}/2 \longleftarrow 0.$$

Consider the inclusion $\text{Dic}_n \subset S^3$ that is given by $\rho \mapsto \cos(\pi/n) + \sin(\pi/n)i$ and $x \mapsto j$.

The elements of Dic_n correspond to the points (in a notational abuse) $\rho^\ell = \cos\left(\frac{\ell\pi}{n}\right) + \sin\left(\frac{\ell\pi}{n}\right)i$, and $\rho^\ell x = \cos\left(\frac{\ell\pi}{n}\right)j + \sin\left(\frac{\ell\pi}{n}\right)k$. Apply the projection p and recall the appropriate trigonometric identities to show that

$$p(\rho^\ell) = \begin{bmatrix} 1 & 0 & 0 \\ 0 & \cos\left(\frac{2\pi\ell}{n}\right) & -\sin\left(\frac{2\pi\ell}{n}\right) \\ 0 & \sin\left(\frac{2\pi\ell}{n}\right) & \cos\left(\frac{2\pi\ell}{n}\right) \end{bmatrix}$$

and

$$p(\rho^\ell x) = \begin{bmatrix} -1 & 0 & 0 \\ 0 & \cos\left(\frac{2\pi\ell}{n}\right) & \sin\left(\frac{2\pi\ell}{n}\right) \\ 0 & \sin\left(\frac{2\pi\ell}{n}\right) & -\cos\left(\frac{2\pi\ell}{n}\right) \end{bmatrix}.$$

The 3-Dimensional Sphere is a Group   179

In particular, $p(x) = T$, and $p(\rho) = S$. So, the projection p induces the projection p' in the seseq that is written above.

Henceforth, the notational abuse $\rho^\ell = \cos\left(\frac{\ell\pi}{n}\right) + \sin\left(\frac{\ell\pi}{n}\right)i$, and $\rho^\ell x = \cos\left(\frac{\ell\pi}{n}\right)j + \sin\left(\frac{\ell\pi}{n}\right)k$ will be maintained. That is, the dicyclic group Dic_n will be identified with its image in S^3. These $4n$ points lie along Hopf link in S^3 that is depicted in Fig. 4.10. But take some care here. If n is odd, the points $\pm i$ are **not** among the points in Dic_n. Nevertheless, there is a cyclic subgroup of order 4 contained within the dicyclic group Dic_n for all n. Let $J = (-1, -j, 1, j)$ denote the ordered subgroup that consists of powers of j.

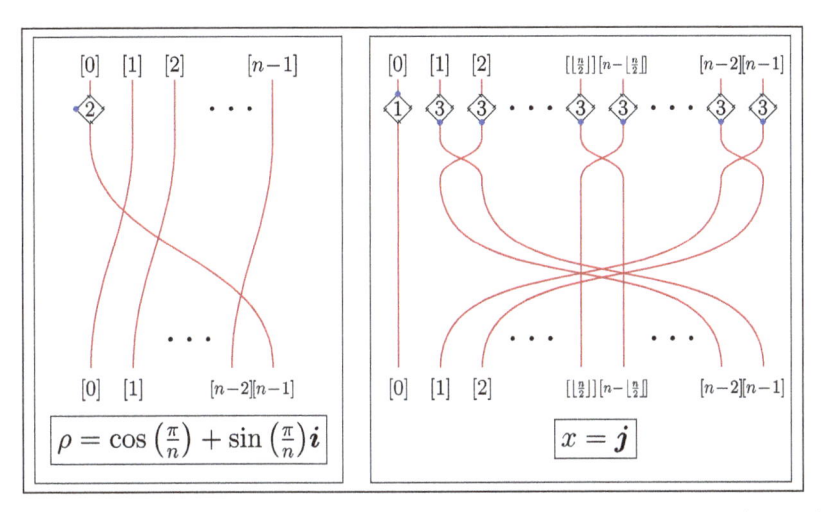

Figure 4.18. The generators of the dicyclic group as strings-with-(mod 4)-quipu.

As in the case of the dihedral group, the base case $n = 3$ for the representation of j may be difficult to discern. So, in Fig. 4.19, the generators of Dic_3 are illustrated.

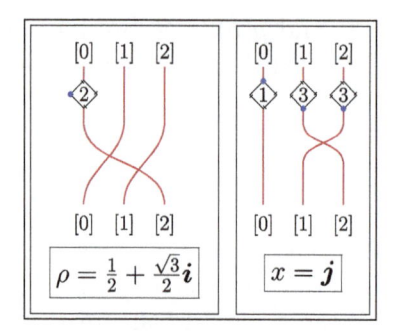

Figure 4.19. The generators of the 12-element dicyclic group Dic$_3$ as strings-with-(mod 4)-quipu.

To justify the n-strings-with-(mod 4)-quipu representations of the generators $\rho = \cos\left(\frac{\pi}{n}\right) + \sin\left(\frac{\pi}{n}\right)\boldsymbol{i}$ and $x = \boldsymbol{j}$ of Dic$_n$ that are depicted in Fig. 4.18, the actions of ρ and $\boldsymbol{j}$ upon the cosets of the ordered subgroup $J = (-1, -\boldsymbol{j}, 1, \boldsymbol{j})$ will be computed.

For $\ell = 0, 1, \ldots, n-1$, let

$$[\ell] = \rho^\ell J = (-\rho^\ell, -\rho^\ell \boldsymbol{j}, \rho^\ell, \rho^\ell \boldsymbol{j}).$$

Note that $[0] = J$. Then for $[\ell] = [0], [1], \ldots, [n-2]$, multiplication by ρ increments the cosets forward: $\rho[\ell] = [\ell + 1]$ and

$$\rho[n-1] = (-\rho^n, -\rho^n \boldsymbol{j}, \rho^n, \rho^n \boldsymbol{j}) = (1, \boldsymbol{j}, -1, -\boldsymbol{j}) = [[0], 2].$$

That is, multiplication by ρ cyclically permutes the cosets, and the last coset is moved to the zeroth coset with a 180° rotation upon it. This gives the left-hand representation in Fig. 4.18.

We compute

$$\boldsymbol{j}[\ell] = (-\boldsymbol{j}\rho^\ell, -\boldsymbol{j}\rho^\ell \boldsymbol{j}, \boldsymbol{j}\rho^\ell, \boldsymbol{j}\rho^\ell \boldsymbol{j}) = (-\rho^{-\ell}\boldsymbol{j}, \rho^{-\ell}, \rho^{-\ell}\boldsymbol{j}, -\rho^{-\ell}).$$

In case $\ell = 0$,

$$\boldsymbol{j}[0] = (-\boldsymbol{j}, 1, \boldsymbol{j}, -1) = [[0], 1].$$

That is, $\boldsymbol{j}$ fixes the coset $[0]$ setwise but rotates its elements by 90° or one step forward. To identify $\boldsymbol{j}[\ell]$, we do a little modular arithmetic. The non-negative representation of $-\ell$, modulo n, is $n - \ell$. Therefore, $\boldsymbol{j}$ projects to the product of transpositions, $(1, n - 1)(2, n - 2) \cdots \left(\lfloor \frac{n}{2} \rfloor, n - \lfloor \frac{n}{2} \rfloor\right)$. See also Fig. 4.16. But each of the cosets is twisted by a 270° rotation or moves 3 steps forward. That is, $\boldsymbol{j}[\ell] = [[n - \ell], 3]$. This gives the representation in the right-hand side of Fig. 4.18.

The projection $D_{2n} \xleftarrow{p'} \mathrm{Dic}_n$ from the dicyclic group to the dihedral group is achieved by ignoring the (mod 4)-quipu upon the strings in either Fig. 4.18 or 4.19 to obtain the diagram that appears in Fig. 4.16 or 4.17.

That the kernel of this projection is $\mathbb{Z}/2$ is less obvious from the illustrations. But observe that $-1 = -1 + 0\boldsymbol{i}$ is also in the dicyclic group, and this element would be illustrated by means of straight strings with the quipu ⌘ upon each. So, in ignoring the quipu, -1 is projected to the same element as 1 is projected, and these are the only two elements that project to 1 in the dihedral group.

4.3.4. Matrices that correspond to the 2-strings-with-quipu representations

In the dicyclic group Dic_n, consider the cyclic group $R = \langle \rho \rangle$, where $\rho = \cos\left(\frac{\pi}{n}\right) + \sin\left(\frac{\pi}{n}\right)\boldsymbol{i}$. The coset $\boldsymbol{j}R$ consists of the vertices of a regular n-gon upon the $(\boldsymbol{j}, \boldsymbol{k})$-circle, and thus, $\mathrm{Dic}_n = R \cup \boldsymbol{j}R$.

Multiplication by ρ rotates the coset $R = \langle R \rangle$ through an angle of $\frac{2\pi}{n}$ in an anti-clockwise direction and rotates $\boldsymbol{j}R$ clockwise. Multiplication by $\boldsymbol{j}$ interchanges the two cosets, but causes a 180° rotation when the right coset $\boldsymbol{j}R$ moves back to the left. These two cosets lie upon a Hopf link in the 3-sphere S^3. In the cases, $n = 3$ and $n = 4$, the situation is illustrated in Fig. 4.20. These two dicyclic groups are isomorphic to subgroups of the binary octahedral group $\widehat{\Sigma}_3$, and the dicyclic group Dic_4 is also a subgroup of the binary tetrahedral group $\widehat{A}_4$.

The two-strings-with-(mod n)-quipu representations of the generators are depicted in Fig. 4.21. In the case of Dic$_4$, the 2-strings representatives are depicted encircling the corresponding octagons in Figs. 4.22 and 4.23.

Since $\rho = e^{(\pi i)/n}$, there is a corresponding matrix representation

$$\rho \leftrightharpoons \begin{bmatrix} e^{(\pi i)/n} & 0 \\ 0 & e^{-(\pi i)/n} \end{bmatrix}, \quad j \leftrightharpoons \begin{bmatrix} 0 & -1 \\ 1 & 0 \end{bmatrix}.$$

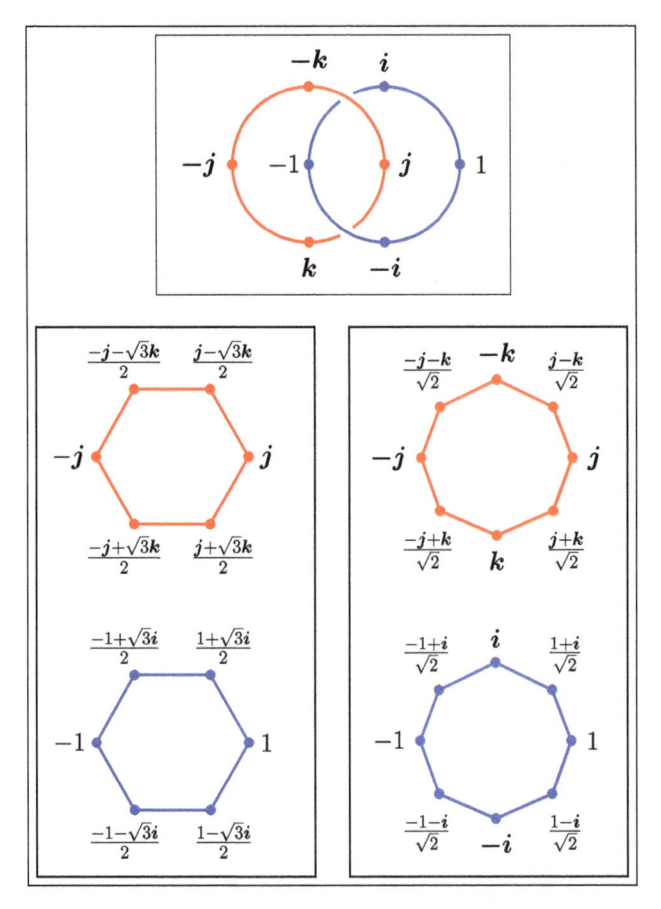

Figure 4.20. The dicyclic groups Dic$_2$, Dic$_3$, and Dic$_4$.

The 3-Dimensional Sphere is a Group 183

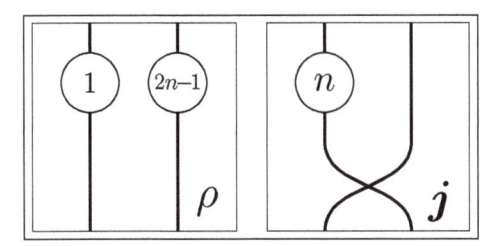

Figure 4.21. The generators for the dicyclic group Dic_n.

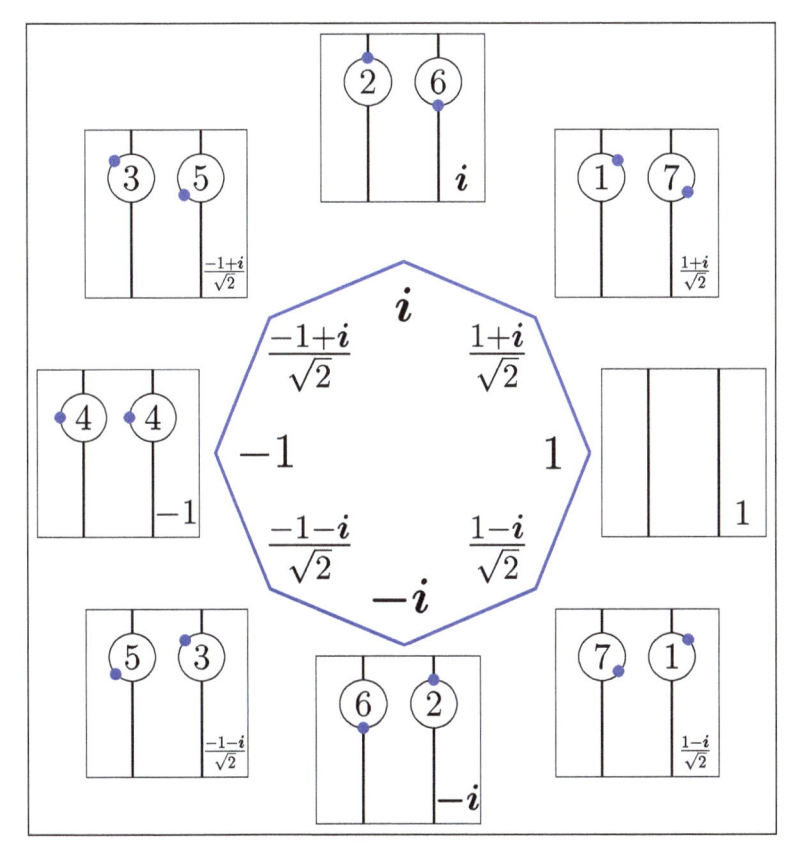

Figure 4.22. The dicyclic group Dic_4: the elements on the $(1, i)$ circle.

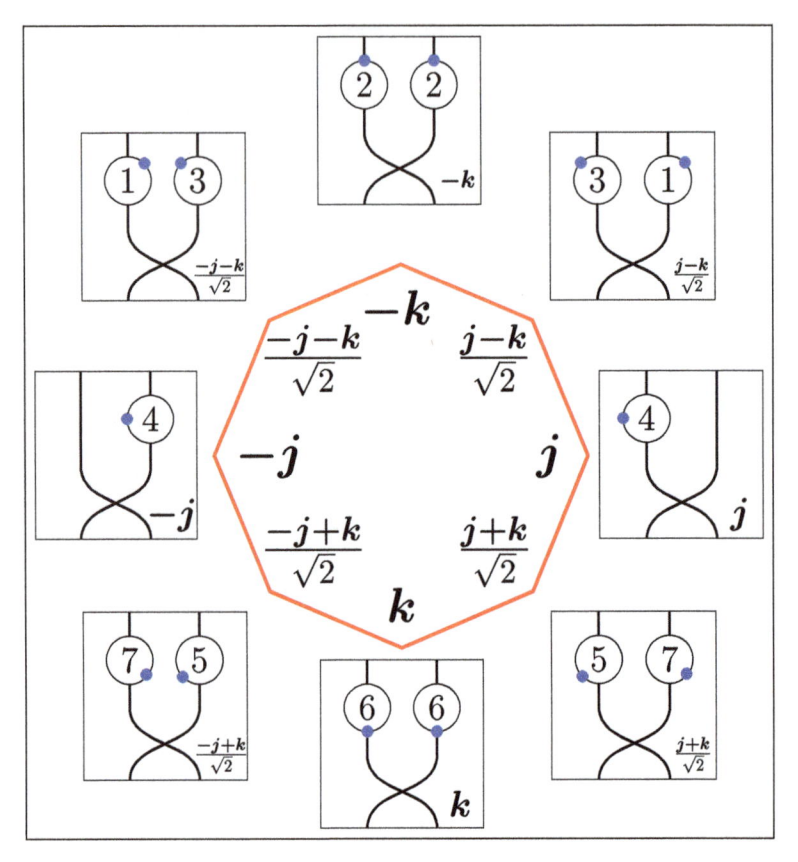

Figure 4.23. The dicyclic group Dic$_4$: the elements on the (j, k) circle.

In the dicyclic group Dic$_n$, the quipu that adorn the pair of strings are (mod $2n$)-quipu. Along the $(1, i)$-circle, proceed anti-clockwise from the identity. A pair of straight stringed (mod $2n$)-quipu appear at the location that corresponds to ρ^ℓ. The sum of the quipu indices adds to 0 (mod $2n$), and the quipu on the left is an ℓ-fold twist.

Along the (j, k)-circle, the pair of strings are always crossed. The quipu indices add to n. The representative of j has a quipu of index of n upon its left string with no quipu on the right. The representatives proceed clockwise from the j (or 3o'clock position) when the k-axis points downward and the j-axis points right from the representative of j. Upon the top of the right string, the indices on the quipu increment by 1, while those on the left decrement by 1; again, these indices are modulo $2n$.

4.3.5. The dicyclic group Dic_2 is the group of quaternions Q_8

Let us consider these n-strings representations in the case that $n = 2$. As mentioned earlier, the dihedral group D_4 is, in fact, the Klein 4-group, $K_4 = \mathbb{Z}/2 \times \mathbb{Z}/2$. Its two-fold extension in S^3 is the dicyclic group Dic_2, where the elements are $\rho^\ell = \cos\left(\frac{\ell\pi}{2}\right) + \sin\left(\frac{\ell\pi}{2}\right)i$, and $\rho^\ell j$, for $\ell = 0, 1$. There is too much obfuscation in these expressions! The dicyclic group $Dic_2 = Q_8$. The correspondence between elements is that $\rho = i$ and $x = j$.

These elements are illustrated in Fig. 4.24 for which the caption describes them as "peculiar." They are different representations of i and j since the ordered subgroup that is used to represent them is the subgroup $J = (-1, -j, 1, j)$. Since the quipu representations of i and j are reversed from the prior representations, the representations of elements k and $-k$ are correspondingly switched.

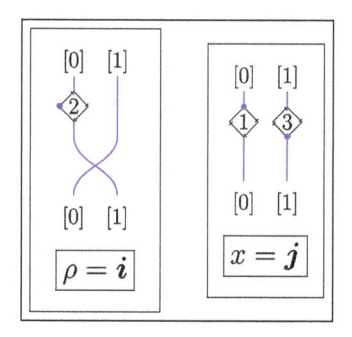

Figure 4.24. A peculiar 2-strings-with-(mod 4)-quipu representation of Q_8.

4.3.6. Describing the projections from the dicyclic groups to the dihedral groups

The projection $D_{2n} \xleftarrow{p'} \text{Dic}_n$ can be visualized at the level of 2-strings-diagrams-with-(mod $2n$)-quipu in a satisfying manner. We hope that illustrating this in the case that $n = 4$ and describing it in general will suffice.

For the description, there are two types of 2-strings-with-(mod $2n$)-quipu diagrams in Dic_n: (1) Those in which the strings do not cross are arranged around the $(1, i)$-circle. Except for the identity element, these have a pair of supplementary quipu. They are supplementary in that the indices in the pair of quipu add to $2n$. (2) The diagrams in which the strings cross are arranged around the (j, k)-circle. The quipu on the strings of a given element are complementary in the sense that the indices add to n. In fact, in both cases, the elements exhaust the different ways the sum of two elements in $\mathbb{Z}/(2n)$ is either $2n$ or n.

The homomorphism is represented at the level of strings with quipu as a reduction of the quipu indices modulo n. To see this, compare the eight representatives in the symmetries of the square D_8 that are depicted in Fig. 2.35, with the 16 representatives in Dic_4 that are depicted in Figs. 4.22 and 4.23. The projection is depicted in Fig. 4.25. Please note that it bears similarity with the squaring function as restricted to the complex circle $S^1 = \{e^{i\theta} : \theta \in [0, 2\pi]\}$. Antipodal points on the 3-sphere S^3 are mapped to the same elements in the group SO(3). The elements on the (j, k)-circle map to the dihedral reflections.

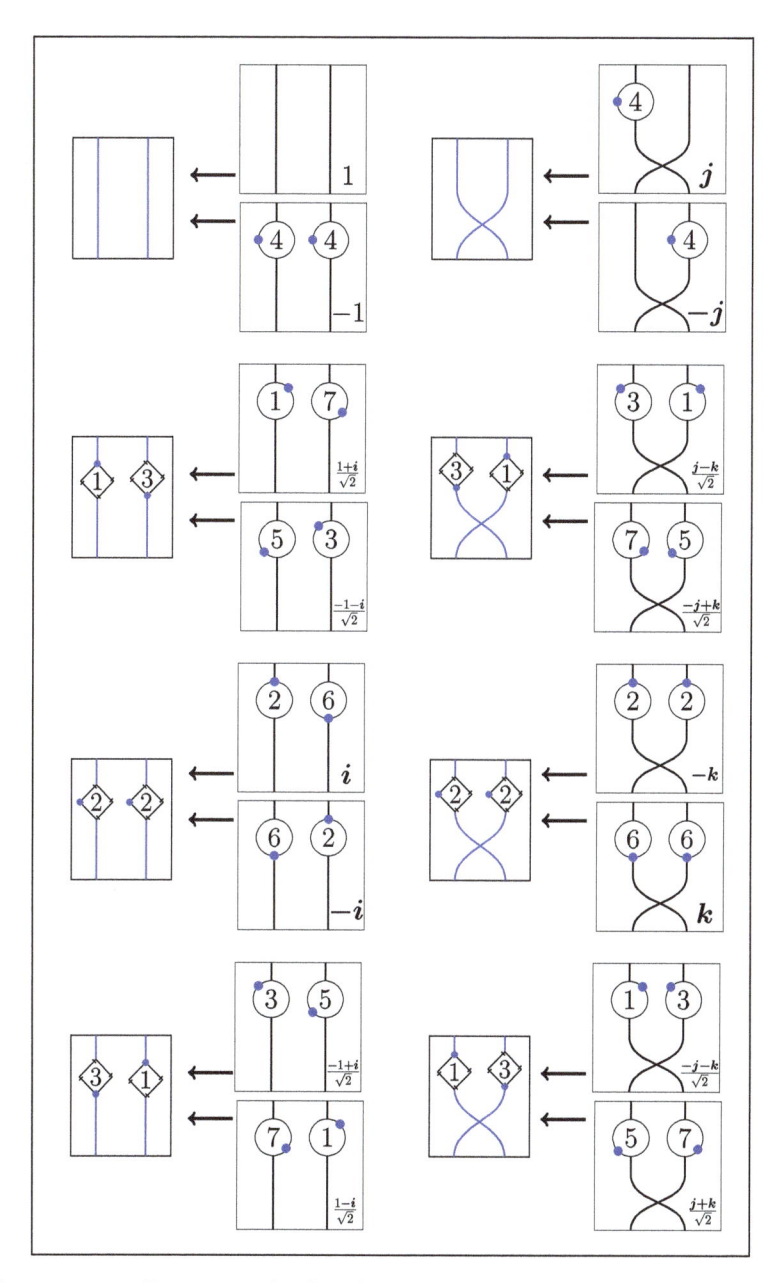

Figure 4.25. Projecting the dicyclic group Dic$_4$ onto the dihedral group D_8.

4.4. Culmination and anticipation

To conclude this chapter, we point out that the only spheres that have a group structure which is compatible with its intrinsic topology are the 0-sphere, S^0, the 1-sphere or circle, S^1, and the 3-sphere that has been the subject of this chapter.

The 7-dimensional sphere has a non-associative multiplication that is defined upon it. This structure is induced by the octonians which form a basis for the corresponding 8-dimensional space in which the 7-sphere surrounds the origin. A discussion of this structure would go far beyond the scope of this book. The possible multiplicative structures on spheres is related to the "Hopf-invariant 1" problem that is discussed, for example, in [4, pp. 427–428].

The dicyclic groups Dic_n are among a short list of subgroups of the 3-dimensional sphere. Others have been mentioned in passing: the binary tetrahedral group $\widetilde{A}_4$ that is a 2-fold extension of the alternating group on 4-elements, the binary octohedral group $\widetilde{\Sigma}_4$ that is a 2-fold extension of the symmetric group Σ_4, and the binary icosahedral group $\widetilde{A}_5$ that is a 2-fold extension of the alternating group A_5. These extensions map, respectively, to the groups of symmetries of the tetrahedron, the octahedron (or cube), and the dodecahedron (or icosahedron).

In subsequent chapters, strings-with-quipu will be used to represent the elements of these more substantial groups. We are describing them as being more substantial since the dicyclic groups cover the symmetries of polygons which are 2-dimensional as opposed to the 3-dimensional regular polyhedra.

In addition, two other related groups will be studied from the point of view of strings-with-quipu. These share some properties with the binary octahedral groups. Their quipu representations will allow us to contrast them.

There are, in fact, many points that the reader can explore, which originate from the discussions presented so far. This section suggests a few of them: classification of division algebra structures upon $\mathbb{R}^n$, possible group actions upon spheres, and H-space structures upon the spheres. Each is a worthwhile journey. But at this junction, we merely choose to explore our own village, and we will avoid the superhighways that lead to more advanced mathematical knowledge.

Chapter 5

Extensions of the Permutation Group Σ_4

This chapter presents the quipu descriptions of four groups:

1. The symmetric group Σ_4 was originally depicted in Fig. 2.3. It is the model case for the remaining three groups of study. Its 3-strings-with-(mod 2)-quipu description that was derived in Chapter 3 will be recalled and compiled in Section 5.1.
2. The group of invertible (2×2)-matrices with entries in the integers modulo 3, $\mathbb{Z}/3$, is denoted by

$$\mathrm{GL}_2(\mathbb{Z}/3) = \left\{ \begin{bmatrix} a & b \\ c & d \end{bmatrix} : a,b,c,d \in \mathbb{Z}/3 \ \& \ ad - bc \neq 0 \in \mathbb{Z}/3 \right\}.$$

3. The group of (2×2)-matrices that have determinant 1 and have entries in the integers modulo 4, $\mathbb{Z}/4$, is denoted by

$$\mathrm{SL}_2(\mathbb{Z}/4) = \left\{ \begin{bmatrix} a & b \\ c & d \end{bmatrix} : a,b,c,d \in \mathbb{Z}/4 \ \& \ ad - bc = 1 \in \mathbb{Z}/4 \right\}.$$

4. The binary octahedral group $\widetilde{\Sigma}_4$ is given via the presentation

$$\widetilde{\Sigma}_4 = \langle a, f : a^3 = f^4 = (af)^2 \rangle.$$

It will be considered as a subgroup of the 3-dimensional sphere, S^3, via $f = (1+i)/\sqrt{2}$ and $a = [1+i+j+k]/2$.

The patient and benevolent experts might be anxious to point out that the last three groups are the non-trivial 2-fold extensions of the symmetric group Σ_4. The trivial extension is $\Sigma_4 \times \mathbb{Z}/2$. Indeed, one such benevolent expert informed us of these facts a few years ago. Moreover, we will, in due course, develop those descriptions and include a definition of "non-trivial 2-fold extension" within the development of the material. But to do so now would put that material a little too far ahead of the narrative.

Let G denote any one of the four groups enumerated above. Each fits into a seseq

$$1 \longleftarrow \Sigma_3 \xleftarrow{p} G \xleftarrow{i} K \longleftarrow 1.$$

In the case of $\mathrm{SL}_2(\mathbb{Z}/4)$, the kernel is $(\mathbb{Z}/2)^3$. In the case of Σ_4, the kernel is the Klein 4-group $K_4 = \mathbb{Z}/2 \times \mathbb{Z}/2$. In the remaining two cases, the kernel is the quaternions, $Q_8 = \{\pm 1, \pm i, \pm j, \pm k\}$.

Each section in this chapter will focus upon one of these groups. String diagrams with quipu will be given for all the cases. Section 5.1 is concerned with the symmetric group Σ_4. It will recapitulate the description that was given in Chapter 3, and it will present an alternative, but related, viewpoint. Section 5.2 will describe the group of invertible (2×2)-matrices over $\mathbb{Z}/3$. Section 5.3 will describe the group of (2×2)-matrices that have determinant 1 over $\mathbb{Z}/4$. Section 5.4 will describe the binary octahedral group. In each section, the group will be described from the point of view of the seseq in which the quotient is the set of permutations of three letters, Σ_3. So, differing 3-strings-with-quipu will be presented. These also will be contrasted among each other.

But the notion of quipu will be generalized to include quipu that are dihedral, quaternionic, or semi-dihedral. The semi-dihedral group of order 16 will be introduced in Section 5.2.2. Furthermore, alternative descriptions will be presented that address the descriptions of the latter three groups as 2-fold extensions of Σ_4.

5.1. The symmetric group Σ_4

The 24 elements of the symmetric group Σ_4 are illustrated alongside a set of 3-strings representatives in Figs. 5.1–5.3.

The columns of these figures correspond to cosets of the Klein 4-group, K_4, which is a subgroup of Σ_4. In Fig. 5.1, the subgroup K_4 and its coset $(12)K_4$ are depicted, and the union of these two cosets form a group H that is isomorphic to the dihedral group D_8.

Extensions of the Permutation Group Σ_4

The elements in Fig. 5.2 form the coset $(123)H$, and those in Fig. 5.3 form the coset $(132)H$.

The figures indicate the seseq

$$1 \longleftarrow \Sigma_3 \xleftarrow{p} \Sigma_4 \xleftarrow{i} K_4 \longleftarrow 1.$$

We hope that the kernel of the projection is apparently K_4 as depicted in the left-hand column of Fig. 5.1. On the other hand, the projection onto Σ_3 is to ignore the (mod 2)-quipu upon the strings.

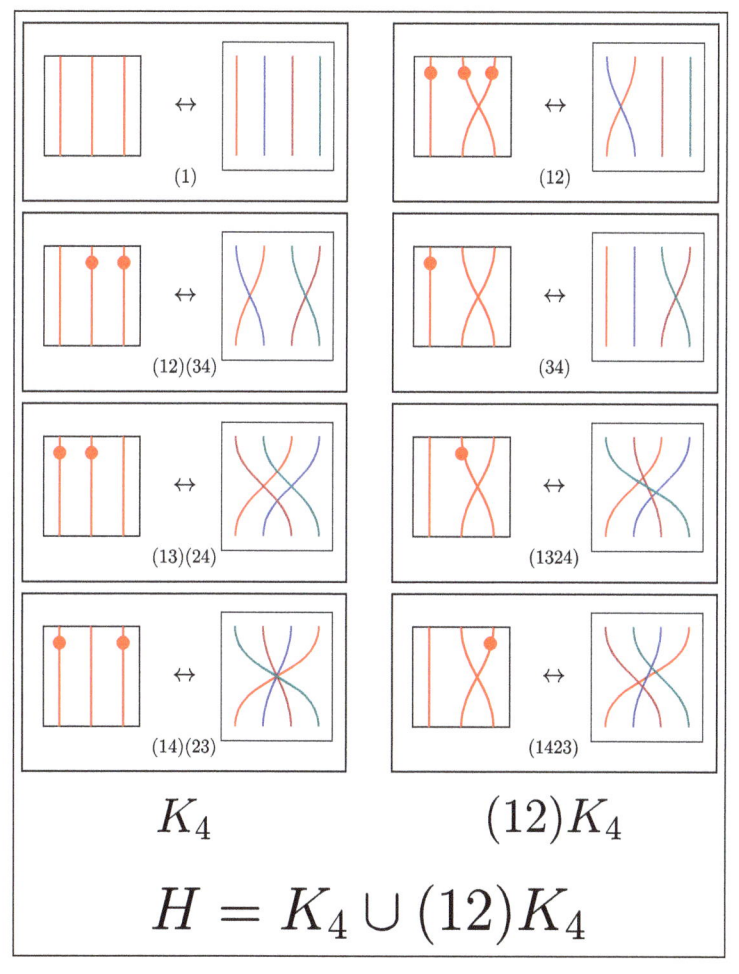

Figure 5.1. Revisiting the 3-strings-with-(mod 2)-quipu description of Σ_4 (part 1).

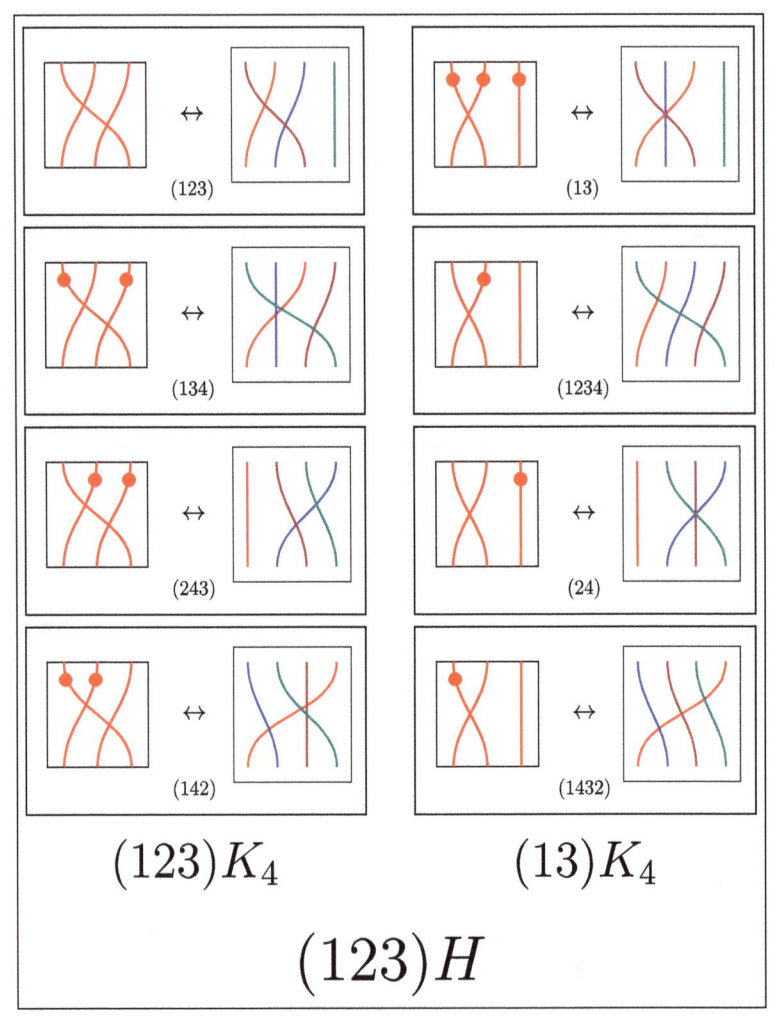

Figure 5.2. Revisiting the 3-strings-with-(mod 2)-quipu description of Σ_4 (part 2).

Extensions of the Permutation Group Σ_4 193

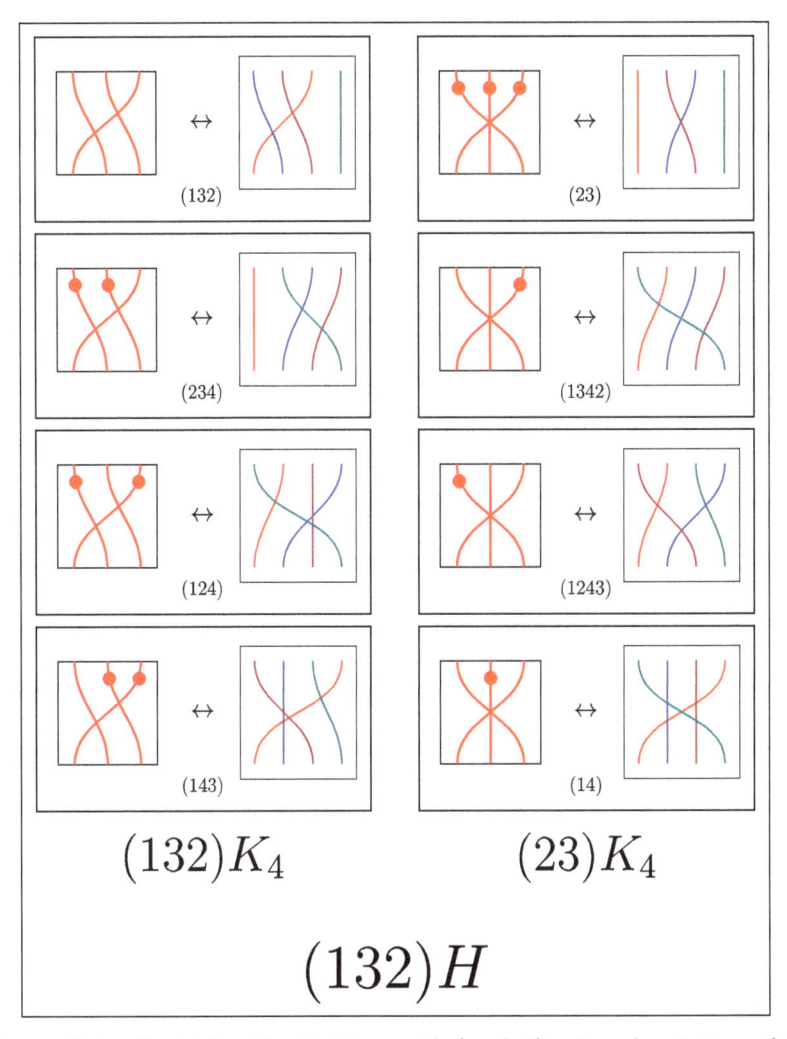

$$(132)K_4 \qquad (23)K_4$$

$$(132)H$$

Figure 5.3. Revisiting the 3-strings-with-(mod 2)-quipu description of Σ_4 (part 3).

Perhaps, from the point of view of the chapter, these figures are sufficient to illustrate the purported extension. However, these 3-strings pictures were obtained via looking at the set of symmetries of the cube C_3. In general, we are also interested in quipu representations of a group wherein the quipu are elements of a subgroup of the given group. So, in the following section and Fig. 5.4, the subgroup H

will be renamed D and reordered so that the quipu which adorn the strings are elements of the dihedral group D_8 that will be depicted as 2-strings-with-(mod 4)-quipu as in Fig. 2.35.

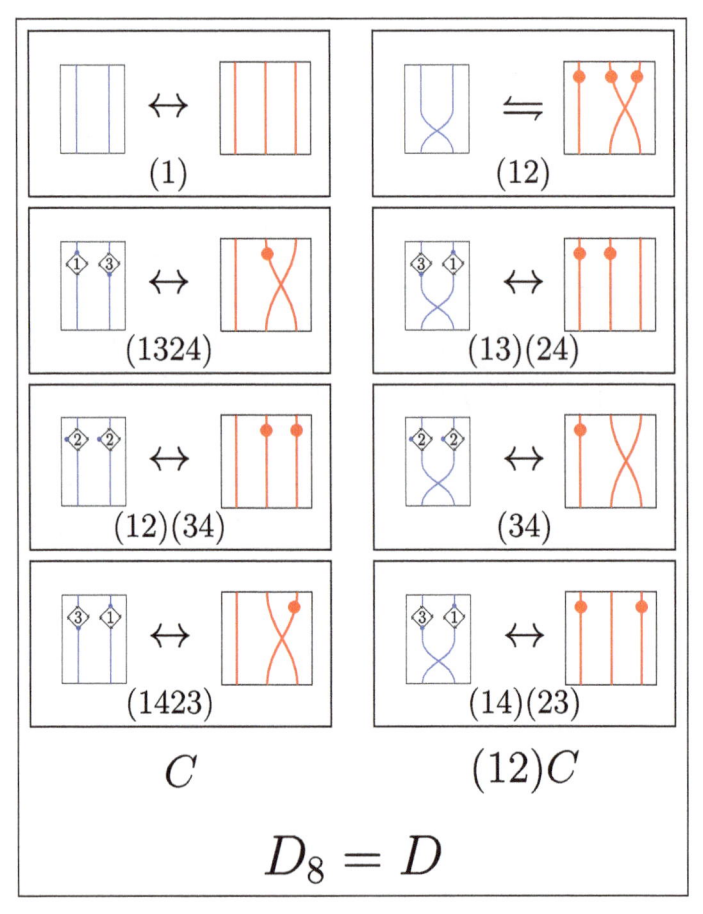

Figure 5.4. The correspondence between elements of D_8 and the subgroup $D \subset \Sigma_4$.

The purpose in establishing these correspondences is to develop a 3-strings-with-dihedral-quipu representation of Σ_4. To that end, let us compile and order the cosets of $D_8 = D$. Let $C = \langle (1324) \rangle$ denote the subgroup that is generated by the 4-cycle (1324). As an ordered set,

$$C = ((1), (1324), (12)(34), (1423)).$$

Then $D = C \cup (12)C$. The symmetric group Σ_4 will be written as a collection of cosets of C. As usual, the order upon C induces orders upon its cosets that are as indicated in the following:

$$D = \begin{bmatrix} C & = ((1), (1324), (12)(34), (1423)), \\ (12)C & = ((12), (13)(24), (34), (14)(23)), \end{bmatrix}$$

$$(123)D = \begin{bmatrix} (123)C & = ((123), (24), (134), (1432)), \\ (13)C & = ((13), (243), (1234), (142)), \end{bmatrix}$$

$$(132)D = \begin{bmatrix} (132)C & = ((132), (1243), (234), (14)), \\ (23)C & = ((23), (124), (1342), (143)). \end{bmatrix}$$

Compute the actions of the generators (12), (23), and (34) upon these cosets while asking, "How are the cosets permuted among themselves, and how are the elements in the cosets transformed?" The former action will be an element in Σ_3; the latter is an action of D_8:

$$(12)D = \begin{bmatrix} ((12), (13)(24), (34), (14)(23)), \\ ((1), (1324), (12)(34), (1423)), \end{bmatrix}$$

$$(12)(123)D = (23)D = \begin{bmatrix} ((23), (124), (1342), (143)), \\ ((132), (1243), (234), (14)), \end{bmatrix}$$

$$(12)(132)D = (13)D = \begin{bmatrix} ((13), (243), (1234), (142)), \\ ((123), (24), (134), (1432)). \end{bmatrix}$$

So, (12) switches the components of D but leaves them in order. It interchanges the cosets $(123)D$ and $(132)D$, and within both cosets, it interchanges the ordered cosets of C. An illustration of the 3-strings-with-dihedral quipu representation of the generator $(12) \in \Sigma_4$ is presented in Fig. 5.5. In addition, this illustration will present the representations of (23) and (34). First, the results of computations of these elements upon the cosets of D will be articulated and analyzed:

$$(23)D = \begin{bmatrix} ((23), (124), (1342), (143)), \\ ((132), (1243), (234), (14))), \end{bmatrix}$$

$$(23)(123)D = (13)D = \begin{bmatrix} ((13), (243), (1234), (142)), \\ ((123), (24), (134), (1432)), \end{bmatrix}$$

$$(23)(132)D = (12)D = \begin{bmatrix} ((12), (13)(24), (34), (14)(23)), \\ ((1), (1324), (12)(34), (1423)). \end{bmatrix}$$

So, (23) interchanges the cosets D and $(132)D$. The generator (23) fixes $(123)D$ set-wise. In each coset, (23) transposes the cosets of the cyclic subgroup, but maintains their order (see Fig. 5.5):

$$(34)D = \begin{bmatrix} ((34),(14)(23),(12),(13)(24)), \\ ((12)(34),(1423),(1),(1324))) , \end{bmatrix}$$

$$(34)(123)D = (1243)D = \begin{bmatrix} ((1243),(234),(14),(132)), \\ ((143),(23),(124),(1342)), \end{bmatrix}$$

$$(34)(132)D = (1432)D = \begin{bmatrix} ((1432),(123),(24),(134)), \\ ((243),(1234),(142),(13)). \end{bmatrix}$$

The coset D is fixed set-wise by (34), but the cosets of C are transposed, and each is rotated two steps forward. The cosets $(123)D$ and $(132)D$ are interchanged set-wise. The cosets of C are correspondingly switched, but the coset $(123)C$ is sent to the coset $(132)C$ with a quarter rotation among its elements since $(34)(132) = (1432)$. Similarly, $(34)(123)(12) = (143)$, and this is the last element of the coset $(132)(12)C$. So, (34) sends the coset $(123)(12)C$ to the coset $(132)(12)C$ with a three-quarter's rotation. Similar verbalizations can be articulated for the actions of (34) upon the cosets $(132)C$ and $(132)(12)C$. Figure 5.5 depicts the actions.

Before articulating all of the elements in Σ_4 by using 3-strings-with-dihedral-quipu, we want to demonstrate how we prove these representatives satisfy the relations $(12)(23)(12) = (23)(12)(23)$, $(23)(34)(23) = (34)(23)(34)$, and that $(12)(34) = (34)(12)$. The process involves juxtaposing the 3-strings diagrams, migrating the quipu up to the top of the corresponding strings, and subsequently multiplying the quipu by an analogous process. The blue-stringed quipu are juxtaposed, the (mod 4)-quipu migrate upward along the corresponding strings, and then the indices in the smaller quipu are multiplied as integers modulo 4.

Figures 5.6 and 5.7 contain the outlines of these computations. At the level of the transpositions, one computes that $(12)(23)(12) = (13)$, and $(23)(34)(23) = (24)$. These results can be determined fairly easily by means of the standard 4-strings diagrams that represent

Extensions of the Permutation Group Σ_4

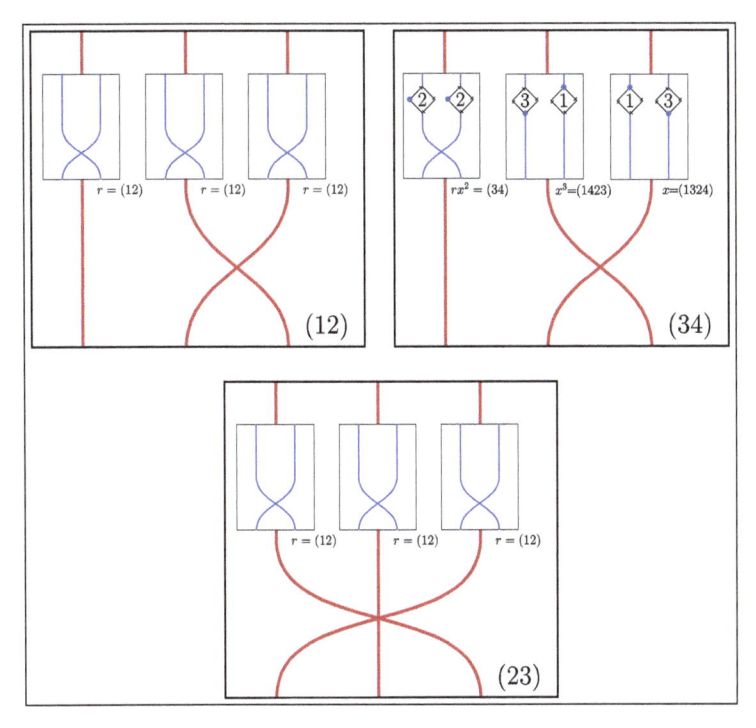

Figure 5.5. Writing the generators (12), (23) and (34) using dihedral quipu.

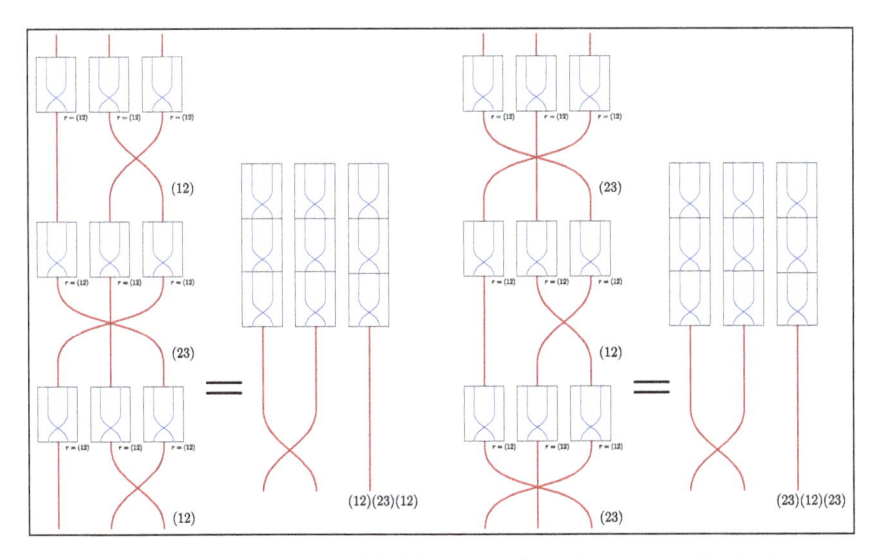

Figure 5.6. The relation $(12)(23)(12) = (23)(12)(23)$ using dihedral quipu.

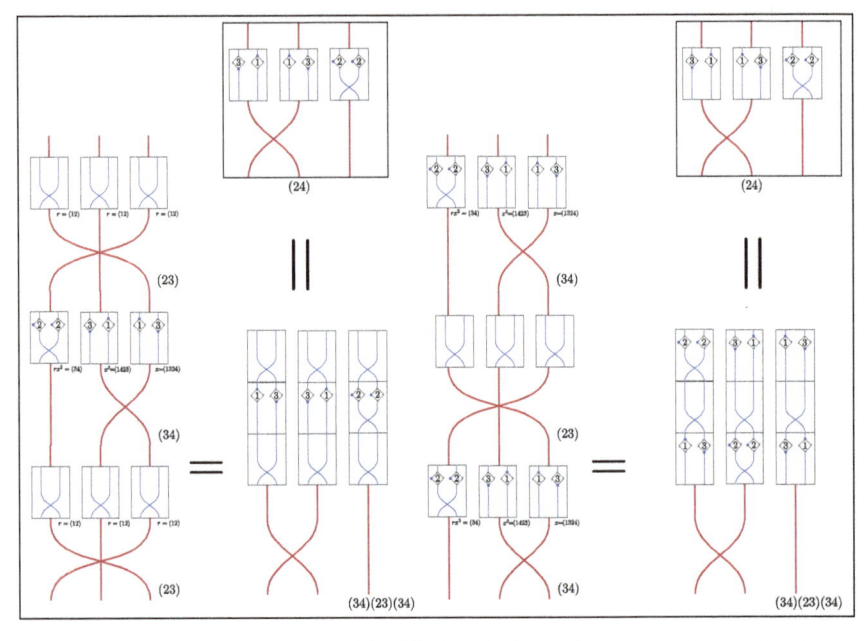

Figure 5.7. The relation $(23)(34)(23) = (34)(23)(34)$ using dihedral quipu.

the permutations in Σ_4 as represented in Figs. 5.1–5.3. See also the illustration that immediately precedes Fig. 2.3.

Figure 5.8 illustrates that the 3-strings-with-dihedral quipu representations of (12) and (34) commute. Similarly, Fig. 5.9 illustrates that the representatives of (13) and (24) commute. In this way, we have demonstrated that the defining relations of Σ_4 hold among the representatives that have been presented.

Exercise 12. Use the relationships among the permutations and, in particular, the 4-strings presentation of Figs. 5.1–5.3 (or Fig. 2.3) to catalog all the 3-strings-with-dihedral representatives that appear in Figs. 5.10–5.12. Note that these are arranged as cosets of the cyclic and dihedral subgroup D.

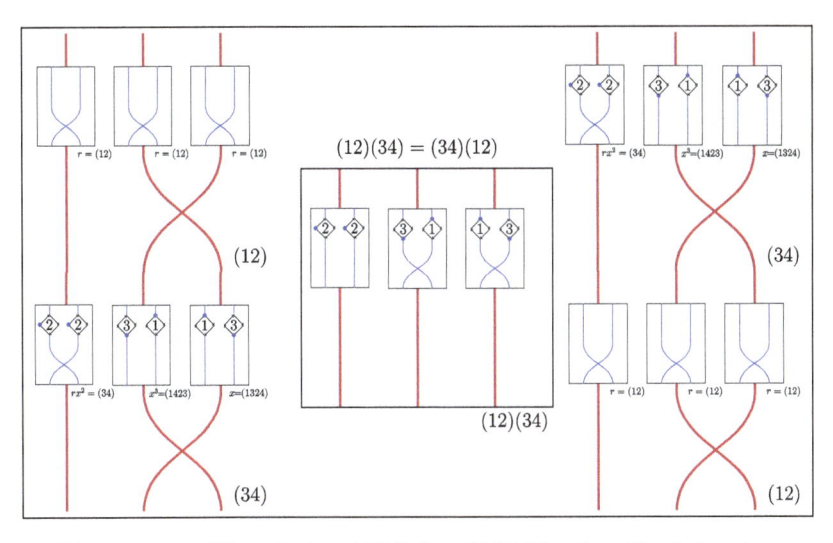

Figure 5.8. The relation $(12)(34) = (34)(12)$ using dihedral quipu.

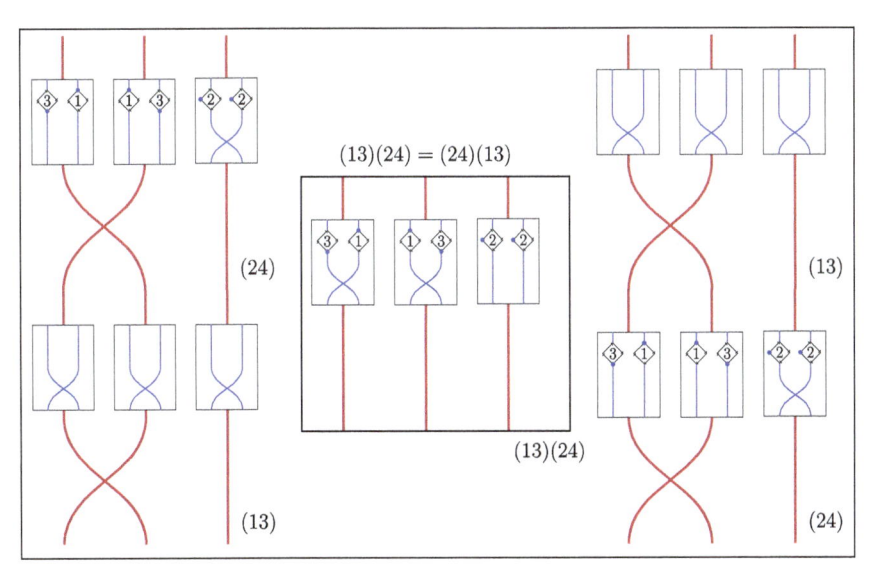

Figure 5.9. The relation $(13)(24) = (24)(13)$ using dihedral quipu.

5.1.1. The cosets of the dihedral group

The subgroup $D \subset \Sigma_4$ is isomorphic to the dihedral group D_8. It is decomposed as a cyclic group C of order 4 and a coset $(12)C$. But these will be considered as subgroups of Σ_4. As indicated in Exercise 12, all 24 of the elements in the symmetric group Σ_4 can be written as 3-strings-with-dihedral-quipu diagrams. These will be arranged in columns that correspond to the cosets of C. Figure 5.10 contains the elements of D, Fig. 5.11 contains the elements of $(123)D$,

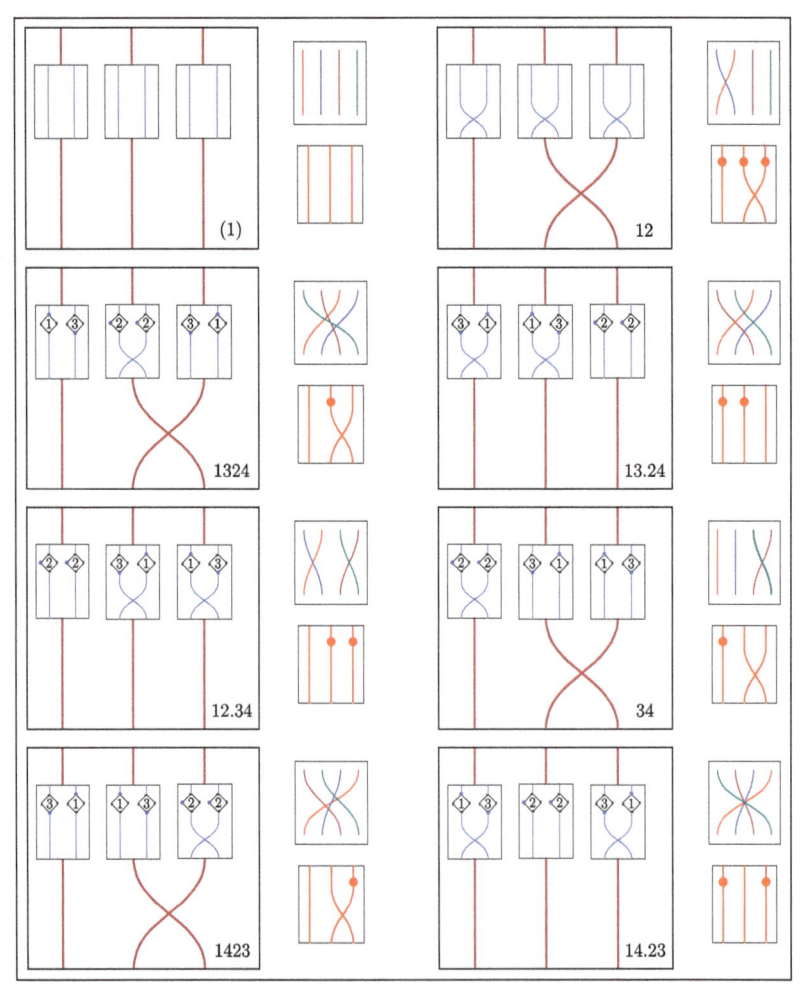

Figure 5.10. 3-strings-with-dihedral-quipu representatives of elements in Σ_4: the subgroup D.

Extensions of the Permutation Group Σ_4 201

and Fig. 5.12 contains the elements of $(132)D$. Each group element is represented as a 3-strings-with-dihedral-quipu, a 3-strings-with-(mod 2)-quipu, and in its original 4-strings notation.

We observe from the figures that the reflections in the dihedral group D_8 map to the non-trivial modulo 2 quipu.

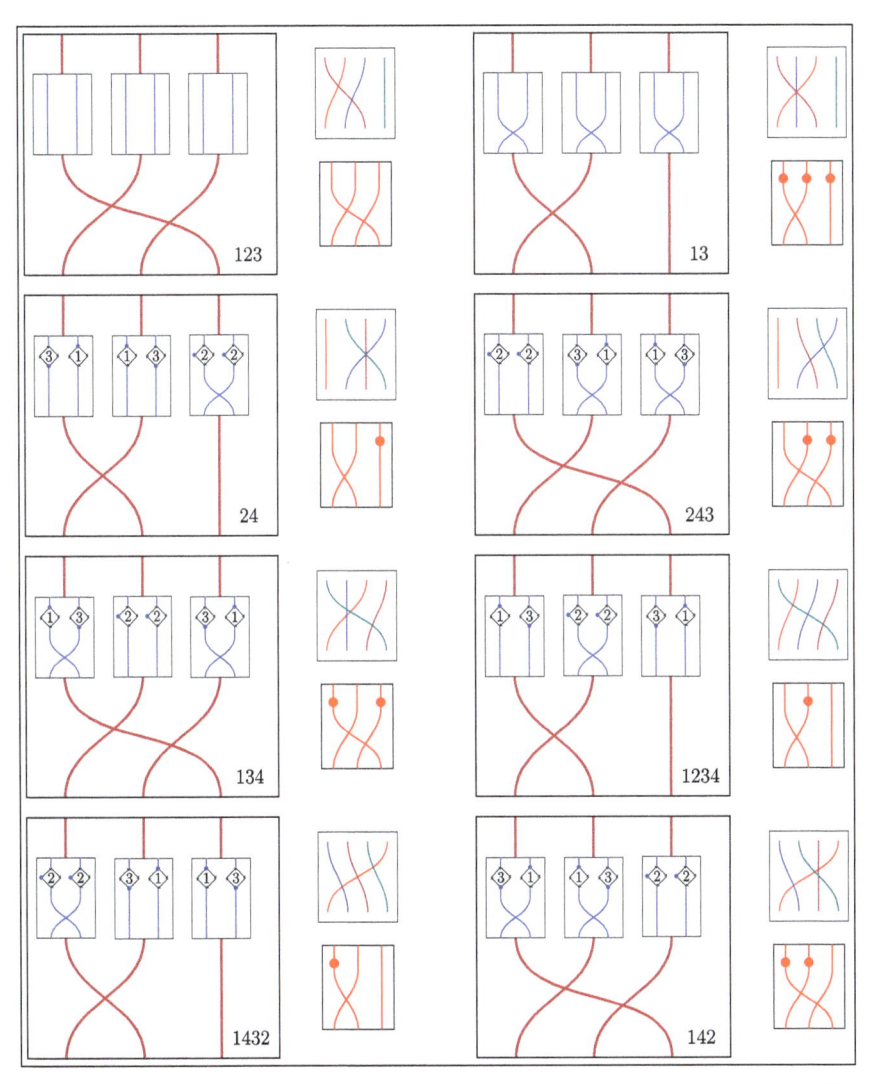

Figure 5.11. 3-strings-with-dihedral-quipu representatives of elements in Σ_4: the coset $(123)D$.

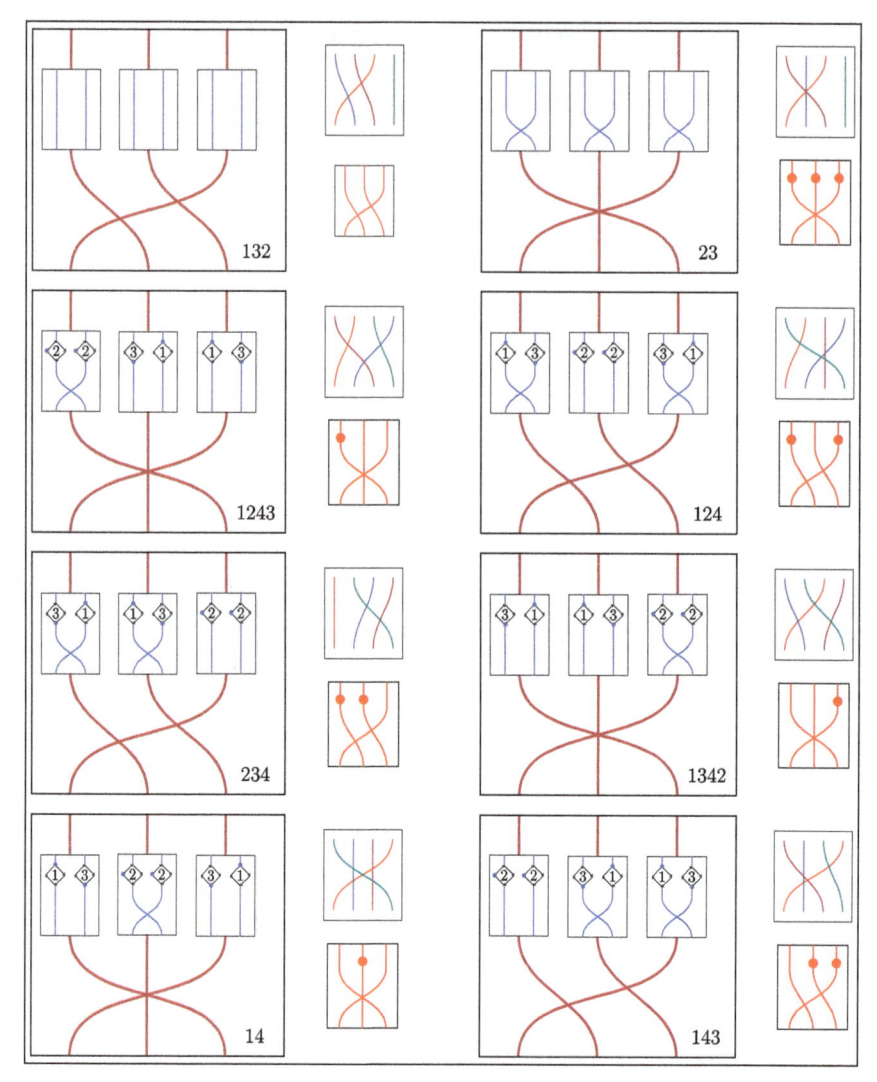

Figure 5.12. 3-strings-with-dihedral-quipu representatives of elements in Σ_4: the coset $(132)D$.

5.2. The group $GL_2(\mathbb{Z}/3)$

The integers modulo 3, $\mathbb{Z}/3$, has both a multiplicative and an additive structure. For only a few moments, the elements will be denoted by 0, 1, and 2 since these are the canonical representatives of the cosets of

$3\mathbb{Z} \subset \mathbb{Z}$. The additive and multiplicative structures are summarized in the following tables:

+	0	1	2
0	0	1	2
1	1	2	0
2	2	0	1

·	0	1	2
0	0	0	0
1	0	1	2
2	0	2	1

The non-zero elements (1 and 2) are multiplicatively invertible and each is its own inverse. Multiplication distributes over addition — $a(b+c) = ab + ac$ and $(a+b)c = ac + bc$ — since this property holds within the integers. Moreover, both the additive and multiplicative structures are associative and commutative. We summarize these properties by saying that the integers modulo 3 forms a *(finite) field*: a set upon which two associative, commutative binary operations, $+$ and $\cdot$, are defined; there is an identity element for each — 0 is the additive identity and 1 is the multiplicative identity — the non-zero elements are multiplicatively invertible, all the elements are additively invertible, and the distributive properties hold.

Since $\mathbb{Z}/3$ is a field,[1] the cartesian product $\mathbb{Z}/3 \times \mathbb{Z}/3$ is a vector space. Or more to the point, the set of (2×2)-matrices that have non-zero determinants and whose entries are taken to be integers modulo 3,

$$\mathrm{GL}_2(\mathbb{Z}/3) = \left\{ \begin{bmatrix} a & b \\ c & d \end{bmatrix} : a, b, c, d \in \mathbb{Z}/3 \ \& \ ad - bc \neq 0 \in \mathbb{Z}/3 \right\},$$

is the group of non-singular transformations of the vector space $(\mathbb{Z}/3)^2$. The matrices are *non-singular* in the sense that the image of any non-zero vector under a matrix transformation is a non-zero vector. No two vectors in the domain share the same image, and therefore, the non-zero vectors in $(\mathbb{Z}/3)^2$ are permuted by any matrix in $\mathrm{GL}_2(\mathbb{Z}/3)$.

It is far more convenient to represent the integers modulo 3 as the elements of the set $\{-1, 0, 1\}$, where, of course, $2 \equiv -1 \mod 3$. Moreover, the elements in the vector space $(\mathbb{Z}/3)^2$ can be represented

---

[1] Some notational purists would prefer that it be written as $\mathbb{F}_3$.

in a 4-dimensional, yet geometric, fashion as the points on the cartesian product of two triangles: $\Delta \times \Delta$. The edges of this 4-dimensional figure are projected into the plane and the vertices are labeled in Fig. 5.13.

Some of the modular arithmetic can be summarized by the relationships $1 + 1 = -1$ and $(-1) \cdot (-1) = 1 = (-1) + (-1)$. Since

$$\begin{bmatrix} a & b \\ c & d \end{bmatrix} \begin{bmatrix} 0 \\ 0 \end{bmatrix} = \begin{bmatrix} 0 \\ 0 \end{bmatrix},$$

the (2×2)-matrices can be understood by means of how they permute the non-zero vectors in $(\mathbb{Z}/3)^2$. Again, because these matrices are non-singular, the eight non-zero vectors are permuted among each other by means of the matrix action.

In the following table, these vectors are written as rows and labeled as elements of the set $\{0, 1, 2, 3, 4, 5, 6, 7\}$. The choice of labeling is a bit peculiar except that antipodal points have labels that differ by 4.

$0 \leftrightarrow (1, -1)$	$4 \leftrightarrow (-1, 1)$
$1 \leftrightarrow (0, -1)$	$5 \leftrightarrow (0, 1)$
$2 \leftrightarrow (1, 1)$	$6 \leftrightarrow (-1, -1)$
$3 \leftrightarrow (-1, 0)$	$7 \leftrightarrow (1, 0)$

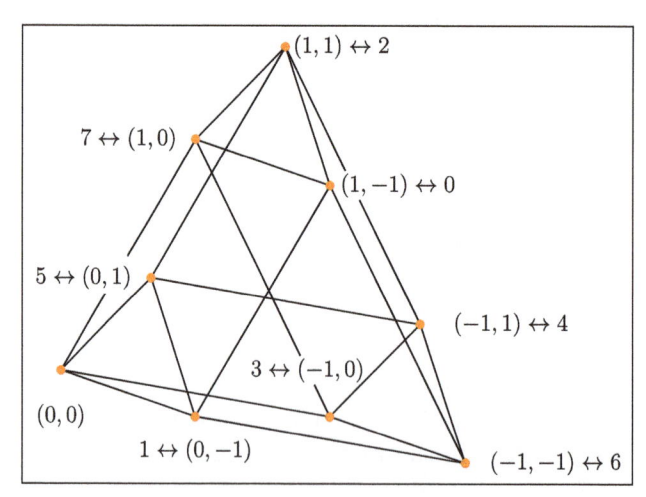

Figure 5.13. The projection of $\Delta \times \Delta = (\mathbb{Z}/3)^2$ and the labels on the points of the vector space.

Extensions of the Permutation Group Σ_4 205

In this section, a 4-strings-with-(mod 2)-quipu description of the group $GL_2(\mathbb{Z}/3)$ will be presented. Then a 3-strings description will be presented in which the quipu are elements of a 16-element group that is called the semi-dihedral group. But even before we get started with these, we'd like to count the number of elements in $GL_2(\mathbb{Z}/3)$.

Consider a matrix of the form $\begin{bmatrix} a & b \\ c & d \end{bmatrix}$ in which the entries $a, b, c, d \in \{0, -1, 1\}$. There are $3^4 = 81$ possible matrices. But only 48 of these has a determinant $ad - bc$ that is non-zero over $\mathbb{Z}/3$. It is not so difficult to list all the possibilities and then eliminate those that are singular, but instead we examine an easier enumeration.

There are nine products of the form bc. Of these, five are 0, and the other four are ± 1. These are tabulated as the multiplication table that appears at the beginning of this section, but with the representatives 0 and ± 1:

$b \cdot c$	0	-1	1
0	0	0	0
-1	0	1	-1
1	0	-1	1

The same census applies to the products ad. So, there are 20 matrices (5×4) in which $bc = 0$ — five of these — and ad is non-zero — four of these. Similarly, there are 20 (4×5) in which $bc \neq 0$ and $ad = 0$. Now, if both $ad, bc \neq 0$, we choose them to have different values so that $ad - bc \neq 0$. There are two ways in which $ad = 1$, namely, $a = d = 1$ and $a = d = -1$. Similarly, there are two ways in which $bc = -1$, namely, $b = -1$ while $c = 1$, or $b = 1$ while $c = -1$. Symmetrically, there are four ways for which $bc = 1$ and $ad = -1$. So, there are eight (2×2)-matrices over $\mathbb{Z}/3$ that have all their entries being non-zero and still have non-zero determinants. The group $GL_2(\mathbb{Z}/3)$ has 48 elements.

Within the section, these 48 matrices will be listed and their correspondences with strings-with-quipu will be made explicit.

5.2.1. 4-strings-with-(mod 2)-quipu

To enumerate the 4-strings-with-(mod 2)-quipu descriptions of the elements in the group $GL_2(\mathbb{Z}/3)$, we first examine the actions of the matrices upon the eight non-zero points in $\Delta \times \Delta$. The computation will begin by examining the actions of three of the matrices upon these vertices that are depicted and labeled in Fig. 5.13. Every matrix in the group $\mathrm{GL}_2(\mathbb{Z}/3)$ can be written as some product of these three elements.

We also recall that since the group is linear, the action of a matrix M upon a vector $\vec{v}$ also determines the value $M(-\vec{v}) = -M(\vec{v})$. Since $4, 5, 6, 7$ are antipodal to $0, 1, 2, 3$, respectively, then we can determine the permutation actions of the elements by merely computing the matrix products with certain (2×4)-matrices.

To be fair to the reader, there is a bit of struggle among the correspondences. So, we write

$$\begin{bmatrix} 0 & 1 & 2 & 3 & 4 & 5 & 6 & 7 \\ 1 & 0 & 1 & -1 & -1 & 0 & -1 & 1 \\ -1 & -1 & 1 & 0 & 1 & 1 & -1 & 0 \end{bmatrix}$$

to indicate the labels upon the vectors. We compute the following:

$$\begin{bmatrix} 0 & 1 \\ -1 & 1 \end{bmatrix} \cdot \begin{bmatrix} 1 & 0 & 1 & -1 \\ -1 & -1 & 1 & 0 \end{bmatrix} = \begin{bmatrix} -1 & -1 & 1 & 0 \\ 1 & -1 & 0 & 1 \end{bmatrix}$$

$$\leftrightarrow (04)(163527),$$

$$\begin{bmatrix} -1 & -1 \\ 0 & -1 \end{bmatrix} \cdot \begin{bmatrix} 1 & 0 & 1 & -1 \\ -1 & -1 & 1 & 0 \end{bmatrix} = \begin{bmatrix} 0 & 1 & 1 & 1 \\ 1 & 1 & -1 & 0 \end{bmatrix}$$

$$\leftrightarrow (37)(056412),$$

$$\begin{bmatrix} 0 & 1 \\ 1 & 1 \end{bmatrix} \cdot \begin{bmatrix} 1 & 0 & 1 & -1 \\ -1 & -1 & 1 & 0 \end{bmatrix} = \begin{bmatrix} -1 & -1 & 1 & 0 \\ 0 & -1 & -1 & -1 \end{bmatrix}$$

$$\leftrightarrow (03164752).$$

Please while doing these matrix calculations, recall that the entries are always reduced modulo 3 to one of $-1, 0, 1$.

When we, the authors, first encountered the index 2 subgroup $\mathrm{SL}_2(\mathbb{Z}/3)$ that consists of matrices of determinant 1, we noted that there was a "ribbon-like" presentation when the antipodal points were arranged to be adjacent. The first two entries above, $a = \begin{bmatrix} 0 & 1 \\ -1 & 1 \end{bmatrix}$ and $b = \begin{bmatrix} -1 & -1 \\ 0 & -1 \end{bmatrix}$, are also elements of $\mathrm{SL}_2(\mathbb{Z}/3)$, by virtue of their determinants, and these are illustrated in Fig. 5.14 along with the other element above. The illustrations indicate their ribbon-like nature and their associated presentation as 4-strings-with-(mod 2)-quipu.

The elements $a = [1 + i + j + k]/2$ and $b = [1 + i + j - k]/2$ that are indicated as the top right and bottom right vertices of the outermost square in Fig. 4.12 correspond to the matrices in $\mathrm{SL}_2(\mathbb{Z}/3)$, as defined in the prior paragraph. Figure 5.14 also indicates this correspondence.

The notation 123[34] indicates that the 4-strings-with-(mod 2)-quipu diagram that represents b corresponds to the 3-cycle $(123) \in \Sigma_4$ (where parenthesis have been omitted) and that there are quipu at the top of the third and fourth strings. Let us explain this new notation a bit further. In Fig. 5.16, the element a^2 is illustrated with the caption 234[0]. The [0] indicates that none of the strings has a (mod 2)-quipu adorning them. An asterisk indicates the complement. So, in Fig. 5.14, the 2-strings-with-(mod 2)-quipu diagram that represents a is denoted as 234[*0] to indicate that **all** the strings have quipu. In this way, b could also be represented as 123[*12], but that notation is not as economical as 123[34]. In general, when two strings have (mod 2)-quipu, the indices at the top of the strings that hold the quipu will be written. Often, either one string or three strings will have quipu. For example, 1423[3] is written for the remaining element in Fig. 5.14. Its negative will be written as 1423[*3] to indicate that the strings 1, 2, and 4 have (mod 2)-quipu upon them.

It may be a good moment to perform four direct and algebraic calculations. First, at the level of matrices in $\mathrm{GL}_2(\mathbb{Z}/3)$, we compute

$$a \cdot b = \begin{bmatrix} 0 & 1 \\ -1 & 1 \end{bmatrix} \cdot \begin{bmatrix} -1 & -1 \\ 0 & -1 \end{bmatrix} = \begin{bmatrix} 0 & -1 \\ 1 & 0 \end{bmatrix}$$

and

$$b \cdot a = \begin{bmatrix} -1 & -1 \\ 0 & -1 \end{bmatrix} \cdot \begin{bmatrix} 0 & 1 \\ -1 & 1 \end{bmatrix} = \begin{bmatrix} 1 & 1 \\ 1 & -1 \end{bmatrix}.$$

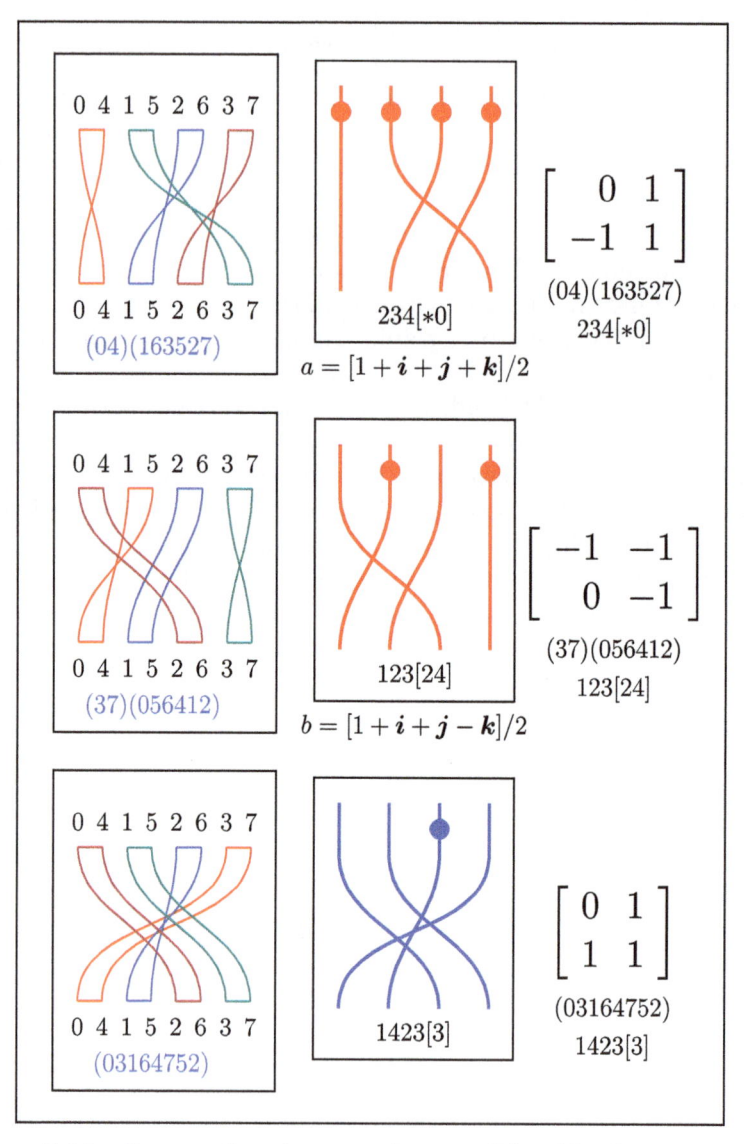

Figure 5.14. Representing three matrices as ribbon-like diagrams and as 4-strings-with-(mod 2)-quipu.

Meanwhile, we compute the corresponding products within S^3:

$a \cdot b$	$1/2$	$(1/2)i$	$(1/2)j$	$-(1/2)k$
$1/2$	$1/4$	$(1/4)i$	$(1/4)j$	$-(1/4)k$
$(1/2)i$	$(1/4)i$	$-(1/4)$	$(1/4)k$	$(1/4)j$
$(1/2)j$	$(1/4)j$	$-(1/4)k$	$-(1/4)$	$-(1/4)i$
$(1/2)k$	$(1/4)k$	$(1/4)j$	$-(1/4)i$	$(1/4)$

$b \cdot a$	$(1/2)$	$(1/2)i$	$(1/2)j$	$(1/2)k$
$(1/2)$	$(1/4)$	$(1/4)i$	$(1/4)j$	$(1/4)k$
$(1/2)i$	$(1/4)i$	$-(1/4)$	$(1/4)k$	$-(1/4)j$
$(1/2)j$	$(1/4)j$	$-(1/4)k$	$-(1/4)$	$(1/4)i$
$-(1/2)k$	$-(1/4)k$	$-(1/4)j$	$(1/4)i$	$(1/4)$

So, $a \cdot b = j$ while $b \cdot a = i$. At the matrix level, let us just show that each corresponding product squares to $\begin{bmatrix} -1 & 0 \\ 0 & -1 \end{bmatrix}$:

$$\begin{bmatrix} 0 & -1 \\ 1 & 0 \end{bmatrix} \begin{bmatrix} 0 & -1 \\ 1 & 0 \end{bmatrix} = \begin{bmatrix} -1 & 0 \\ 0 & -1 \end{bmatrix}$$

and

$$\begin{bmatrix} 1 & 1 \\ 1 & -1 \end{bmatrix} \begin{bmatrix} 1 & 1 \\ 1 & -1 \end{bmatrix} = \begin{bmatrix} -1 & 0 \\ 0 & -1 \end{bmatrix}.$$

We are obliged to repeat an admonition. All of the above calculations are presented for the reader's convenience. However, it is our expectation that you replicate the calculations for yourself. For example, in the matrix, calculations align pencils along the appropriate row and column of the matrices, multiply the entries pair by pair and add the results. Again, recall the entries are in $\mathbb{Z}/3$. The results of the calculations are replicated within Fig. 5.15 using 4-strings-with-(mod 2)-quipu.

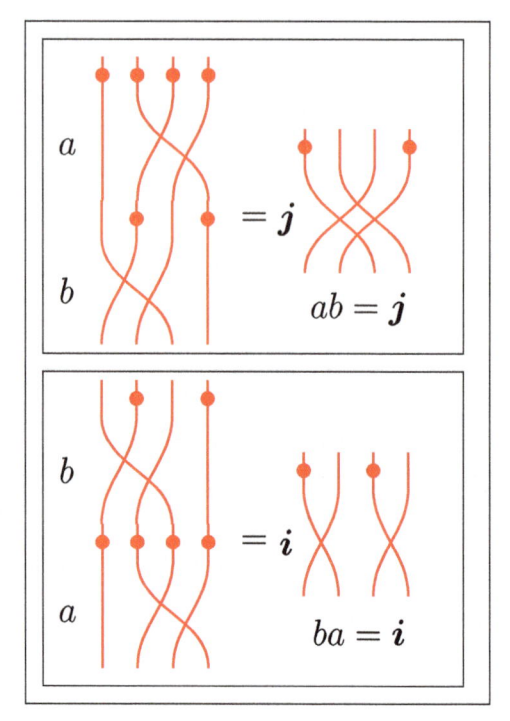

Figure 5.15. 4-strings-with-(mod 2)-quipu that indicate $a \cdot b = j$ and $b \cdot a = i$.

Table 4.1 can be used to demonstrate that when the element $a = [1 + i + j + k]/2 \in S^3$ is represented in quaternionic form, its square is $a^2 = [-1 + i + j + k]/2$. While you are reviewing that section, please also observe that $a^2 = -a^{-1}$, for the inverse of any element $s = [w + xi + yj + zk]/2$ is given as $s^{-1} = [w - xi - yj - zk]/2$.

Meanwhile, also recall the recipe, "switch the diagonals, change the signs of the off-diagonals, and divide through by the determinant" that gives the inverse of a (2×2)-matrix. This process also works over $\mathbb{Z}/3$. So, the formula

$$\begin{bmatrix} a & b \\ c & d \end{bmatrix}^{-1} = \frac{1}{ad - bc} \begin{bmatrix} d & -b \\ -c & a \end{bmatrix}$$

holds for matrices in $GL_2(\mathbb{Z}/3)$.

Exercise 13. Verify that the matrix and quaternionic representations of the 4-strings-with-(mod 2)-quipu depictions of the powers of a that are indicated in Fig. 5.16 are correct.

$$\begin{bmatrix} 0 & 1 \\ -1 & 1 \end{bmatrix} \cdot \begin{bmatrix} 0 & 1 \\ -1 & 1 \end{bmatrix} = \begin{bmatrix} -1 & 1 \\ -1 & 0 \end{bmatrix}$$

$= a^2$

243[0]

$a^2 = [-1 + \boldsymbol{i} + \boldsymbol{j} + \boldsymbol{k}]/2$

$$\begin{bmatrix} 0 & 1 \\ -1 & 1 \end{bmatrix} \cdot \begin{bmatrix} -1 & 1 \\ -1 & 0 \end{bmatrix} = \begin{bmatrix} -1 & 0 \\ 0 & -1 \end{bmatrix}$$

$= -1$

1.2.3.4[*0]

$a^3 = -1$

$$\begin{bmatrix} 0 & -1 \\ 1 & -1 \end{bmatrix}$$

$a^4 = -a$
$= [-1 - \boldsymbol{i} - \boldsymbol{j} - \boldsymbol{k}]/2$

234[0]

$$\begin{bmatrix} 1 & -1 \\ 1 & 0 \end{bmatrix}$$

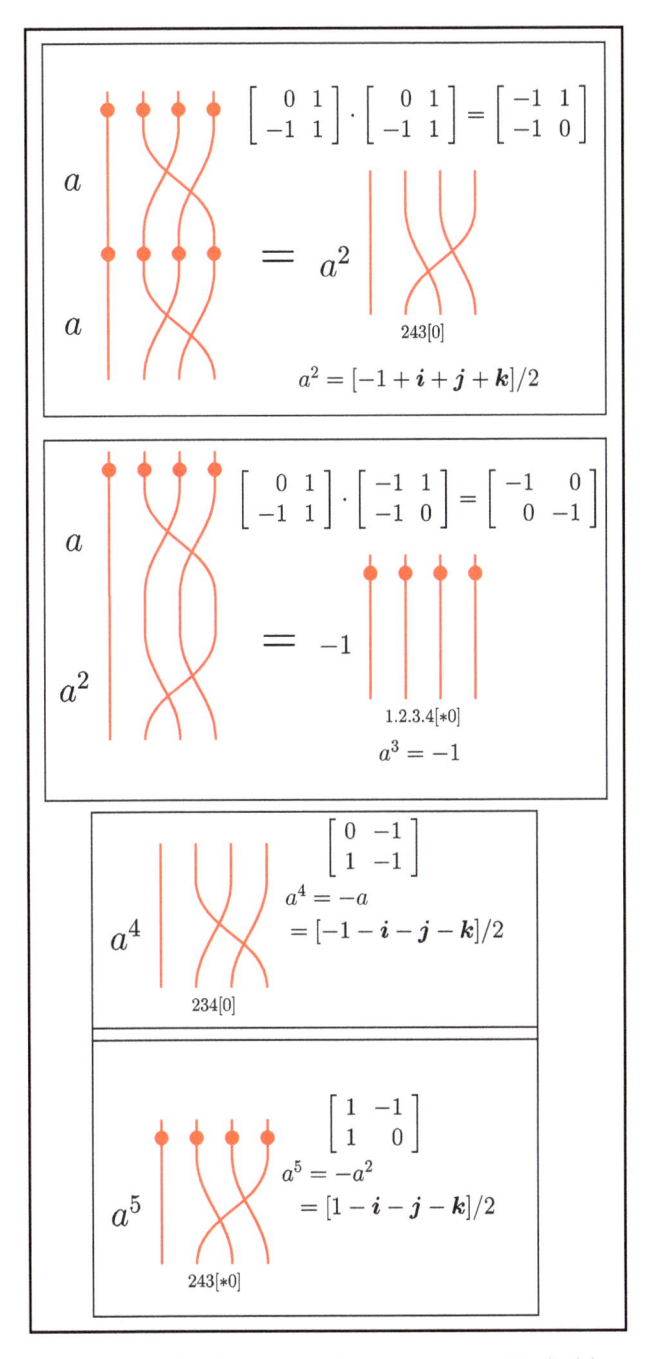

$a^5 = -a^2$
$= [1 - \boldsymbol{i} - \boldsymbol{j} - \boldsymbol{k}]/2$

243[*0]

Figure 5.16. Powers of the element $a \in \mathrm{GL}_2(\mathbb{Z}/3)$.

Exercise 14. Consider the matrix $g = \begin{bmatrix} 0 & 1 \\ 1 & 1 \end{bmatrix} \in \mathrm{GL}_2(\mathbb{Z}/3)$. Verify that

$$\begin{bmatrix} 0 & 1 \\ 1 & 1 \end{bmatrix} \cdot \begin{bmatrix} 0 & 1 \\ 1 & 1 \end{bmatrix} = \begin{bmatrix} 1 & 1 \\ 1 & -1 \end{bmatrix} \leftrightarrow i$$

and that

$$\begin{bmatrix} 0 & 1 \\ 1 & 1 \end{bmatrix} \cdot \begin{bmatrix} 1 & 0 & 1 & -1 \\ -1 & -1 & 1 & 0 \end{bmatrix} = \begin{bmatrix} -1 & -1 & 1 & 0 \\ 0 & -1 & -1 & -1 \end{bmatrix}$$
$$\leftrightarrow (03164752) \leftrightarrow 1423[2].$$

Show that the 4-strings-with-(mod 2)-quipu representation for g is $1423[2]$, as depicted at the northeast vertex in Fig. 5.17.

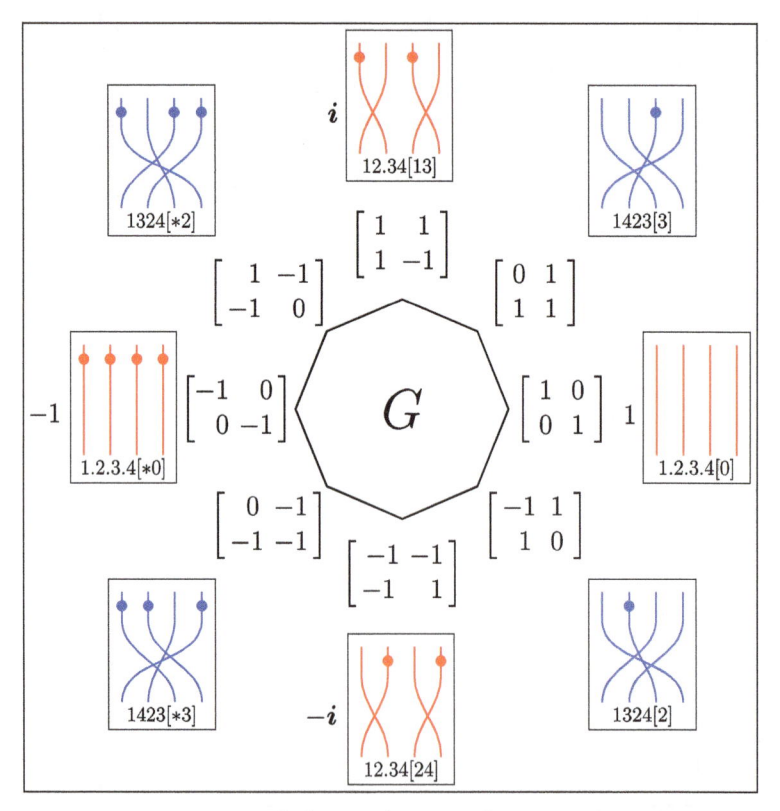

Figure 5.17. 4-strings-with-(mod 2)-quipu for a cyclic subgroup $G \subset \mathrm{GL}_2(\mathbb{Z}/3)$.

Extensions of the Permutation Group Σ_4 213

Let $G = \langle g \rangle$ denote the subgroup of $\mathrm{GL}_2(\mathbb{Z}/3)$ that consists of powers of g. Verify that the elements of G are as indicated in Fig. 5.17.

The subgroup $G \subset \mathrm{GL}_2(\mathbb{Z}/3)$ has order 8, and so, there are 6 cosets of G in the larger group. As it happens, $G \cup jG$ is isomorphic to a group that is known as the semi-dihedral group of order 16. The cosets G, jG, aG, ajG, bG, and bjG are depicted in Figs. 5.17–5.22. You are urged to verify that the representations are correct.

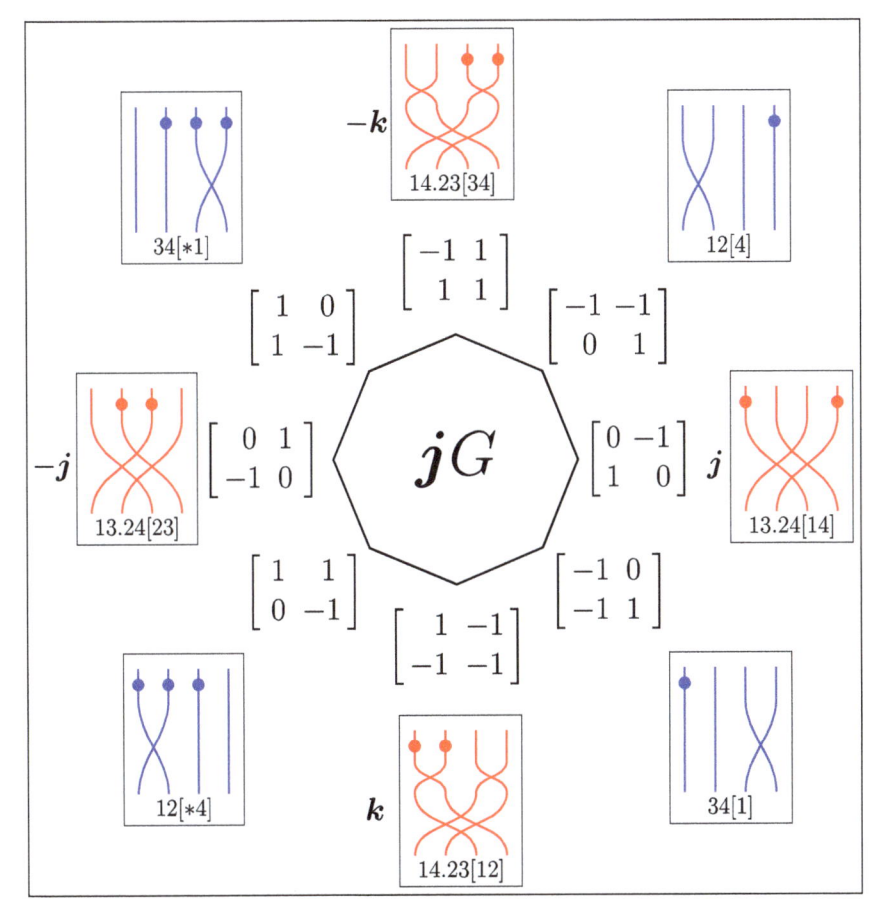

Figure 5.18. 4-strings-with-(mod 2)-quipu for the coset $jG \subset \mathrm{GL}_2(\mathbb{Z}/3)$.

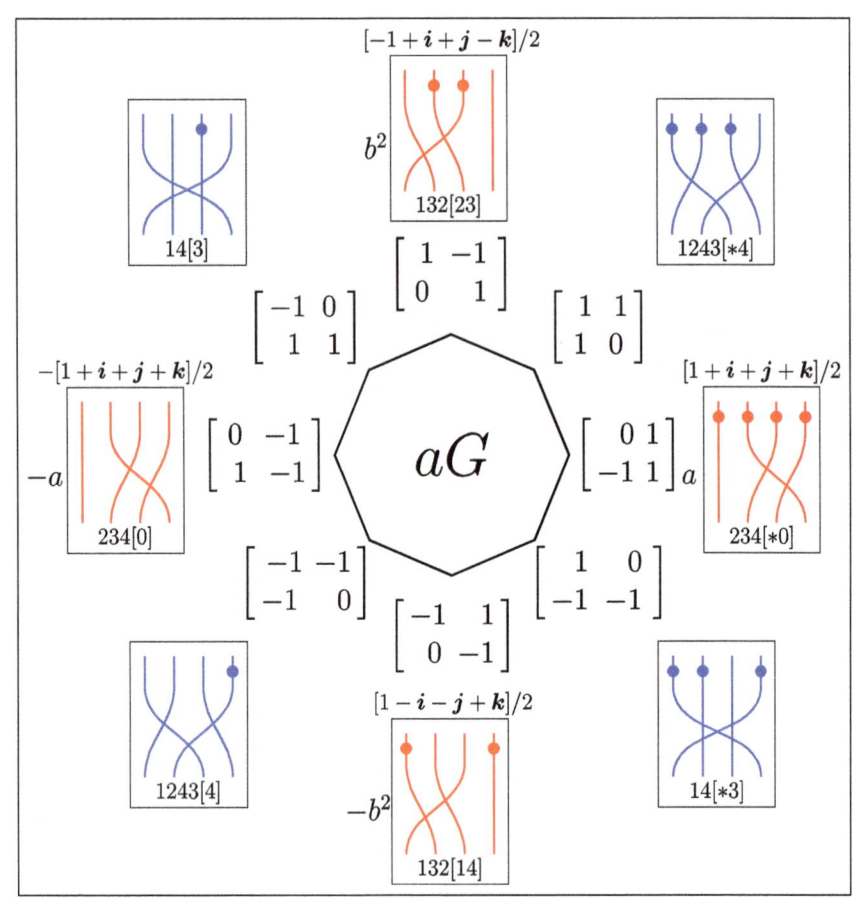

Figure 5.19. 4-strings-with-(mod 2)-quipu for the coset $aG \subset \mathrm{GL}_2(\mathbb{Z}/3)$.

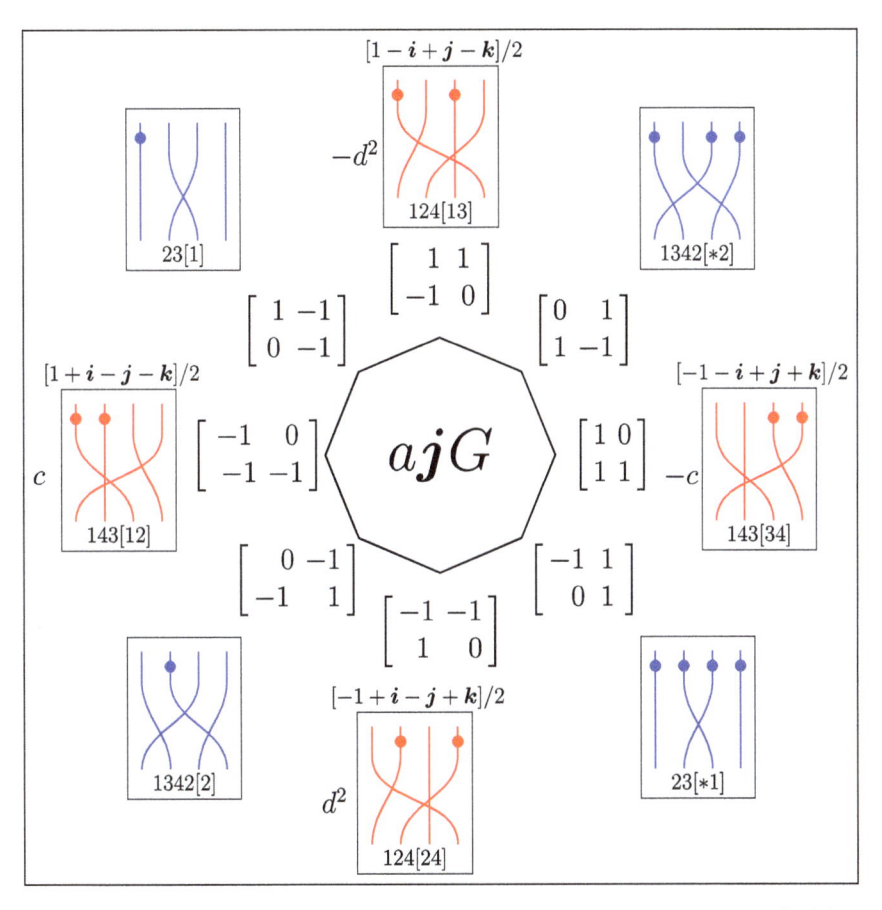

Figure 5.20. 4-strings-with-(mod 2)-quipu for the coset $ajG \subset \mathrm{GL}_2(\mathbb{Z}/3)$.

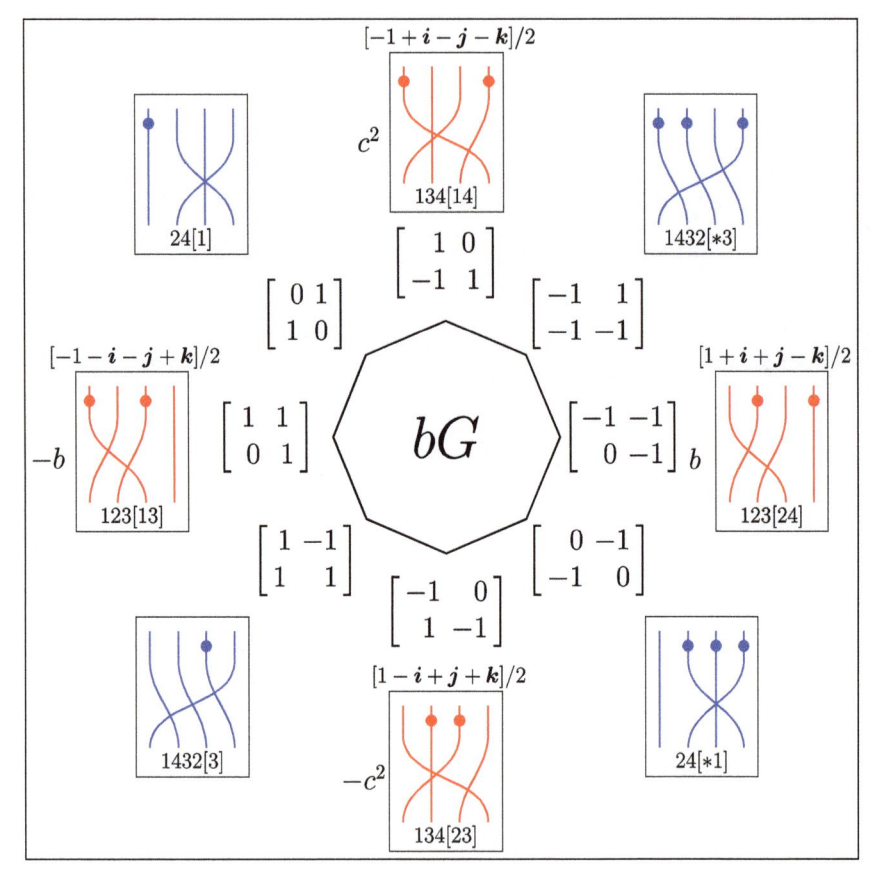

Figure 5.21. 4-strings-with-(mod 2)-quipu for the coset $bG \subset \mathrm{GL}_2(\mathbb{Z}/3)$.

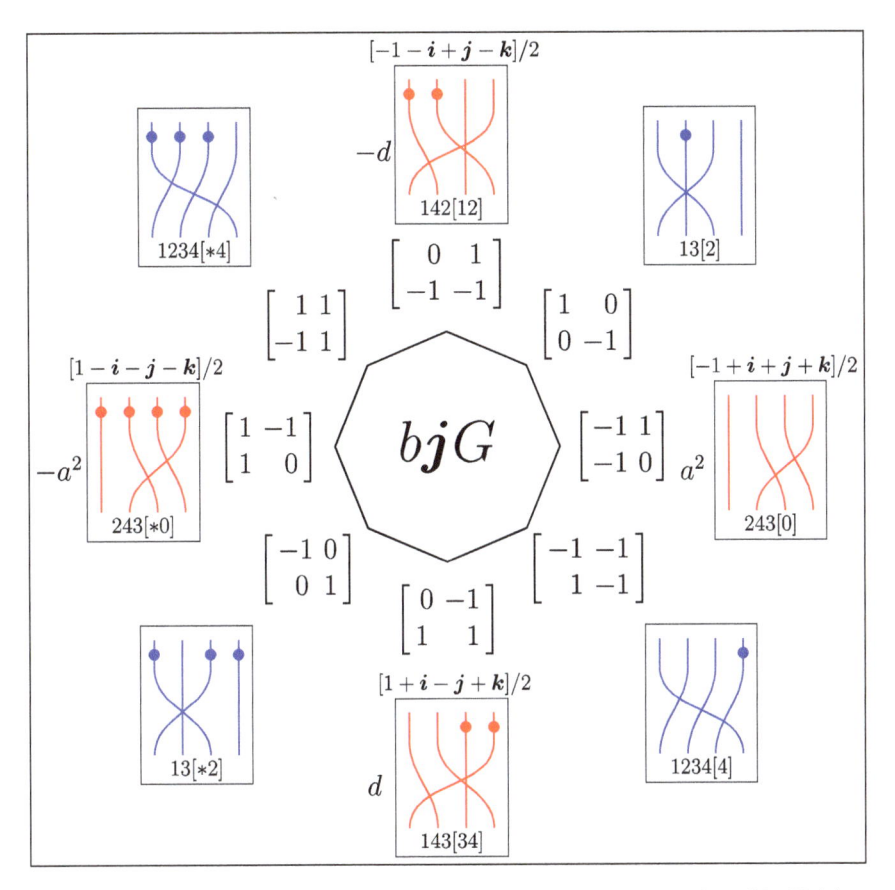

Figure 5.22. 4-strings-with-(mod 2)-quipu for the coset $bjG \subset \mathrm{GL}_2(\mathbb{Z}/3)$.

The 4-strings-with-(mod 2)-quipu diagrams are colored red if the determinant of the corresponding matrix is $+1$ and they are colored blue if the determinant is -1.

5.2.2. The semi-dihedral group

The subgroup $G \cup jG$, that will be used to construct the quipu for the 3-strings depiction of $\mathrm{GL}_2(\mathbb{Z}/3)$, can be generated by the two elements $g = \begin{bmatrix} 0 & 1 \\ 1 & 1 \end{bmatrix}$ and $j = \begin{bmatrix} 0 & -1 \\ 1 & 0 \end{bmatrix}$. The element g is depicted at the northeast vertex in Fig. 5.24 as a 2-strings-with-(mod 8)-quipu diagram, and the element that represents j in a 2-strings-with-(mod 8)-quipu diagram appears at the eastern position in Fig. 5.25. Let us explain how these representations and the remaining elements in these two figures were obtained.

The subgroup and the coset will be ordered as

$$G = (1, g, g^2 = i, g^3, g^4 = -1, g^5 = -g, g^6 = -i, g^7 = -g^3)$$
$$= (1, g, i, ig, -1, -g, -i, -ig)$$

and

$$jG = (j, jg, -k, -kg, -j, -jg, k, kg).$$

Since $j \cdot j = -1$, the result of multiplying the cosets by j is as follows:

$$j \cdot G = jG = [jG, 0], \quad j \cdot jG = [G, 4].$$

That is, j interchanges the cosets but rotates G four steps forward. Meanwhile, $gG = [G, 1]$ since G is a cyclic group and multiplication by g increments the elements forward one place. In Fig. 5.23, the product gj is computed by means of the 4-strings-with-(mod 2)-quipu representatives. The figure also includes the matrix calculation. In this way, $gj = jg^3$. Therefore, $g \cdot jG = [jG, 3]$. That is, multiplication by g fixes the cosets set-wise and increments G one step forward while it increments the elements of jG three steps forward.

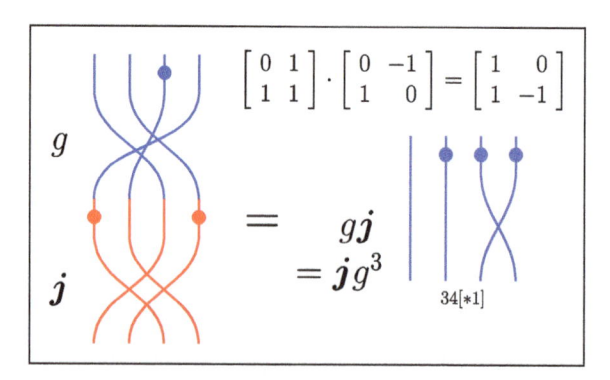

Figure 5.23. The product $g \cdot j$.

Extensions of the Permutation Group Σ_4 219

There is a groupprops.subwiki that describes the "semi-dihedral group of order 16" via the presentation

$$\mathrm{SD}_{16} = \langle g, x : g^8 = x^2 = 1, xgx = g^3 \rangle.$$

In the notation that we have adopted, the element x can be taken to be jg, for $x^2 = jg \cdot jg = j \cdot jg^4 = -g^4 = 1$ while $xgx = jggjg = jijg = jkg = ig = g^3$. These relations can also be verified by using the 2-strings-with-(mod 8)-quipu diagrams which, by the way, was how the authors originally did the calculations.

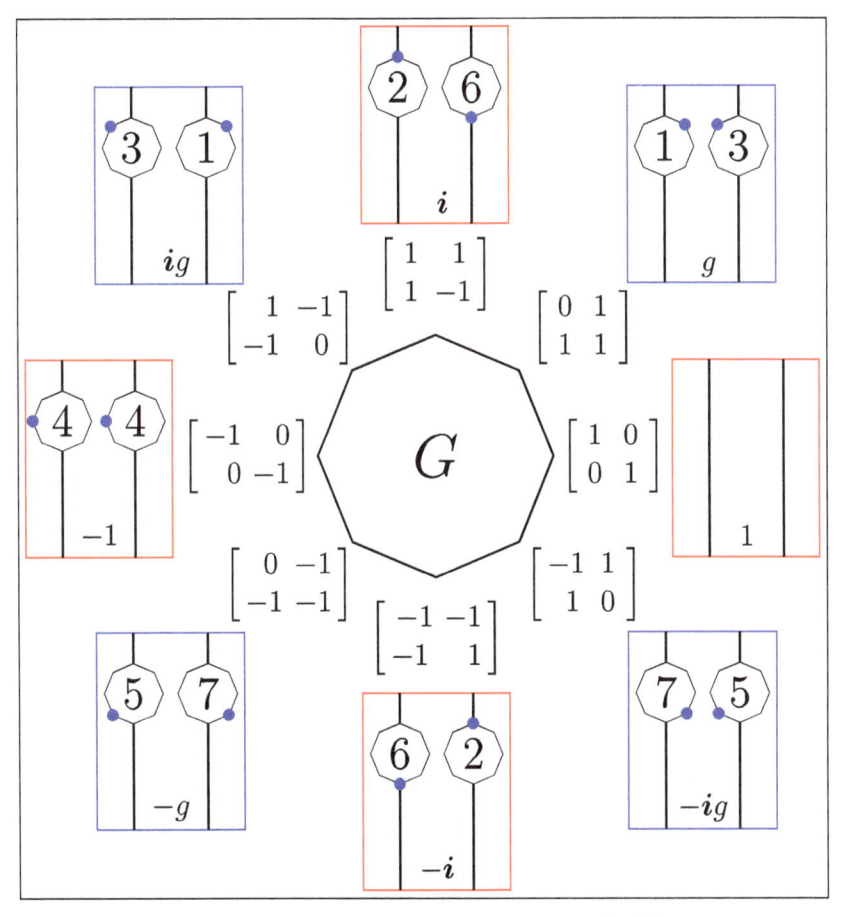

Figure 5.24. Alternate representation of $G \cup jG \subset \mathrm{GL}_2(\mathbb{Z}/3)$: part 1, the cyclic group G.

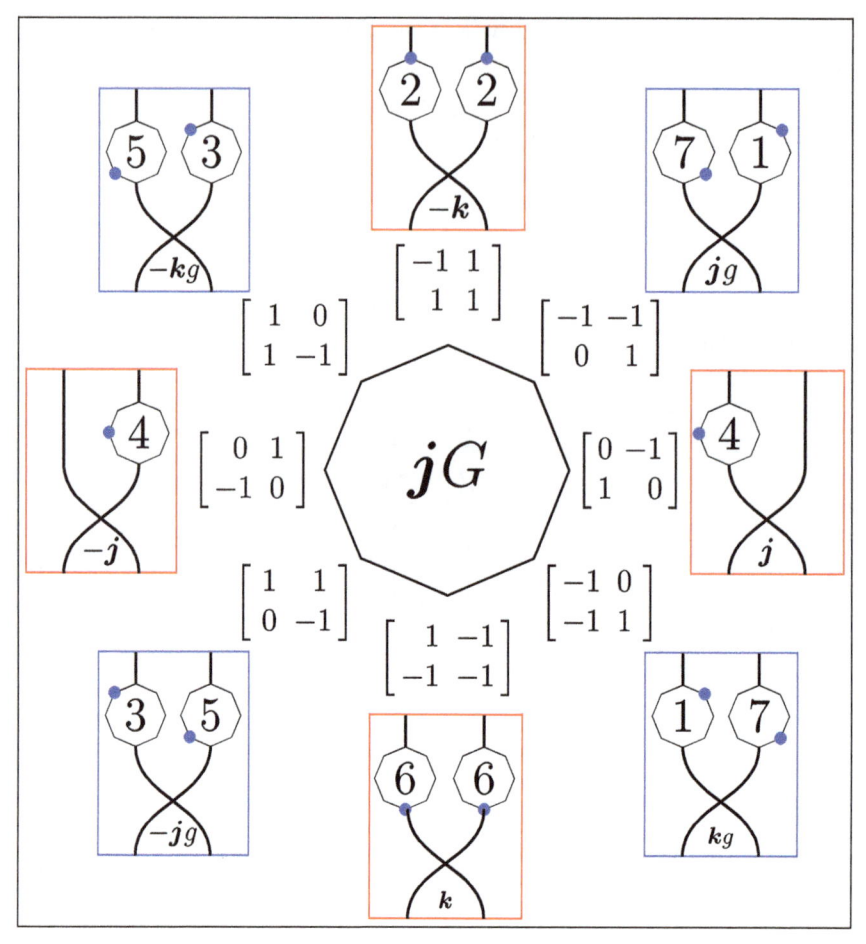

Figure 5.25. Alternate representation of $G \cup jG \subset \mathrm{GL}_2(\mathbb{Z}/3)$: part 2, the coset jG.

Figures 5.24 and 5.25 indicate 2-strings-with-(mod 8)-quipu for the elements of the semi-dihedral group $G \cup jG$. At the cardinal directions E, W, N, S in the diagrams, the elements in the quaternion group $Q_8 = \{\pm 1, \pm i, \pm j, \pm k\}$ appear. It is worth noting that the prior 2-strings-with-(mod 4)-quipu descriptions from Fig. 4.15 can be obtained from the 2-strings-with-(mod 8)-quipu representations by dividing the index in the quipu from the latter by 2.

5.2.3. A 3-strings representation of $GL_2(\mathbb{Z}/3)$

In order to illustrate the elements of $GL_2(\mathbb{Z}/3)$ as 3-strings-with-semi-dihedral-quipu, we will first compute the actions of the elements a, b, and g upon the ordered set of ordered cosets

$$((G, jG), (aG, ajG), (bG, bjG)).$$

Some of these actions can be determined by means of prior computations, and others are virtually tautological.

Let's consider the tautological actions first. Well,

$$a \cdot \begin{cases} G = aG, \\ jG = ajG, \end{cases} \text{ and } b \cdot \begin{cases} G = bG, \\ jG = bjG. \end{cases}$$

We previously determined that

$$g \cdot \begin{cases} G = [G, 1], \\ jG = [jG, 3]. \end{cases}$$

In order to compute the action of a upon aG, we computed a^2 by using its 4-strings-with-(mod 2)-quipu representation in Fig. 5.16. In Fig. 5.26, the identity $b \cdot j = a^2 = 243[0]$ is illustrated. We can conclude that $a \cdot aG = [bjG, 0]$ since that element is in the eastern most position of the octagon illustrated in Fig. 5.22.

Figure 5.15 indicates that $a \cdot b = j$. From this identity, it follows that $a \cdot bG = [jG, 0]$. Furthermore, $a \cdot bjG = j \cdot jG = -G = [G, 4]$. So, to determine the remaining action of a, we illustrate that $a^2 \cdot j = -b = 123[13]$ in Fig. 5.26. Thus, we conclude that $a \cdot ajG = [bG, 4]$.

Let us summarize. Label the coset $[0] = (G, jG)$, $[A] = (aG, ajG)$, and $[B] = (bG, bjG)$. The element a acts on the set of cosets as the 3-cycle $([0], [A], [B])$. With more specificity,

$$a \cdot \begin{cases} G = [aG, 0], \\ jG = [ajG, 0], \\ \hline aG = [bjG, 0], \\ ajG = [bG, 4], \\ \hline bG = [jG, 0], \\ bjG = [G, 4]. \end{cases}$$

Before this is illustrated as 3-strings-with-quipu, let's determine the actions of b and g.

Figure 5.15 indicates that $b \cdot a = i = 12.34[13]$. This element is located at the northern vertex of Fig. 5.17. So, $b \cdot aG = [G, 2]$. In a similar fashion, $b \cdot aj = i \cdot j = k$ implies that $b \cdot ajG = [jG, 6]$ since k is located at the south vertex of jG illustrated in Fig. 5.18.

Figure 5.26 computes that $b^2 = 132[23]$, and this element is located at the northern vertex of aG, as illustrated in Fig. 5.19. We also determine in Fig. 5.26 that $b^2j = d^2 = 124[24]$, and this element appears at the southern vertex of ajG in Fig. 5.20. Thus, $bjG = [ajG, 6]$.

The element b corresponds to the 3-cycle $([0], [B], [A])$. In detail,

$$b \cdot \begin{cases} G = [bG, 0], \\ jG = [bjG, 0], \\ \hline aG = [G, 2], \\ ajG = [jG, 6], \\ \hline bG = [aG, 2], \\ bjG = [ajG, 6]. \end{cases}$$

There are a few remaining calculations needed to determine the action of g upon aG, ajG, bG, and bjG, namely, Fig. 5.27 indicates that $ga = 1432[3*]$ and this element is at the northeast position of bG indicated in Fig. 5.21.

In addition, $gaj = 1234[*4]$ which is located at the northwest vertex of bjG. The identities $gb = 1342[2]$ and $gbj = 14[3]$ are also computed in Fig. 5.27. The former is located at the southwest of ajG (Fig. 5.20) and the latter at the northwest vertex of aG (Fig. 5.19).

Therefore, g acts at the course level as the transposition $([A], [B])$. Specifically,

$$g \cdot \begin{cases} G = [G, 1], \\ jG = [jG, 3], \\ \hline aG = [bG, 1], \\ ajG = [bjG, 3], \\ \hline bG = [ajG, 5], \\ bjG = [aG, 3]. \end{cases}$$

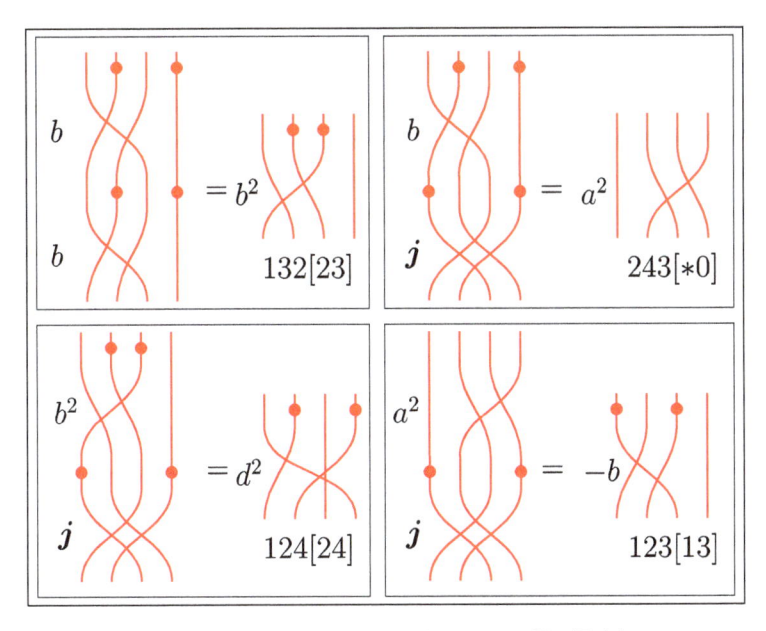

Extensions of the Permutation Group Σ_4 223

Figure 5.26. Some calculations in $SL_2(\mathbb{Z}/3)$.

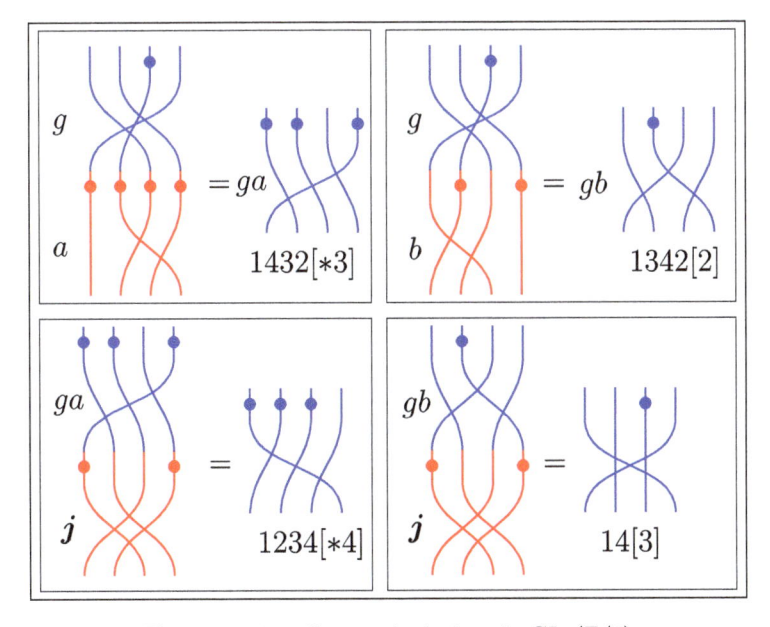

Figure 5.27. Some calculations in $GL_2(\mathbb{Z}/3)$.

The reader may feel more comfortable in performing the matrix calculations that correspond to those illustrated in Figs. 5.26 and 5.27. If so, please do. Also note that the determinants of the corresponding matrices are color encoded upon the 4-strings. Blue strings indicate a (-1)-determinant, and red strings indicate a $(+1)$ determinant.

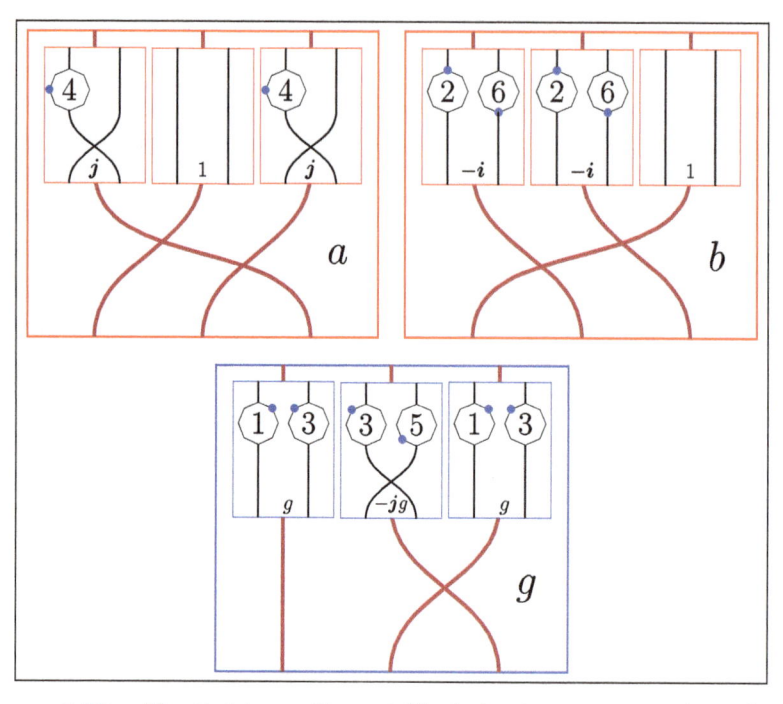

Figure 5.28. The 3-strings-with-semi-dihedral-quipu representatives of a, b, and g.

The 3-strings-with-semi-dihedral-quipu representations of the elements in $\mathrm{GL}_2(\mathbb{Z}/3)$ are compiled, coset by coset, in Figs. 5.29–5.34. The first columns are compilations of the elements in the binary tetrahedral group $\mathrm{SL}_2(\mathbb{Z}/3)$. The quipu upon these elements are always found within the group of unit quaternions $Q_8 = \{\pm 1, \pm \boldsymbol{i}, \pm \boldsymbol{j}, \pm \boldsymbol{k}\}$.

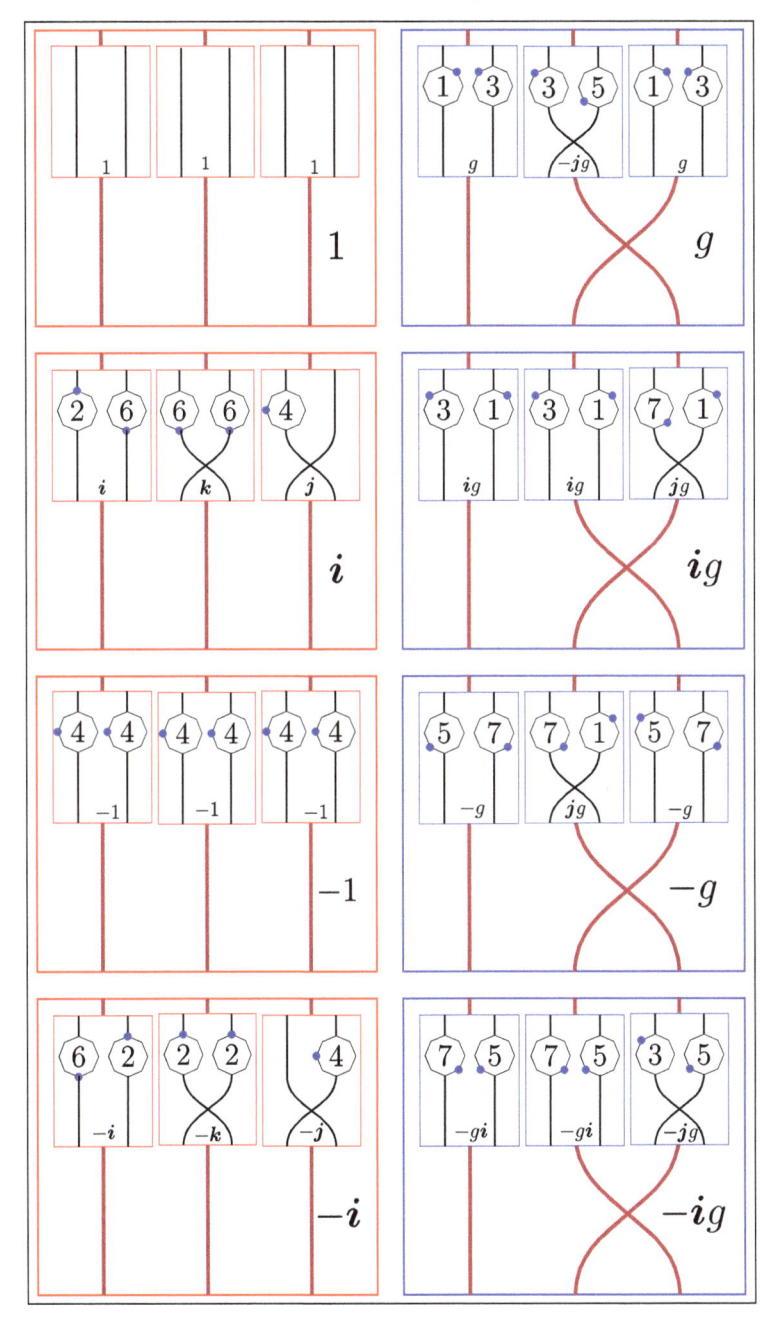

Figure 5.29. 3-strings-with-semi-dihedral-quipu for the subgroup $G \subset \mathrm{GL}_2(\mathbb{Z}/3)$.

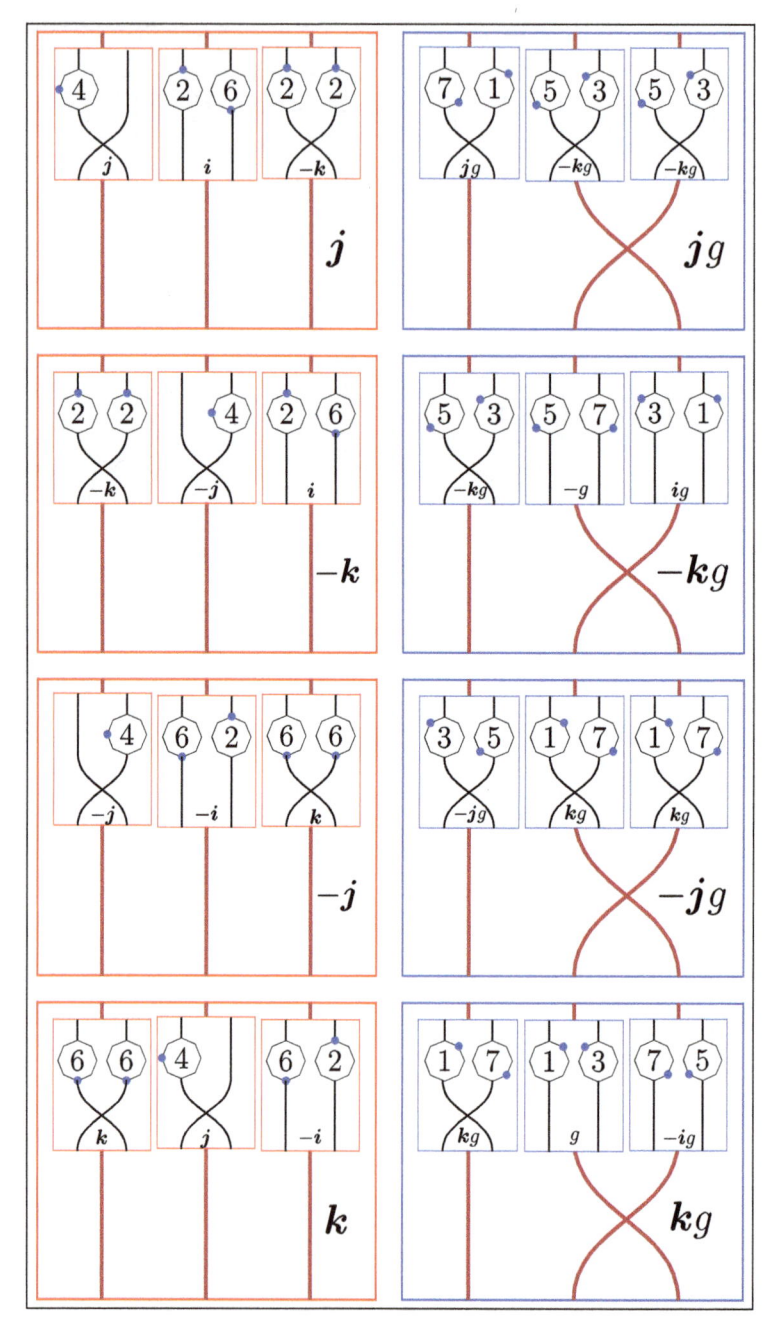

Figure 5.30. 3-strings-with-semi-dihedral-quipu for the coset $jG \subset \mathrm{GL}_2(\mathbb{Z}/3)$.

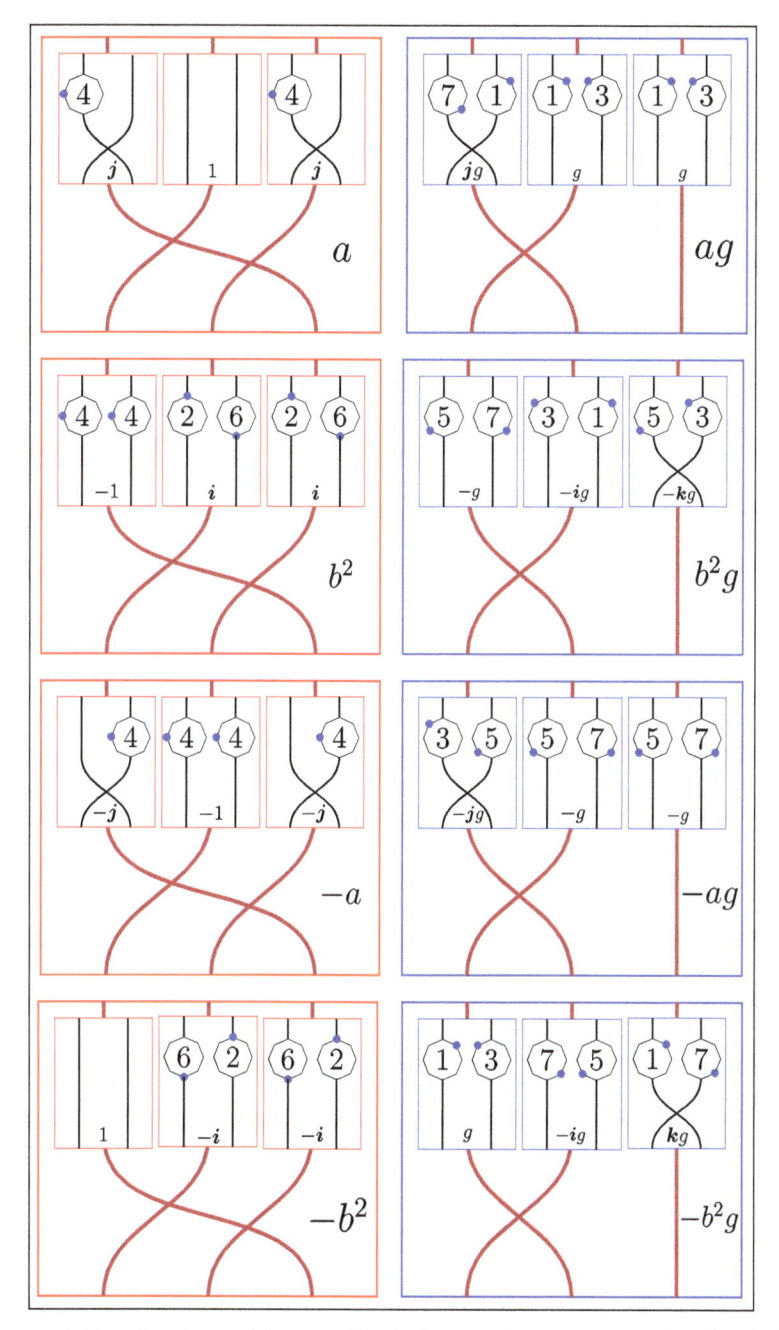

Figure 5.31. 3-strings-with-semi-dihedral-quipu for the coset $aG \subset \mathrm{GL}_2(\mathbb{Z}/3)$.

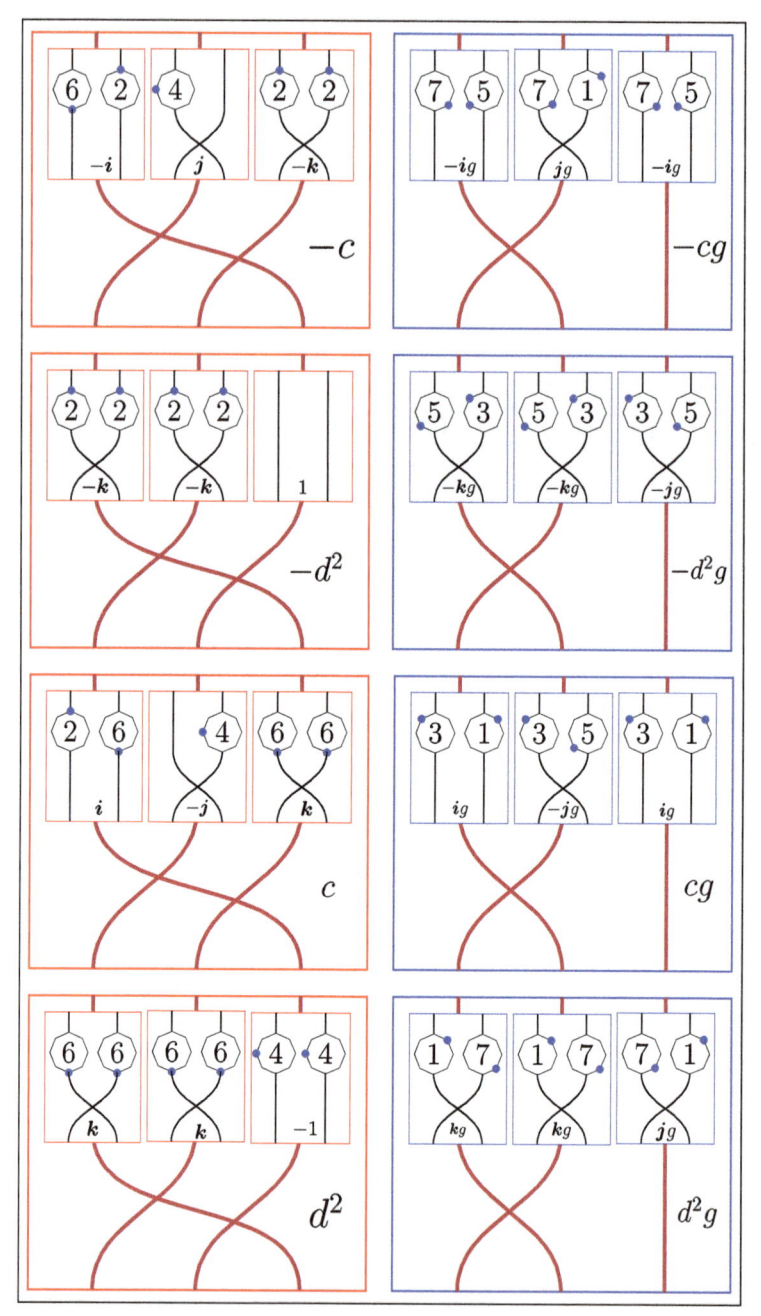

Figure 5.32. 3-strings-with-semi-dihedral-quipu for the coset $ajG \subset \mathrm{GL}_2(\mathbb{Z}/3)$.

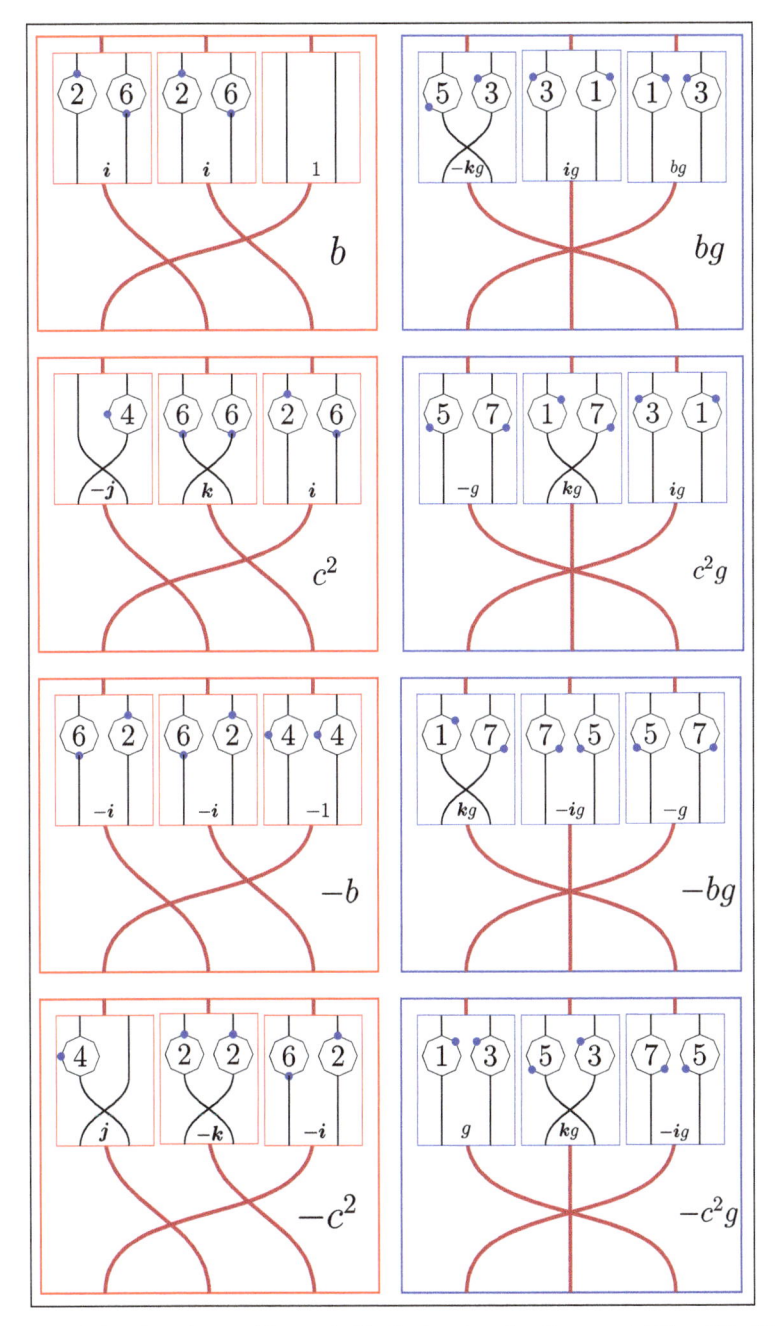

Figure 5.33. 3-strings-with-semi-dihedral-quipu for the coset $bG \subset \mathrm{GL}_2(\mathbb{Z}/3)$.

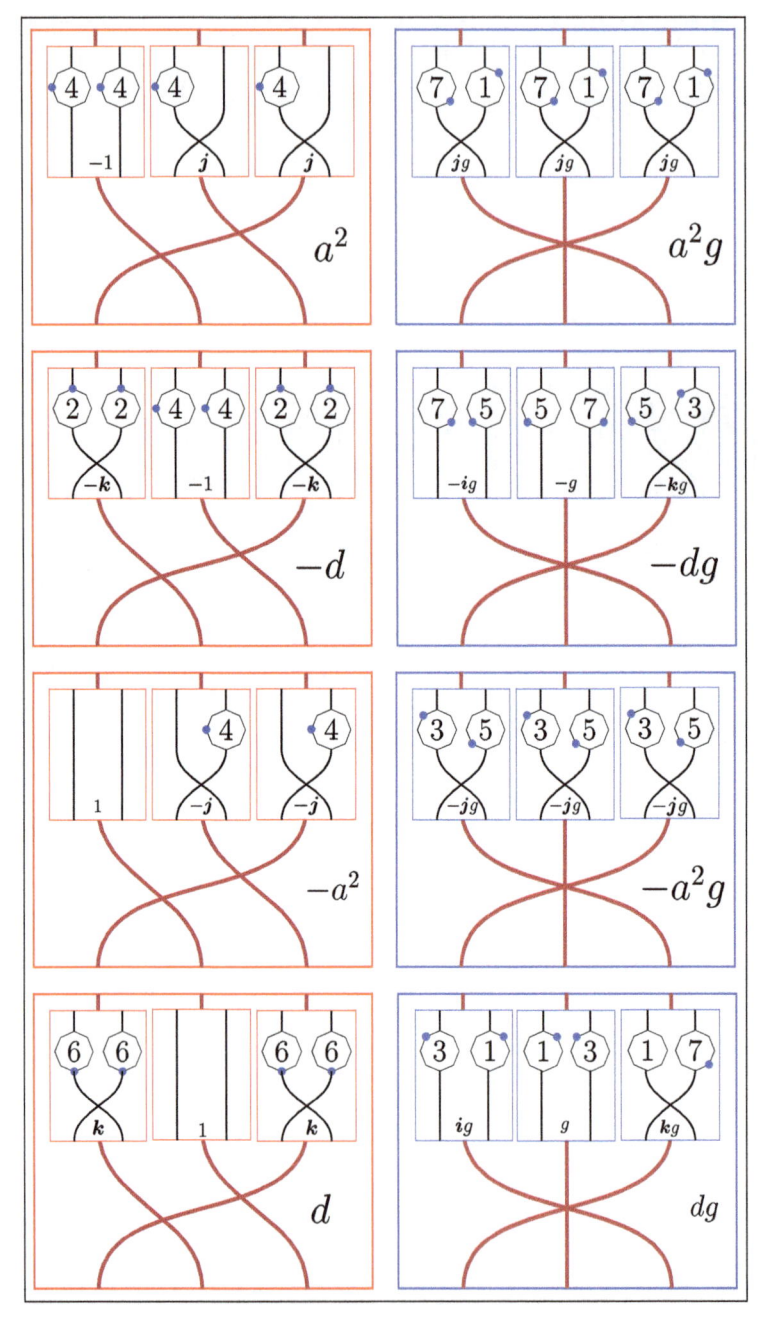

Figure 5.34. 3-strings-with-semi-dihedral-quipu for the coset $bjG \subset \mathrm{GL}_2(\mathbb{Z}/3)$.

Recall that Σ_3 denotes the permutation group upon 3-elements and that Q_8 denotes the quaternions. There is a seseq

$$1 \longleftarrow \Sigma_3 \xleftarrow{q} \mathrm{GL}_2(\mathbb{Z}/3) \xleftarrow{i} Q_8 \longleftarrow 1$$

that is analogous to the seseq

$$1 \longleftarrow \Sigma_3 \xleftarrow{q} \Sigma_4 \xleftarrow{i} K_4 \longleftarrow 1.$$

5.2.4. Peculiar correspondences

In this section, two projections from the group of invertible (2×2)-matrices over $\mathbb{Z}/3$ $(\mathrm{GL}_2(\mathbb{Z}/3))$ to the symmetric group Σ_4 will be discussed and compared.

The first of these is fairly easy to understand. In Figs. 5.17–5.22, the matrices in $\mathrm{GL}_2(\mathbb{Z}/3)$ are demonstrated to correspond to 4-strings-with-(mod 2)-quipu that adorn the vertices of octagonally depicted cosets of the cyclic subgroup G. For example, $g = \begin{bmatrix} 0 & 1 \\ 1 & 1 \end{bmatrix}$ corresponds to the diagram labeled 1423[3]. By ignoring the quipu (dot) upon the third string, we can map the element g to the 4-cycle in Σ_4 as follows: $g \mapsto (1432)$. This process yields the surjective homomorphism q_1 in the seseq

$$1 \longleftarrow \Sigma_4 \xleftarrow{q_1} \mathrm{GL}_2(\mathbb{Z}/3) \xleftarrow{i} \mathbb{Z}/2 \longleftarrow 1.$$

The kernel of the homomorphism is illustrated in Fig. 5.17 as $\pm I$ — the identity matrix and its negative. These lie at the east and west vertices of the octagon that represents the cyclic subgroup G.

That representation was dependent upon the choice of labels upon the non-zero points in $\mathbb{Z}/3 \times \mathbb{Z}/3$, as depicted in Fig. 5.13. The significant feature about that choice is that antipodal points in $\mathbb{Z}/3 \times \mathbb{Z}/3 = \Delta \times \Delta$ have labels that differ by 4. Any choice in which the labels upon antipodal points are congruent modulo 4 would be a good labeling.

The authors now ask the reader to hold that thought for a little while.

Next, we'd like to ask you to recall the process that was depicted in Figs. 5.10–5.12. Therein the dihedral quipu project to (mod 2)-quipu by sending reflections to (-1) and sending rotations to $(+1)$. This projection is a surjective homomorphism whose kernel is isomorphic to $\mathbb{Z}/4$ — it being the subgroup of rotations: $\langle x \rangle = \{x^\ell : \ell = 0, 1, 2, 3\}$. Thus, there is a seseq

$$1 \longleftarrow \mathbb{Z}/2 \overset{p}{\longleftarrow} D_8 \overset{i}{\longleftarrow} \mathbb{Z}/4 \longleftarrow 1.$$

Figures 5.10 and 5.11 indicate that the depiction of the symmetric group Σ_4 as 3-strings-with-(mod 2)-quipu is induced from the depiction as 3-strings-with-D_8-quipu via the projection $\mathbb{Z}/2 \overset{p}{\longleftarrow} D_8$.

Our immediate goal is to emulate that relationship, but now within the context of the semi-dihedral group, $SD_{16} = G \cup jG$. The elements in G act analogously to the rotations because their 2-strings-with-(mod 8)-quipu representatives do not involve a pair of crossing strings. The elements in jG act analogously to the reflections.

Therefore, there is a surjective homomorphism $\mathbb{Z}/2 \overset{p'}{\longleftarrow} SD_{16}$ whose kernel is the cyclic subgroup G. Specifically, the elements in G get mapped to $(+1)$, and the elements in jG get mapped to (-1). The map can be envisioned by examining the crossings between the 2-strings representations in Figs. 5.24 and 5.25. Then any element in jG projects to a dot upon a string. In this way, p' extends to $GL_2(\mathbb{Z}/3)$, and there is another seseq

$$1 \longleftarrow \Sigma_4 \overset{q_2}{\longleftarrow} GL_2(\mathbb{Z}/3) \overset{i}{\longleftarrow} \mathbb{Z}/2 \longleftarrow 1$$

because the elements of Σ_4 can be represented as 3-strings-with-(mod 2)–quipu.

In Figs. 5.35–5.40, the correspondences among different representations of half of the elements of $GL_2(\mathbb{Z}/3)$ are compiled. If $x \in GL_2(\mathbb{Z}/2)$, then $\pm x$ project to the same point in Σ_4 under either of the projections q_1 or q_2 that appear in seseqs above. In the compilation, the 3-strings-with-(mod 2)-quipu representatives of the elements of Σ_3 are included and labeled via the correspondences given in Figs. 5.10–5.12. The 4-strings-with-(mod 2)-quipu are relabeled alphabetically.

Let us recap. The figures indicate that there are two possible projections to the symmetric group: (1) ignore the quipu upon the 4-strings-with-(mod 2)-quipu or (2) map the elements in G to $(+1)$ and those in jG to (-1) to create 3-strings-with-(mod 2)-quipu depictions. Then employ the previous correspondences between elements in Σ_4 and the 3-strings-with-(mod 2)-quipu representatives. These projections disagree!

To resolve the disagreement, we will use the compilation of the figures to determine that there is a relabeling of the non-zero elements in $\Delta \times \Delta$. Such a relabeling involves finding a correspondence between the elements of $\{p, q, r, s\}$ and $\{1, 2, 3, 4\}$. Here, we will only provide an outline of the relabeling. We will not implement it!

We can use Figs. 5.35–5.40 to reconcile the disparity between the labels p, q, r, s in the 4-strings pictures and the labels $1, 2, 3, 4$ in the 3-strings representatives of the elements in Σ_4. By using the elements that project to 3-cycles, we can identify the label upon the fixed string. For example, the element a projects to the 3-cycle (134) in Fig. 5.37 and also to (qrs). Thus, p corresponds to 2. The element b projects to (123) in Fig. 5.39 and also to (pqr). So, s corresponds to 4. The elements $-c$ and c project to (243) and (psr) in Fig. 5.38. So, the fixed string indicates that q corresponds to 1. The elements $-d$ and d project to (124) and to (psq). So, the fixed string r corresponds to 3.

Now, it is a bookkeeping exercise to see that these relabelings are consistent throughout the remaining elements.

It is fair to ask then: "Why not go back through the text and interchange the labels on the non-zero vectors in $\Delta \times \Delta$ and redraw the corresponding 4-strings diagrams?" Well, as it happens, the current labels on the 4-strings diagrams coincide with some labelings that will occur in future sections in the text. Moreover, the 4-strings-with-(mod 2)-quipu representations for the unit quaternions $Q_8 = \{\pm 1, \pm i, \pm j, \pm k\}$ would become discombobulated. Finally, the process of relabeling is fraught with the potential for typographical errors and incomplete edits.

To conclude this section, we will merely let the peculiar correspondences remain as they are. For those who are cognizant, we remark that the two projections q_1 and q_2 onto the symmetric group Σ_4 are conjugate.

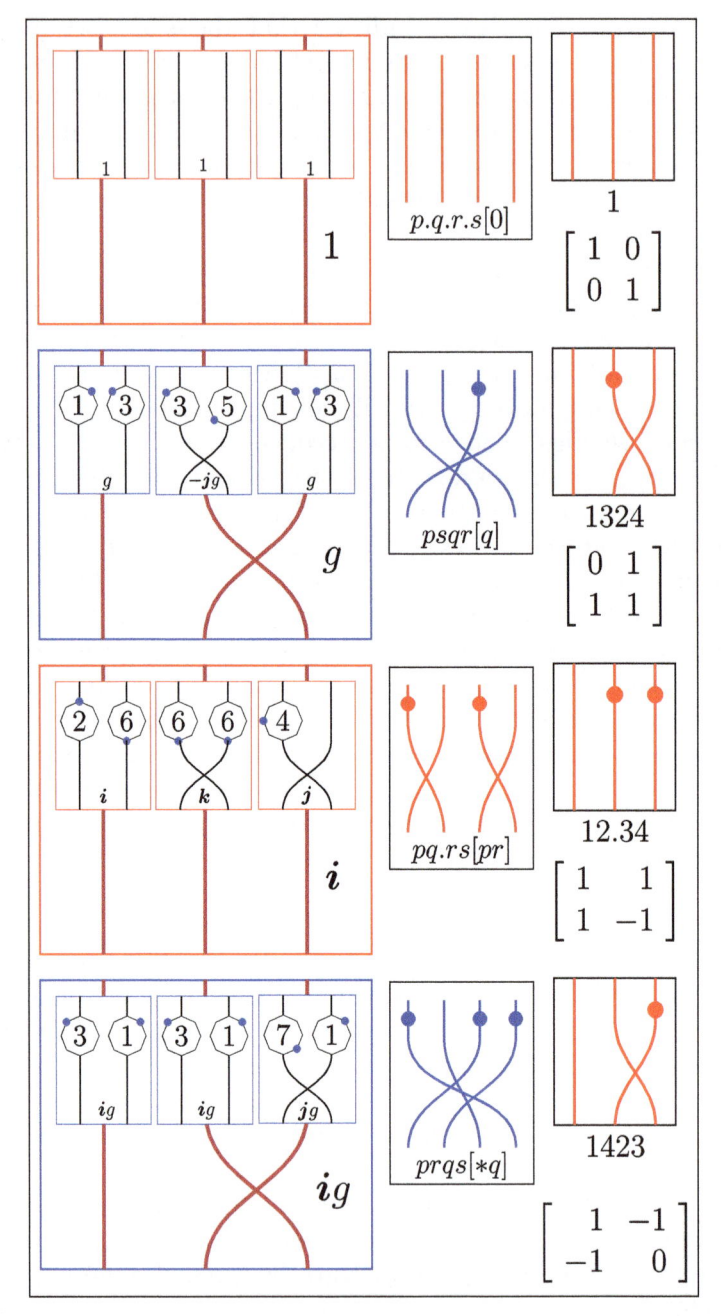

Figure 5.35. Representing half the elements in $G \subset \mathrm{GL}_2(\mathbb{Z}/3)$.

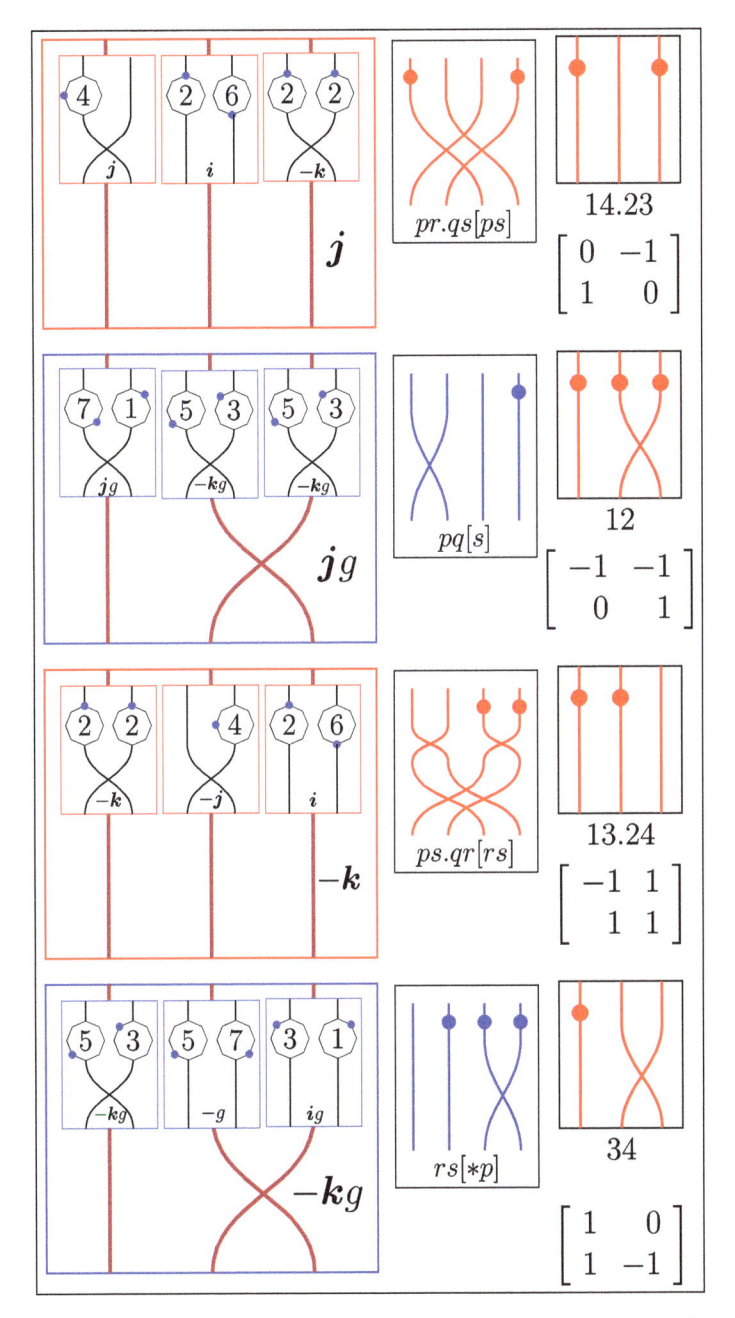

Figure 5.36. Representing half the elements in $jG \subset \mathrm{GL}_2(\mathbb{Z}/3)$.

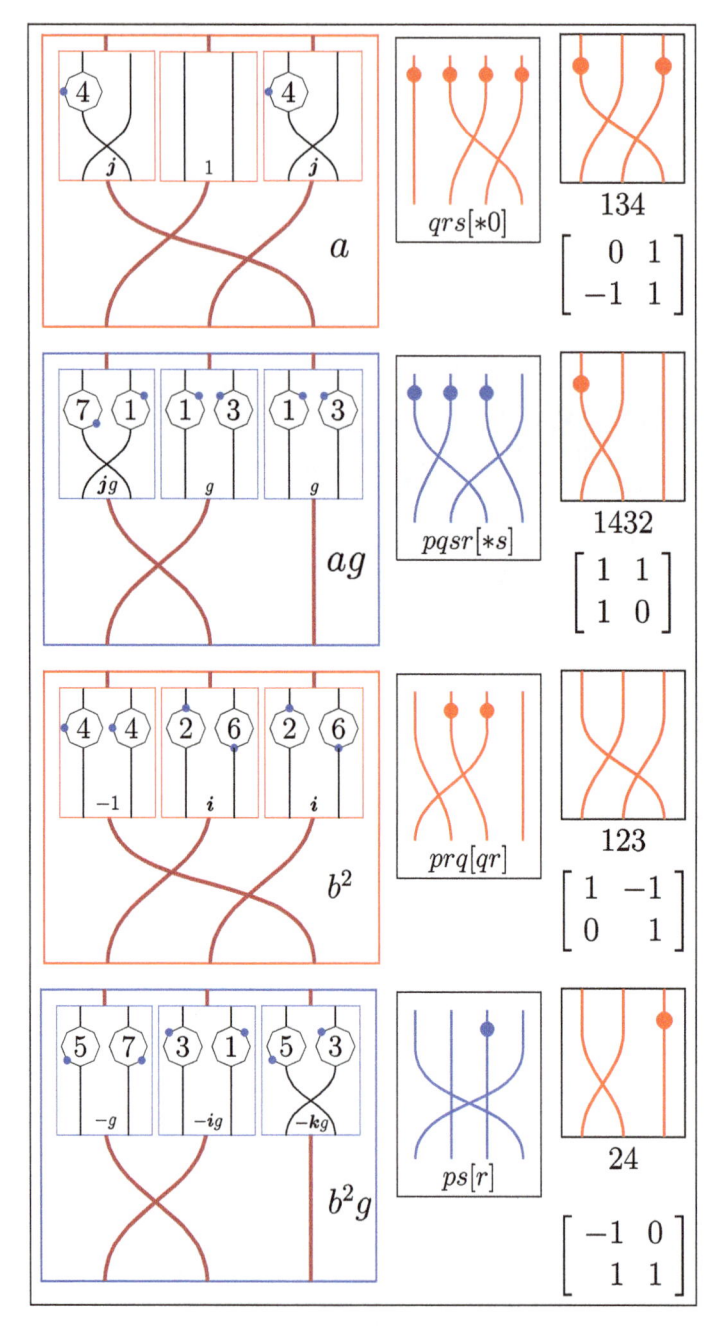

Figure 5.37. Representing half the elements in $aG \subset \mathrm{GL}_2(\mathbb{Z}/3)$.

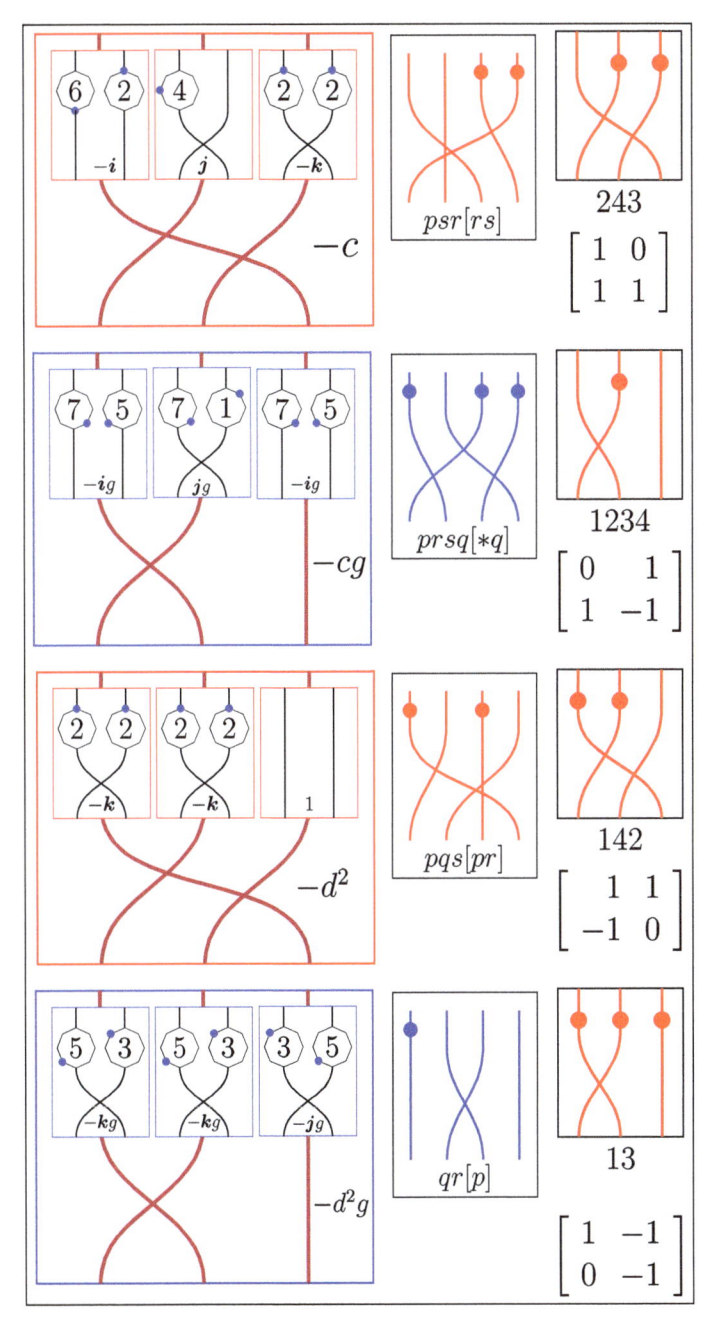

Figure 5.38. Representing half the elements in $ajG \subset \mathrm{GL}_2(\mathbb{Z}/3)$.

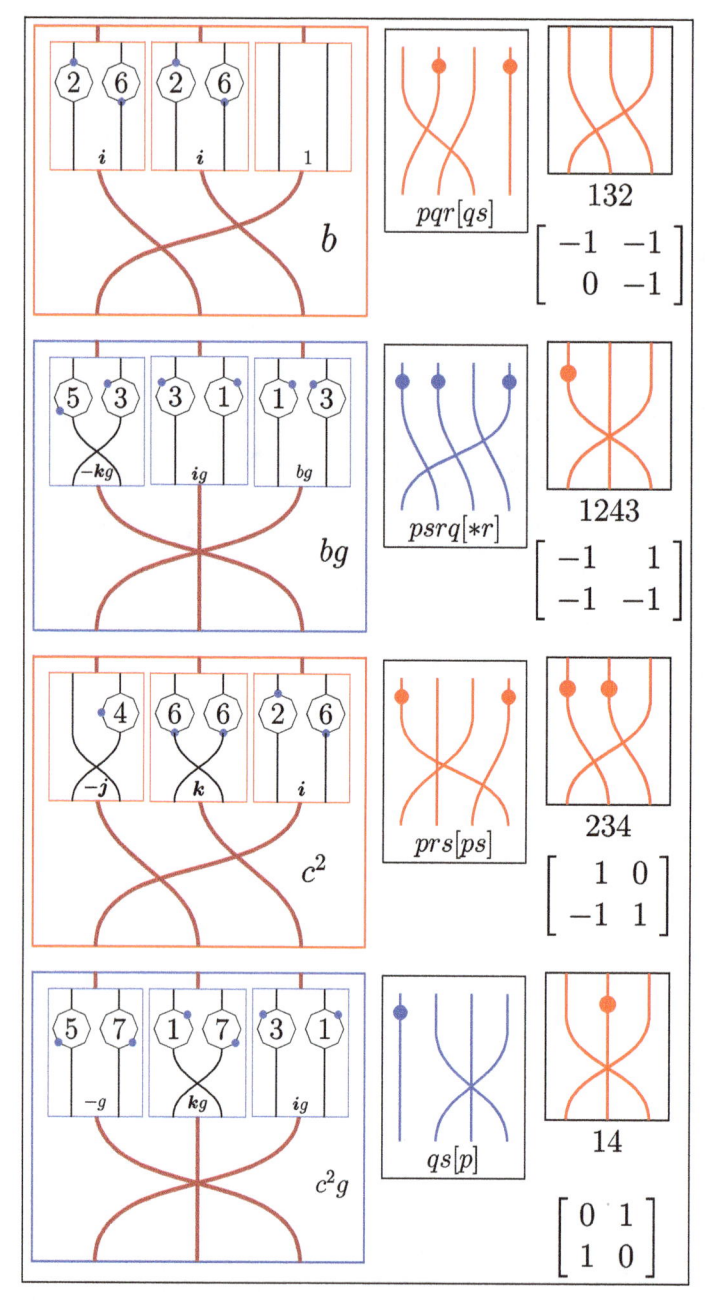

Figure 5.39. Representing half the elements in $bG \subset \mathrm{GL}_2(\mathbb{Z}/3)$.

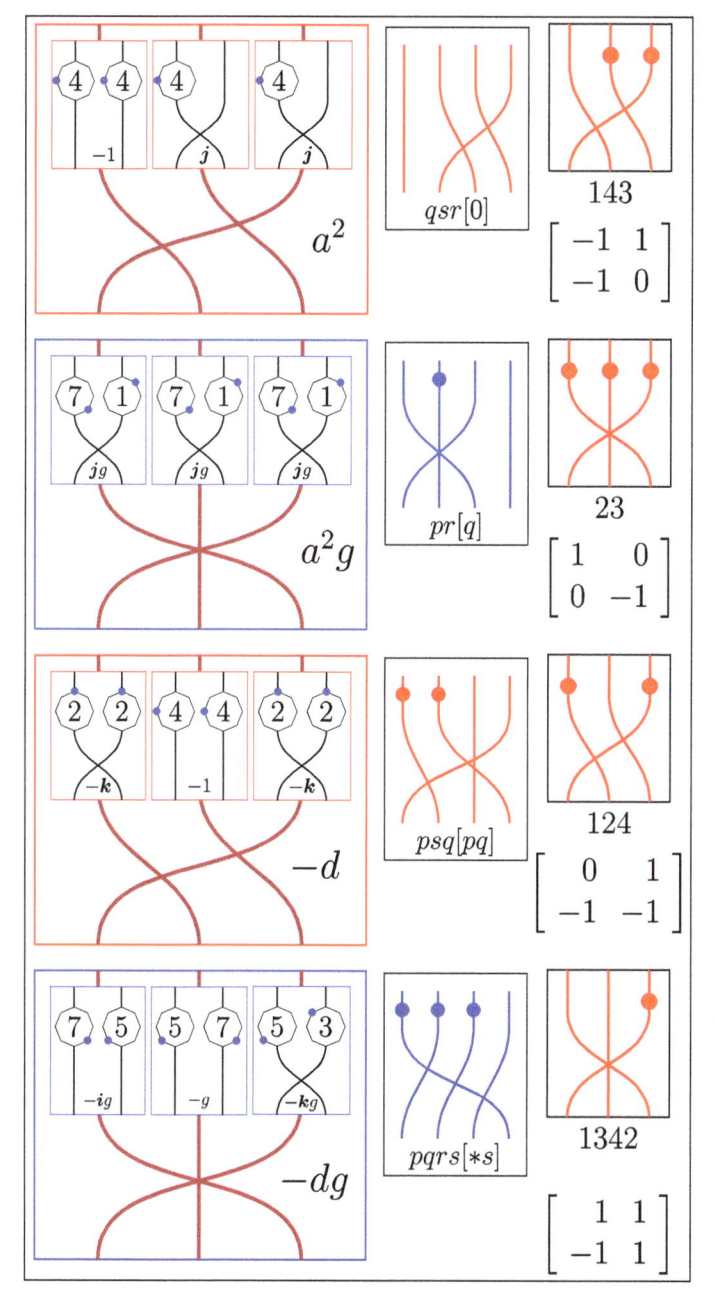

Figure 5.40. Representing half the elements in $bjG \subset \mathrm{GL}_2(\mathbb{Z}/3)$.

5.3. The group $SL_2(\mathbb{Z}/4)$

The elements of $\mathbb{Z}/4$ will be represented by $-1, 0, 1, 2$. The addition and multiplication tables are presented:

$r+c$	0	1	2	-1
0	0	1	2	-1
1	1	2	-1	0
2	2	-1	0	1
-1	-1	0	1	2

$r \cdot c$	0	1	2	-1
0	0	0	0	0
1	0	1	2	-1
2	0	2	0	2
-1	0	-1	2	1

Structurally, $\mathbb{Z}/4$ is said to be a *ring*. This is a set upon which there are two operations, $+$ and $\cdot$, which are associative and commutative. Multiplication distributes over additions ($a \cdot (b+c) = a \cdot b + a \cdot c$ and $(a+b) \cdot c = a \cdot c + b \cdot c$), there is an additive identity 0, and every element has an additive inverse. In $\mathbb{Z}/4$, the element 2 is its own additive inverse since $2 + 2 = 0$, the additive inverse of $+1$ is -1, and vice versa. There is no assertion about multiplicative inverses, and since $2 \cdot 2 = 0$, they don't always exist. Yet, 1 is a multiplicative identity in that $1 \cdot a = a$.

A pair of (2×2)-matrices whose entries are in $\mathbb{Z}/4$ can be multiplied in the usual manner. The determinant of a (2×2)-matrix $M = \begin{bmatrix} a & b \\ c & d \end{bmatrix}$ whose entries are in $\mathbb{Z}/4$ is still given by the expression $\det M = ad - bc$.

An element M of $SL_2(\mathbb{Z}/4)$ is one for which the determinant, $\det M = 1$. The multiplicative inverse is given as

$$M^{-1} = \begin{bmatrix} d & -b \\ -c & a \end{bmatrix}.$$

The recipe "switch the diagonals and change the signs of the off diagonals" applies because $\det M = 1$. But be aware that $-2 = 2 \in \mathbb{Z}/4$. So, the set

$$SL_2(\mathbb{Z}/4) = \left\{ \begin{bmatrix} a & b \\ c & d \end{bmatrix} : a, b, c, d \in \mathbb{Z}/4 \ \& \ ad - bc = 1 \in \mathbb{Z}/4 \right\}$$

is a group under matrix multiplication.

If you write down the 128 possible values for a, b, c, d, it is possible to filter out those for which $ad - bc \neq 1$. There are 48 matrices that remain. These will be enumerated in due course. That filtering process is a bit tedious, but not impossible. The clever way to determine the 48 matrices in $SL_2(\mathbb{Z}/4)$ is described subsequently.

Consider the entries in the multiplication table that appears above. There are eight 0s, two 1s, four 2s, and two (-1)s. These are the possible values for the products ad and bc that are involved in computing the determinant. Suppose that $ad = 0$. Then for $ad - bc = 1$, the value of bc must be (-1). There are $8 \times 2 = 16$ possible combinations of this form. Suppose that $ad = 1$. Then if $ad - bc = 1$, the value of bc must be 0. This gives another $2 \times 8 = 16$ possibilities. Suppose that $ad = 2$. Then if $ad - bc = 1$, the value of bc must be 1. There are $4 \times 2 = 8$ possibilities. Finally, suppose that $ad = -1$. Then $bc = 2$ gives that $ad - bc = -1 - 2 = -3 = 1 \in \mathbb{Z}/4$. There are $2 \times 4 = 8$ possibilities. Then $16 + 16 + 8 + 8 = 48$.

Exercise 15. Let

$$A = \begin{bmatrix} 1 & 2 \\ 0 & 1 \end{bmatrix}, \quad B = \begin{bmatrix} 1 & 0 \\ 2 & 1 \end{bmatrix}, \quad \text{and} \quad C = \begin{bmatrix} 1 & 2 \\ 2 & 1 \end{bmatrix}.$$

Compute directly that when the matrix entries are considered as elements in $\mathbb{Z}/4$, each squares to the identity, and the product of any two of them is the third. Explicitly, $A^2 = B^2 = C^2 = \begin{bmatrix} 1 & 0 \\ 0 & 1 \end{bmatrix} = I$ while $AB = BA = C$, $AC = CA = B$, and $BC = CB = A$. Conclude that the set $\{A, B, C, I\}$ is isomorphic to the Klein 4-group K_4.

Exercise 16. Let

$$\rho = \begin{bmatrix} 2 & 1 \\ 1 & -1 \end{bmatrix} \quad \text{and} \quad J = \begin{bmatrix} 0 & -1 \\ 1 & 0 \end{bmatrix}.$$

- Show that $\rho^6 = \begin{bmatrix} 1 & 0 \\ 0 & 1 \end{bmatrix} = I$.
- Show that $\rho^3 = J^2 = \begin{bmatrix} -1 & 0 \\ 0 & -1 \end{bmatrix}$.
- Show that $\rho J = J\rho^5$.

Conclude that the subgroup of $SL_2(\mathbb{Z}/4)$ that is generated by ρ and J is isomorphic to the dicyclic group Dic_3 that is of order 12.

Here is a bit of a hint to make the exercise easier.

$$\rho^2 = \begin{bmatrix} 2 & 1 \\ 1 & -1 \end{bmatrix} \cdot \begin{bmatrix} 2 & 1 \\ 1 & -1 \end{bmatrix} = \begin{bmatrix} 1 & 1 \\ 1 & 2 \end{bmatrix} = -\rho^{-1}.$$

Therefore, $\rho^3 = \rho^2 \cdot \rho = -\rho^{-1}\rho = -I$.

The result of the exercise is that the element $J = \begin{bmatrix} 0 & -1 \\ 1 & 0 \end{bmatrix}$ is a poorly disguised version of $\boldsymbol{j}$. So, henceforth, we write the element in quaternionic notation. We will also write $D = Y \cup \boldsymbol{j}Y$ where the cyclic subgroup $Y = \langle \rho \rangle = (I, \rho, \rho^2, -1, -\rho, -\rho^2)$ is ordered as indicated and the coset $\boldsymbol{j}Y = (\boldsymbol{j}, \boldsymbol{j}\rho, \boldsymbol{j}\rho^2, -\boldsymbol{j}, -\boldsymbol{j}\rho, -\boldsymbol{j}\rho^2)$ has the induced ordering.

The representation of ρ and $\boldsymbol{j}$ that appear in Fig. 4.21 when $n = 3$ will be used. In this way, the the elements in $\mathrm{SL}_2(\mathbb{Z}/4)$ will be depicted as 4-strings-with-dicyclic-quipu. The cosets of D are AD, BD, and CD. As an initial step, Fig. 5.41 illustrates the elements in Y, while Fig. 5.42 illustrates the elements in $\boldsymbol{j}Y$ in terms of the dicyclic quipu.

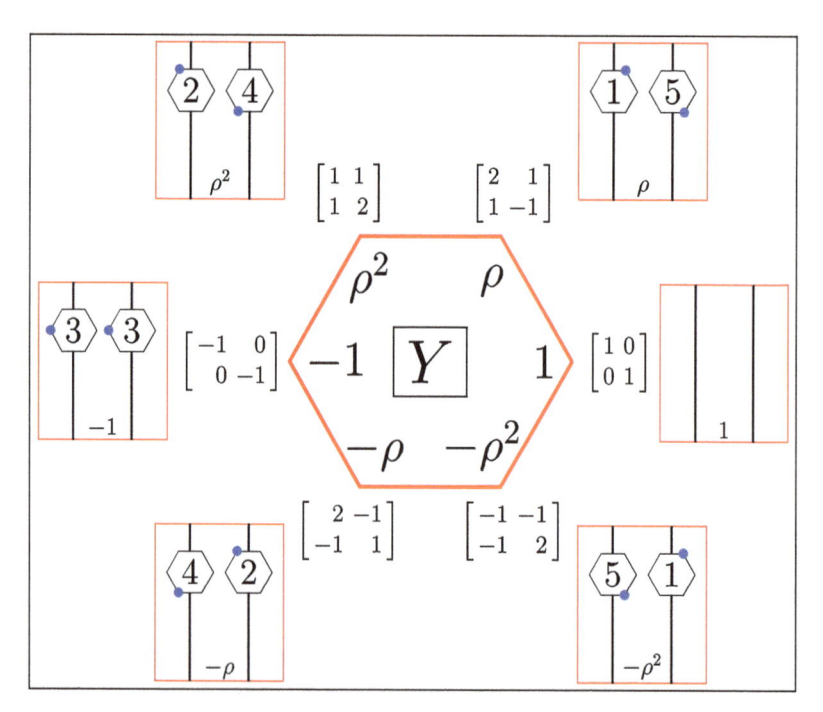

Figure 5.41. The coset $Y \subset D \subset \mathrm{SL}_2(\mathbb{Z}/4)$.

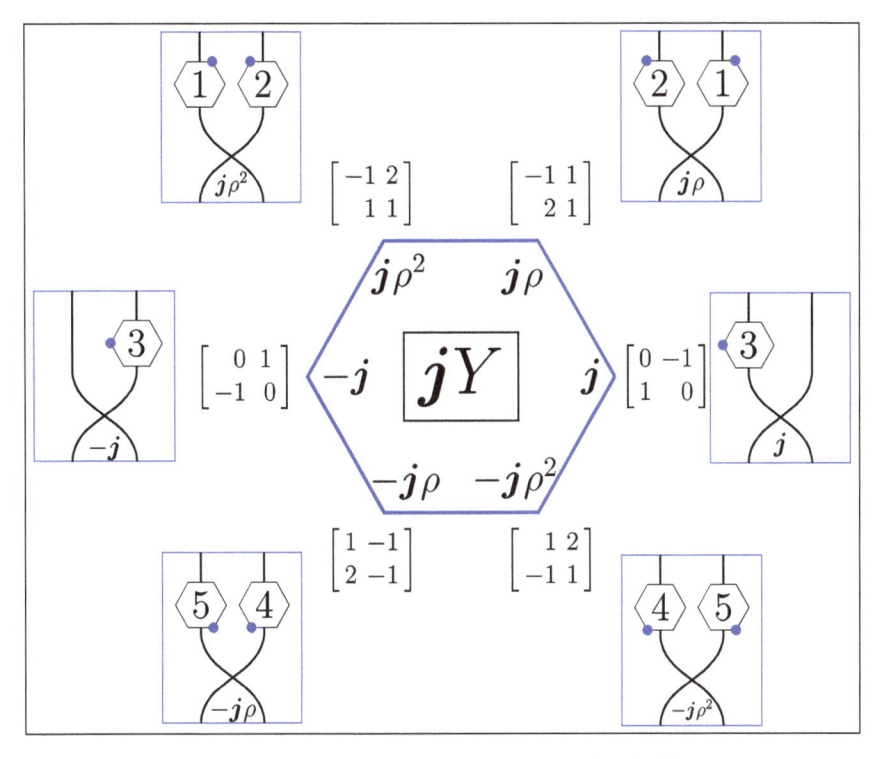

Figure 5.42. The coset $jY \subset D \subset \mathrm{SL}_2(\mathbb{Z}/4)$.

We take time out to apologize for using upper case letters (A–D) for both the matrices in $\mathrm{SL}_2(\mathbb{Z}/4)$ and for the subgroup D. Unfortunately, the notation that we use for the elements in this group of matrices are a hodge-podge. Our goal has been to reserve the lower case letters a, b, c, d to indicate elements in the binary tetrahedral group. Since D is isomorphic to a dicyclic group, powers of ρ and j feel more natural. But we are also in the habit of using upper case letters to denote both matrices and subgroups. Again, we are sorry.

Next, let's suppose that the cosets D, AD, BD, and CD are indeed disjoint. Then we can describe the respective actions of the elements A, B, and C upon these cosets. We write

$$A(D, AD, BD, CD) = (AD, D, CD, BD),$$
$$B(D, AD, BD, CD) = (BD, CD, D, AD),$$

and

$$C(D, AD, BD, CD) = (CD, BD, AD, D).$$

Since $A^2 = B^2 = C^2 = I$, $AB = BA = C$, $AC = CA = B$, and $BC = CB = A$, each acts as an identity on the individual cosets. The 4-strings depictions of the matrices A, B, and C are depicted in Fig. 5.43.

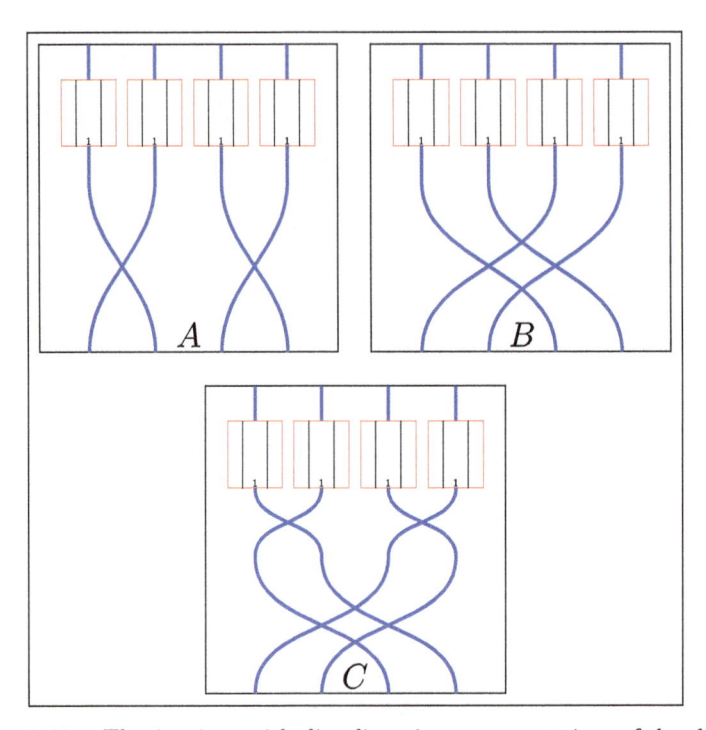

Figure 5.43. The 4-strings-with-dicyclic-quipu representations of the elements $A, B, C \in \mathrm{SL}_2(\mathbb{Z}/4)$.

We will compute the actions of these elements upon the cosets Y and jY so that we have an ordered list of the matrices in each coset. But since $Y = (1, \rho, \rho^2, -1, -\rho, -\rho^2)$ and $jY = (j, j\rho, j\rho^2, -j, -j\rho, -j\rho^2)$, not every matrix product is needed. We first will determine the matrices that represent $A\rho$, $A\rho^2$, Aj, and so forth. Then we will compile the cosets.

Please verify for yourself that the following matrix calculations, in which the entries represent elements of $\mathbb{Z}/4$, are correct:

$$A\rho = \begin{bmatrix} 1 & 2 \\ 0 & 1 \end{bmatrix} \cdot \begin{bmatrix} 2 & 1 \\ 1 & -1 \end{bmatrix} = \begin{bmatrix} 0 & -1 \\ 1 & -1 \end{bmatrix},$$

$$A\rho^2 = \begin{bmatrix} 1 & 2 \\ 0 & 1 \end{bmatrix} \cdot \begin{bmatrix} 1 & 1 \\ 1 & 2 \end{bmatrix} = \begin{bmatrix} -1 & 1 \\ 1 & 2 \end{bmatrix},$$

$$Aj = \begin{bmatrix} 1 & 2 \\ 0 & 1 \end{bmatrix} \cdot \begin{bmatrix} 0 & -1 \\ 1 & 0 \end{bmatrix} = \begin{bmatrix} 2 & -1 \\ 1 & 0 \end{bmatrix},$$

$$Aj\rho = \begin{bmatrix} 1 & 2 \\ 0 & 1 \end{bmatrix} \cdot \begin{bmatrix} -1 & 1 \\ 2 & 1 \end{bmatrix} = \begin{bmatrix} -1 & -1 \\ 2 & 1 \end{bmatrix},$$

$$Aj\rho^2 = \begin{bmatrix} 1 & 2 \\ 0 & 1 \end{bmatrix} \cdot \begin{bmatrix} -1 & 2 \\ 1 & 1 \end{bmatrix} = \begin{bmatrix} 1 & 0 \\ 1 & 1 \end{bmatrix},$$

$$B\rho = \begin{bmatrix} 1 & 0 \\ 2 & 1 \end{bmatrix} \cdot \begin{bmatrix} 2 & 1 \\ 1 & -1 \end{bmatrix} = \begin{bmatrix} 2 & 1 \\ 1 & 1 \end{bmatrix},$$

$$B\rho^2 = \begin{bmatrix} 1 & 0 \\ 2 & 1 \end{bmatrix} \cdot \begin{bmatrix} 1 & 1 \\ 1 & 2 \end{bmatrix} = \begin{bmatrix} 1 & 1 \\ -1 & 0 \end{bmatrix},$$

$$Bj = \begin{bmatrix} 1 & 0 \\ 2 & 1 \end{bmatrix} \cdot \begin{bmatrix} 0 & -1 \\ 1 & 0 \end{bmatrix} = \begin{bmatrix} 0 & -1 \\ 1 & 2 \end{bmatrix},$$

$$Bj\rho = \begin{bmatrix} 1 & 0 \\ 2 & 1 \end{bmatrix} \cdot \begin{bmatrix} -1 & 1 \\ 2 & 1 \end{bmatrix} = \begin{bmatrix} -1 & 1 \\ 0 & -1 \end{bmatrix},$$

$$Bj\rho^2 = \begin{bmatrix} 1 & 0 \\ 2 & 1 \end{bmatrix} \cdot \begin{bmatrix} -1 & 2 \\ 1 & 1 \end{bmatrix} = \begin{bmatrix} -1 & 2 \\ -1 & 1 \end{bmatrix},$$

$$C\rho = \begin{bmatrix} 1 & 2 \\ 2 & 1 \end{bmatrix} \cdot \begin{bmatrix} 2 & 1 \\ 1 & -1 \end{bmatrix} = \begin{bmatrix} 0 & -1 \\ 1 & 1 \end{bmatrix},$$

$$C\rho^2 = \begin{bmatrix} 1 & 2 \\ 2 & 1 \end{bmatrix} \cdot \begin{bmatrix} 1 & 1 \\ 1 & 2 \end{bmatrix} = \begin{bmatrix} -1 & 1 \\ -1 & 0 \end{bmatrix},$$

$$Cj = \begin{bmatrix} 1 & 2 \\ 2 & 1 \end{bmatrix} \cdot \begin{bmatrix} 0 & -1 \\ 1 & 0 \end{bmatrix} = \begin{bmatrix} 2 & -1 \\ 1 & 2 \end{bmatrix},$$

$$Cj\rho = \begin{bmatrix} 1 & 2 \\ 2 & 1 \end{bmatrix} \cdot \begin{bmatrix} -1 & 1 \\ 2 & 1 \end{bmatrix} = \begin{bmatrix} -1 & -1 \\ 0 & -1 \end{bmatrix},$$

$$Cj\rho^2 = \begin{bmatrix} 1 & 2 \\ 2 & 1 \end{bmatrix} \cdot \begin{bmatrix} -1 & 2 \\ 1 & 1 \end{bmatrix} = \begin{bmatrix} 1 & 0 \\ -1 & 1 \end{bmatrix}.$$

These calculations and their consequences are summarized within Figs. 5.45–5.52. A thorough check indicates that the cosets, Y, jY, AY, AjY, BY, BjY, CY, and CjY, are disjoint.

To compute the actions of ρ, j, A, B, and C upon the ordered list of cosets

$$((Y, jY), A(Y, jY), B(Y, jY), C(Y, jY))$$

that are also partitioned into cosets of $D = Y \cup jY$, we assign the following exercise.

Exercise 17. Calculate directly that

$$\rho A = B\rho, \quad \rho A j = -Bj\rho^2,$$
$$\rho B = -C\rho, \quad \rho Bj = Cj\rho^2,$$
$$\rho C = -A\rho, \quad \rho Cj = Aj\rho^2,$$

$$jA = Bj, \quad jAj = -B,$$
$$jB = Aj, \quad jBj = -A,$$
$$jC = Cj, \quad jCj = -C.$$

Conclude that the actions of ρ and j upon the various cosets are as follows:

$$\rho \cdot Y = [Y, 1], \quad \rho \cdot jY = [jY, 5],$$
$$\rho \cdot AY = [BY, 1], \quad \rho \cdot AjY = [BjY, 5],$$
$$\rho \cdot BY = [CY, 4], \quad \rho \cdot BjY = [CjY, 2],$$
$$\rho \cdot CY = [AY, 4], \quad \rho \cdot CjY = [AjY, 2],$$

$$j \cdot Y = [jY, 0], \quad j \cdot jY = [Y, 3],$$
$$j \cdot AY = [BjY, 0], \quad j \cdot AjY = [BY, 3],$$
$$j \cdot BY = [AjY, 0], \quad j \cdot BjY = [AY, 3],$$
$$j \cdot CY = [CjY, 0], \quad j \cdot CjY = [CY, 3].$$

The meaning of the notation $\rho \cdot BY = [CY, 4]$ and $\rho \cdot BjY = [CjY, 2]$, for example, is that ρ takes the coset BY to the coset CY where the elements are cyclically permuted four units, while ρ takes the coset BjY to the coset CjY where the elements are cyclically permuted two units.

The 4-strings-with-dicyclic-quipu diagrams for ρ and j are depicted in Fig. 5.44. From these illustrations and those in Fig. 5.43, all of the elements in $\mathrm{SL}_2(\mathbb{Z}/4)$ can be represented in this 4-strings form, as depicted in Figs. 5.45–5.52.

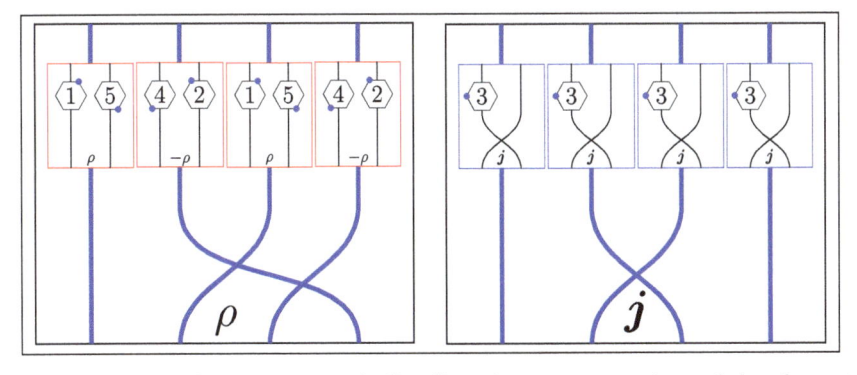

Figure 5.44. The 4-strings-with-dicyclic-quipu representations of the element $\rho, j \in \mathrm{SL}_2(\mathbb{Z}/4)$.

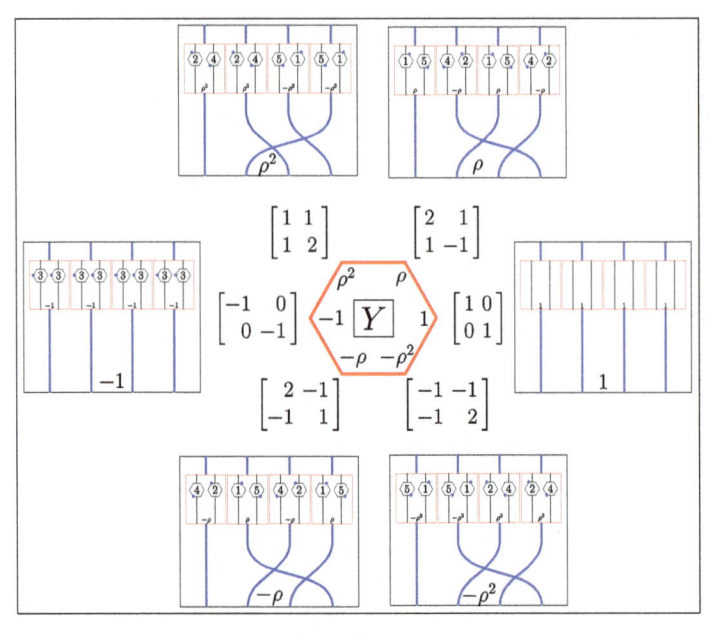

Figure 5.45. 4-strings representatives of the elements in $Y \subset \mathrm{SL}_2(\mathbb{Z}/4)$.

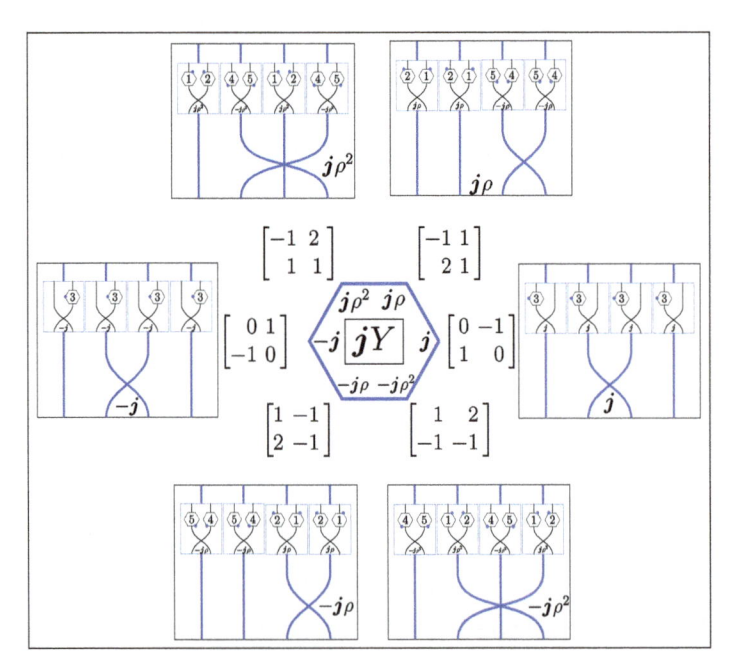

Figure 5.46. 4-strings representatives of the elements in $jY \subset \mathrm{SL}_2(\mathbb{Z}/4)$.

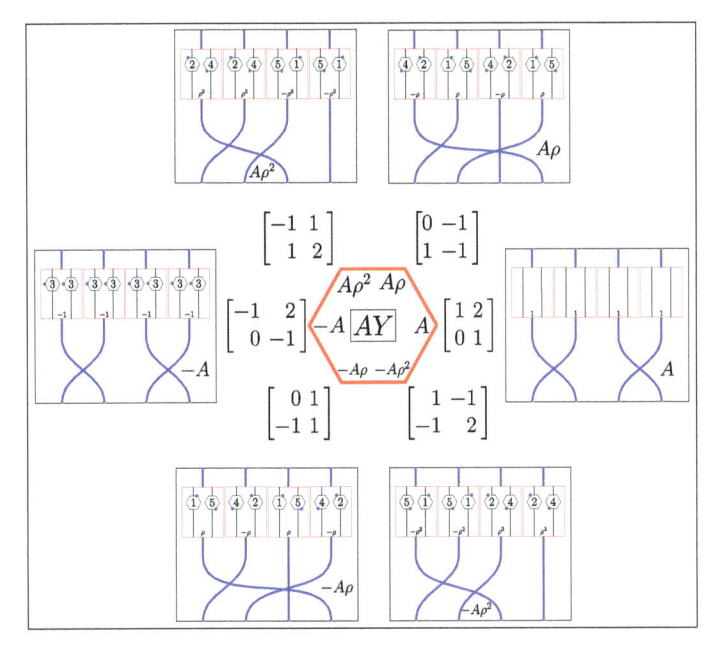

Extensions of the Permutation Group Σ_4 249

Figure 5.47. 4-strings representatives of the elements in $AY \subset \mathrm{SL}_2(\mathbb{Z}/4)$.

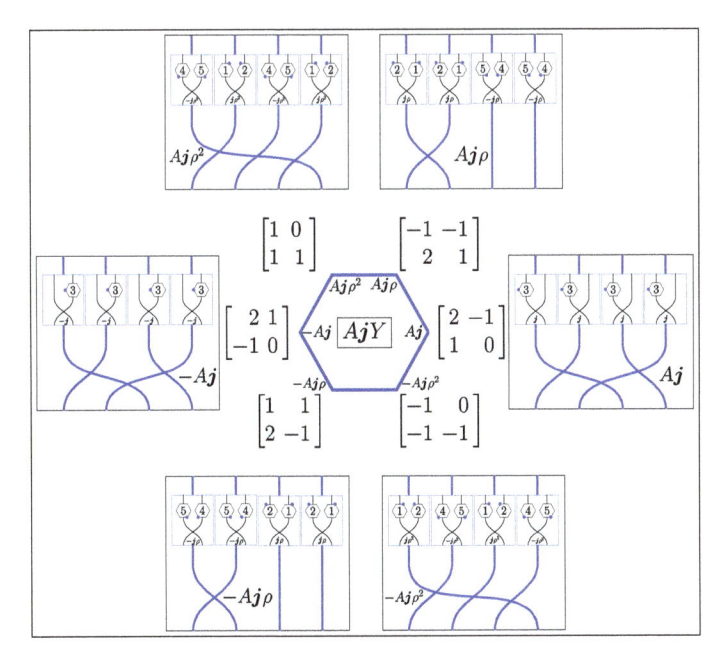

Figure 5.48. 4-strings representatives of the elements in $AjY \subset \mathrm{SL}_2(\mathbb{Z}/4)$.

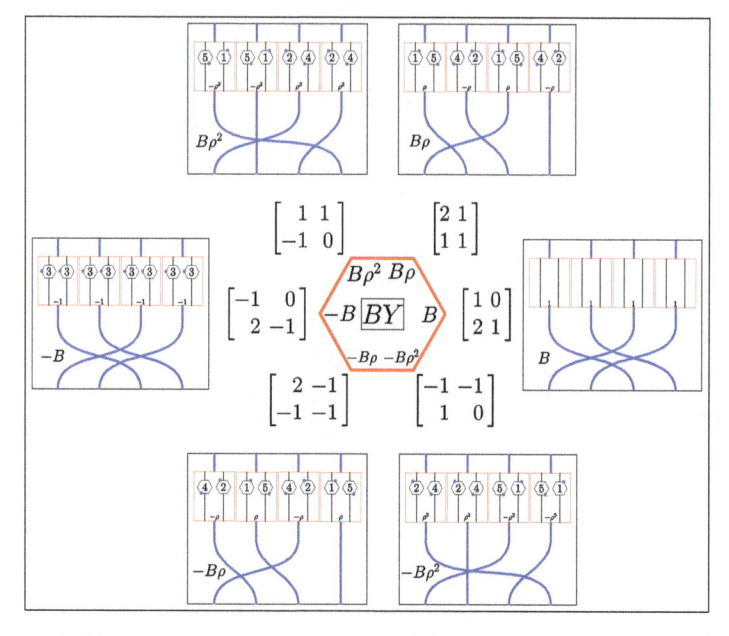

Figure 5.49. 4-strings representatives of the elements in $BY \subset \mathrm{SL}_2(\mathbb{Z}/4)$.

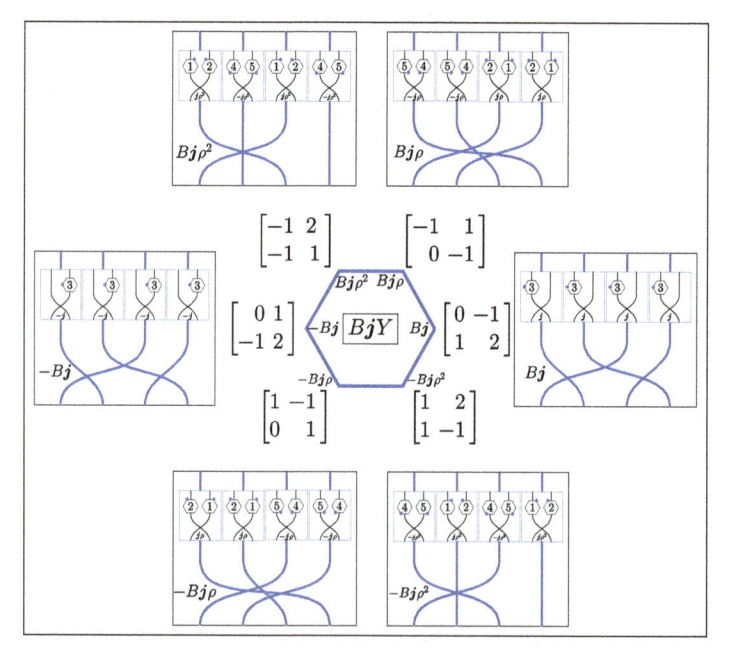

Figure 5.50. 4-strings representatives of the elements in $BjY \subset \mathrm{SL}_2(\mathbb{Z}/4)$.

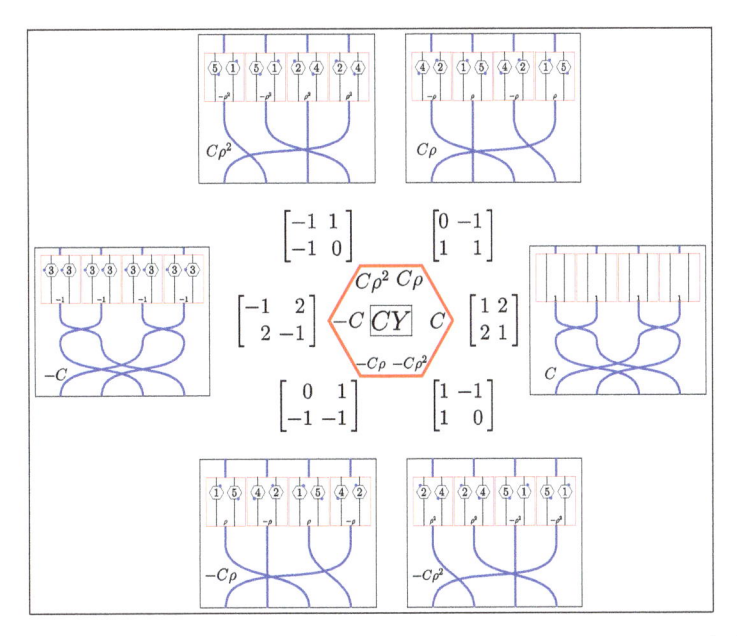

Extensions of the Permutation Group Σ_4 251

Figure 5.51. 4-strings representatives of the elements in $CY \subset \mathrm{SL}_2(\mathbb{Z}/4)$.

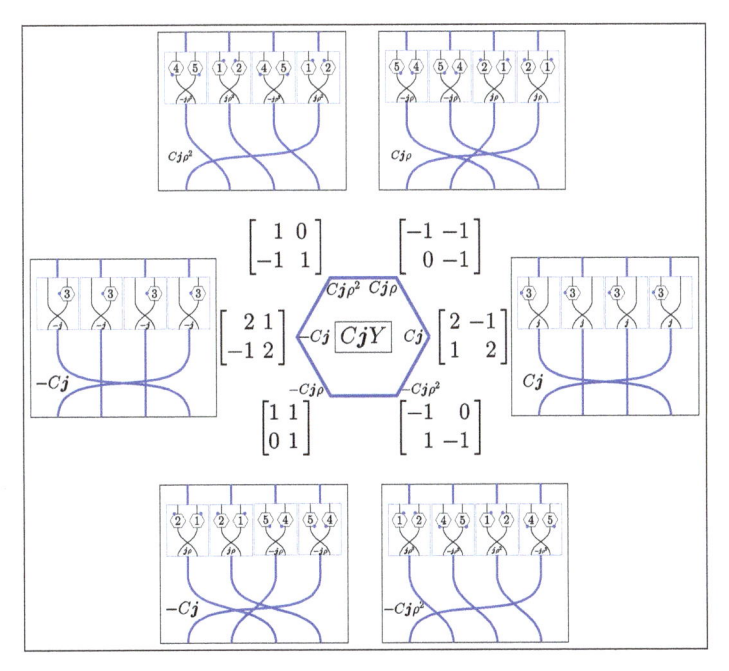

Figure 5.52. 4-strings representatives of the elements in $CjY \subset \mathrm{SL}_2(\mathbb{Z}/4)$.

Let us observe that by ignoring the quipu upon the representations in Figs. 5.45–5.52, a projection onto the symmetric group Σ_4 can be obtained. This yields a seseq

$$1 \longleftarrow \Sigma_4 \xleftarrow{q_1} \mathrm{SL}_2(\mathbb{Z}/4) \xleftarrow{i} \mathbb{Z}/2 \longleftarrow 1,$$

in which the kernel consists of the matrices $\pm I$, where $I = \begin{bmatrix} 1 & 0 \\ 0 & 1 \end{bmatrix}$ is the identity matrix.

5.3.1. A 3-strings representation of $\mathrm{SL}_2(\mathbb{Z}/4)$

In order to construct a 3-strings-with-dihedral-quipu representation of the matrix group $\mathrm{SL}_2(\mathbb{Z}/4)$, we will begin by studying the inverse image, $q_1^{-1}(A_4)$ of the alternating group. By counting the crossings between pairs of blue strings in Figs. 5.45–5.52, one can see that this subgroup of $\mathrm{SL}_2(\mathbb{Z}/4)$ is depicted in the odd-numbered figures (Figs. 5.45, 5.47, 5.49, 5.51) and thus consists of the union of cosets Y, AY, BY, and CY.

The set $\{I, A, B, C\}$ is isomorphic to the Klein 4-group, and more generally, the inverse image of $K_4 \subset \Sigma_4$ is the set $\{\pm I, \pm A, \pm B, \pm C\}$ which, as a group, is isomorphic to $\mathbb{Z}/2 \times \mathbb{Z}/2 \times \mathbb{Z}/2$. We begin the current discussion by means of finding a (mod 2)-quipu representation of this group of order 8.

Recall that K_4 is represented as

$$\left\{ \;\vcenter{\hbox{∥}}\;,\;\vcenter{\hbox{∥}}\;,\;\vcenter{\hbox{⋈}}\;,\;\vcenter{\hbox{⋈}}\;\right\}.$$

The representatives for A, B, and C are interchangeable, and at this moment, we won't make an assignment even though an obvious assignment is determined by the 4-strings pictures thereof.

Instead, we will horizontally juxtapose three of these 2-strings-with-(mod 2)-quipu to represent the elements in $\{\pm 1, \pm A, \pm B, \pm C\}$ and demonstrate in Fig. 5.53 that $A^2 = B^2 = C^2 = 1$,

$AB = BA = C$, $AC = CA = B$, and $BC = CB = A$ in these 6-strings representations.

Let

$$A = \boxed{|\;|\;\times\;\times}\;;\; B = \boxed{\times\;|\;|\;\times}\;;\; C = \boxed{\times\;\times\;|\;|}.$$

The representation of (-1) in the 6-strings-with-(mod 2)-quipu is as follows:

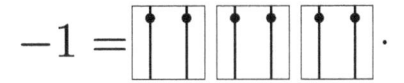

$$-1 = \boxed{\overset{\bullet}{|}\;\overset{\bullet}{|}\;\overset{\bullet}{|}\;\overset{\bullet}{|}\;\overset{\bullet}{|}\;\overset{\bullet}{|}}.$$

We believe that it is kinematically easy to see that this element commutes with the elements A, B, and C. Moreover, to obtain negatives of these, replace those strings that have non-trivial quipu with strings that don't have them and replace strings that don't have them with those that do. See, for example, the 6-strings-with-(mod 2)-quipu representation of $-C$ is presented in the following. Also, the identity, $(+1)$, consists of six straight strings.

We refer to these diagrams as 6-strings-with-(mod 2)-quipu, but we are grouping the strings into three boxes of two each. In this way, the diagrams are also thought of as 3-strings-with-K_4-quipu. While considering the group $\{\pm 1, \pm A, \pm B, \pm C\}$ in and of itself, these K_4-quipu adorn three straight strings that will be omitted from the diagrams.

In the top frame of Fig. 5.53, A^2, B^2, and C^2 are illustrated by vertically juxtaposing the diagrams. That each product is an identity follows since two (mod 2)-quipu upon any one string cancel, and the pairs of crossings between pairs of strings can be eliminated. In the subsequent three panels, each of A, B, and C is illustrated along with the corresponding products BC and CB, AC and CA, and AB and BA from bottom to top, respectively. We hope that the identities $BC = A = CB$, $AC = B = CA$ and $AB = C = BA$ are visually apparent in these representations.

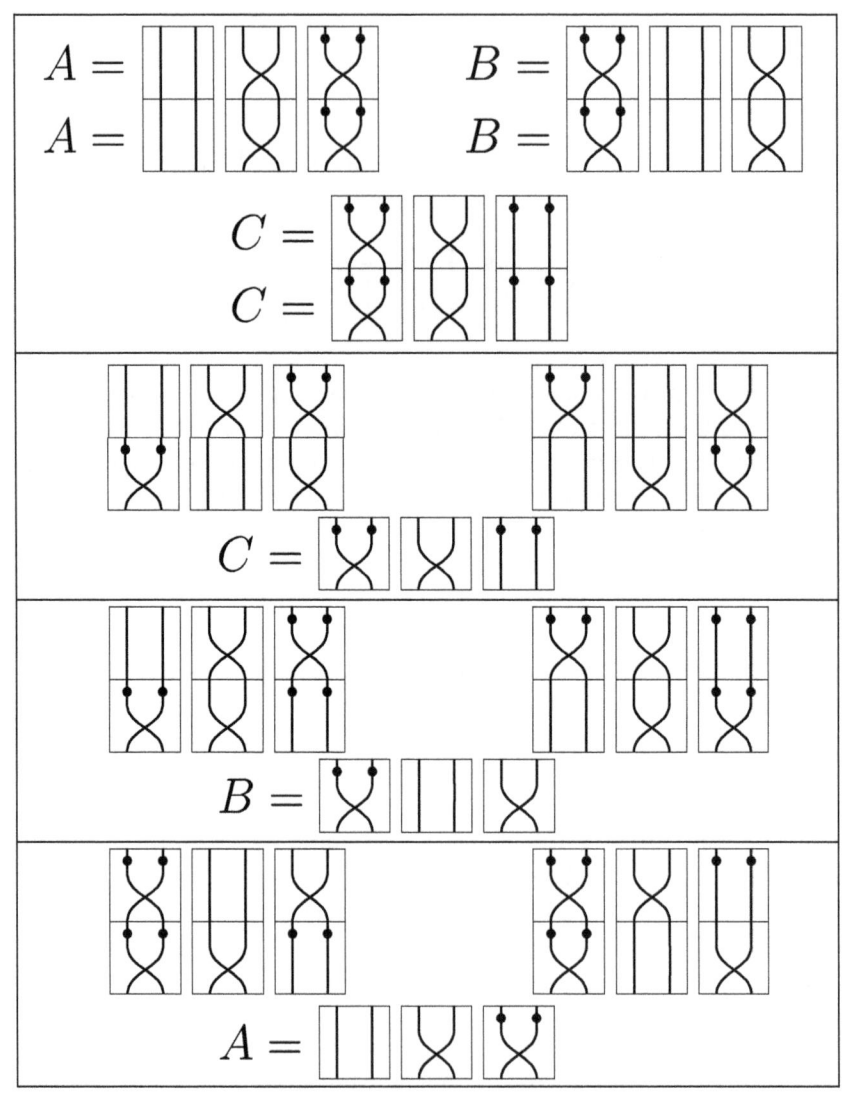

Figure 5.53. Identities among the 6-strings representatives of A, B, and C.

Now, consider the representations of A, B, and $-C$ as 3-strings-with-K_4-quipu that are listed as follows:

$$A = \boxed{||\,\cancel{|}\,\cancel{|}}\,;\; B = \boxed{\cancel{|}\,||\,\cancel{|}}\,;\; -C = \boxed{\cancel{|}\,\cancel{|}\,||}\,.$$

For a more easy textual discussion, let us call the quipu from left to right in the expression of A as $1, X$, and $-X$, so that $A = (1, X, -X)$, $B = (-X, 1, X)$, and $-C = (X, -X, 1)$. These are cyclic permutations of each other. Define a 3-strings-with-K_4-quipu that represents the permutation $A \mapsto B \mapsto -C \mapsto A$ by the following diagram:

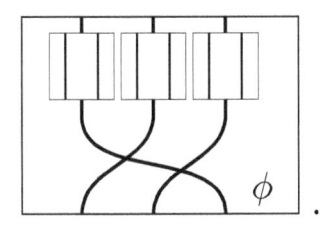

[diagram of braid with label ϕ]

The element ϕ is chosen to represent the matrix $\begin{bmatrix} 0 & 1 \\ -1 & -1 \end{bmatrix} = -C\rho$.

Please recall that $A = \begin{bmatrix} 1 & 2 \\ 0 & 1 \end{bmatrix}$, $B = \begin{bmatrix} 1 & 0 \\ 2 & 1 \end{bmatrix}$, and $-C = \begin{bmatrix} -1 & 2 \\ 2 & -1 \end{bmatrix}$. Then verify, by means of matrix multiplication, that $\phi A = B\phi$, $\phi B = (-C)\phi$, and $\phi(-C) = A\phi$. Now, imagine feeding each of the 3-strings-with-K_4-quipu pictures for A, B, and $(-C)$ into the bottom of the diagram for ϕ. The matrix relations that you were asked to verify are precisely the results of pulling the three K_4-quipu of each of A, B, and $(-C)$ to the top of the diagram of ϕ.

The subgroup $q_1^{-1}(A_4) \subset SL_2(\mathbb{Z}/4)$ is generated by A, B, $-C$, and ϕ. The elements will be listed in Figs. 5.54–5.59. These figures are paired so that Figs. 5.54 and 5.55 represent the subgroup $q_1^{-1}(A_4)$, and each subsequent pair of figures represents a coset. Within the figures, the elements are labeled as matrices and as products that involve ρ or ϕ.

Let $\mathcal{K} = q_1^{-1}(K_4)$ denote the subgroup of $SL_2(\mathbb{Z}/4)$ that consists of the eight elements $\pm I$, $\pm A$, $\pm B$, and $\pm C$. These matrix representation of these elements are found at the horizontal vertices in Figs. 5.45–5.52. Here, they will be represented in their 3-strings-with-K_4-quipu representation that has been outlined above. The nontrivial cosets of $\mathcal{K}$ in $\mathcal{A} = q_1^{-1}(A_4)$ are $\phi \mathcal{K}$ and $\phi^2 \mathcal{K}$. We can also write $\mathcal{K}$ as the union $+K_4 = \{I, A, B, C\}$ and $-K_4 = \{-I, -A, -B, -C\}$, the latter being a coset of the former which is a subgroup of $\mathcal{K}$.

A perceptive observer might note that a 4-strings-with-(mod 2)-quipu representation of $\mathcal{K}$ would have sufficed to distinguish $-A$ from $+A$, and so forth. The first two K_4-quipu, in order, are enough to specify the eight elements in $\mathcal{K}$. However, a thrifty representation of $\mathcal{K}$ would result in an expensive representation of the larger subgroup $\mathcal{A}$ and indeed of the entire group $\mathrm{SL}_2(\mathbb{Z}/4)$. The cliqué, "penny-wise and pound foolish," comes to mind.

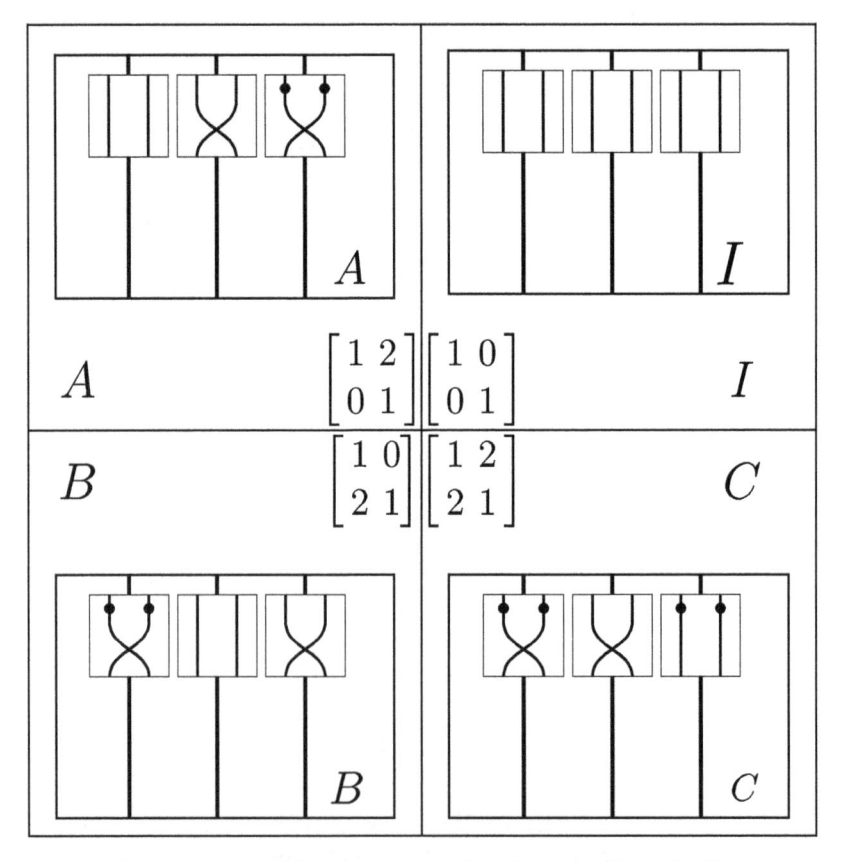

Figure 5.54. The elements in the subgroup $\{I, A, B, C\}$.

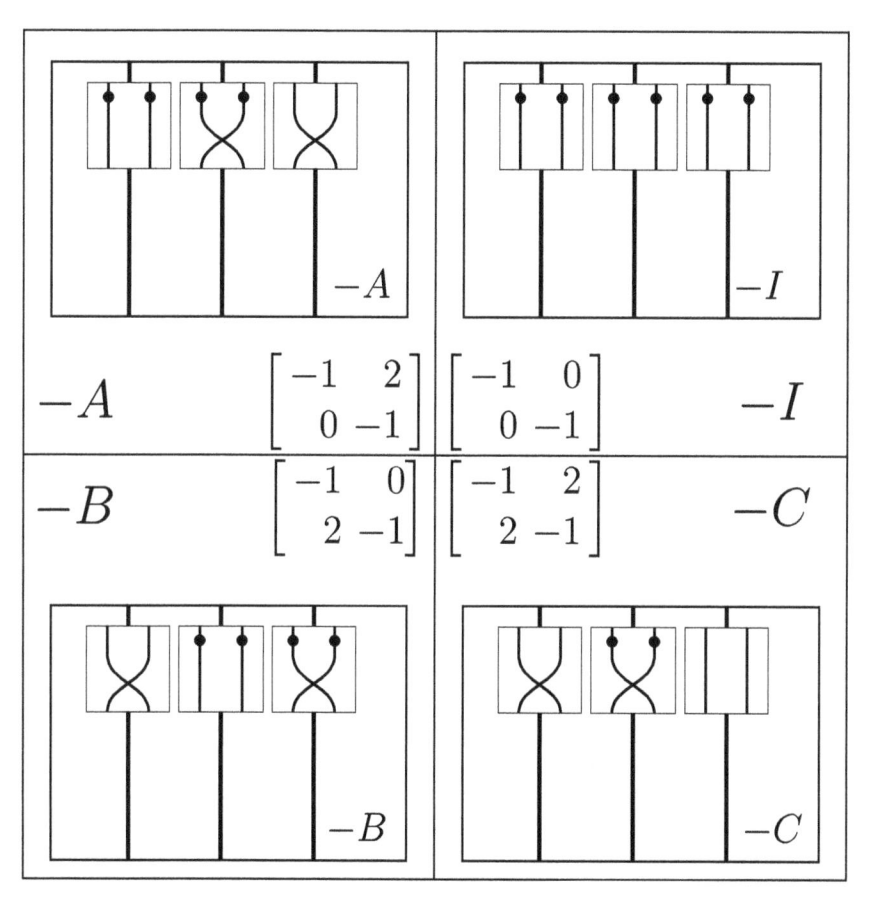

Figure 5.55. The elements in the coset $\{-I, -A, -B, -C\}$.

258   Quipu: Decorated Permutation Representations of Finite Groups

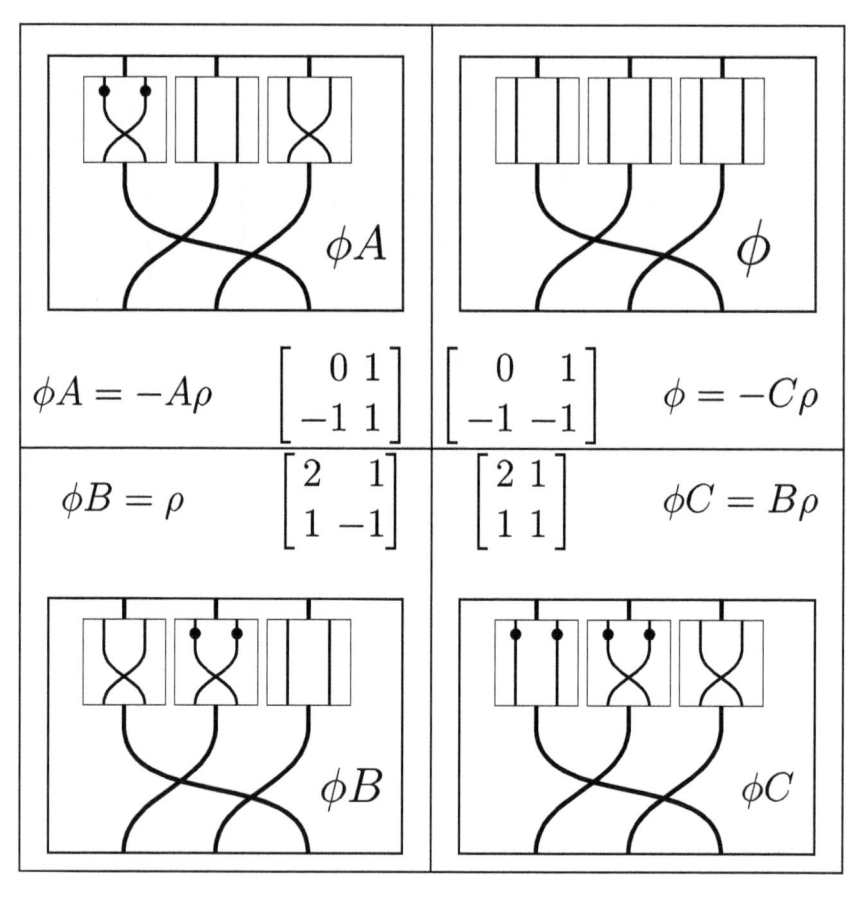

$\phi A = -A\rho$ $\begin{bmatrix} 0 & 1 \\ -1 & 1 \end{bmatrix}$ $\begin{bmatrix} 0 & 1 \\ -1 & -1 \end{bmatrix}$ $\phi = -C\rho$

$\phi B = \rho$ $\begin{bmatrix} 2 & 1 \\ 1 & -1 \end{bmatrix}$ $\begin{bmatrix} 2 & 1 \\ 1 & 1 \end{bmatrix}$ $\phi C = B\rho$

Figure 5.56. The elements in the coset $\{\phi, \phi A, \phi B, \phi C\}$.

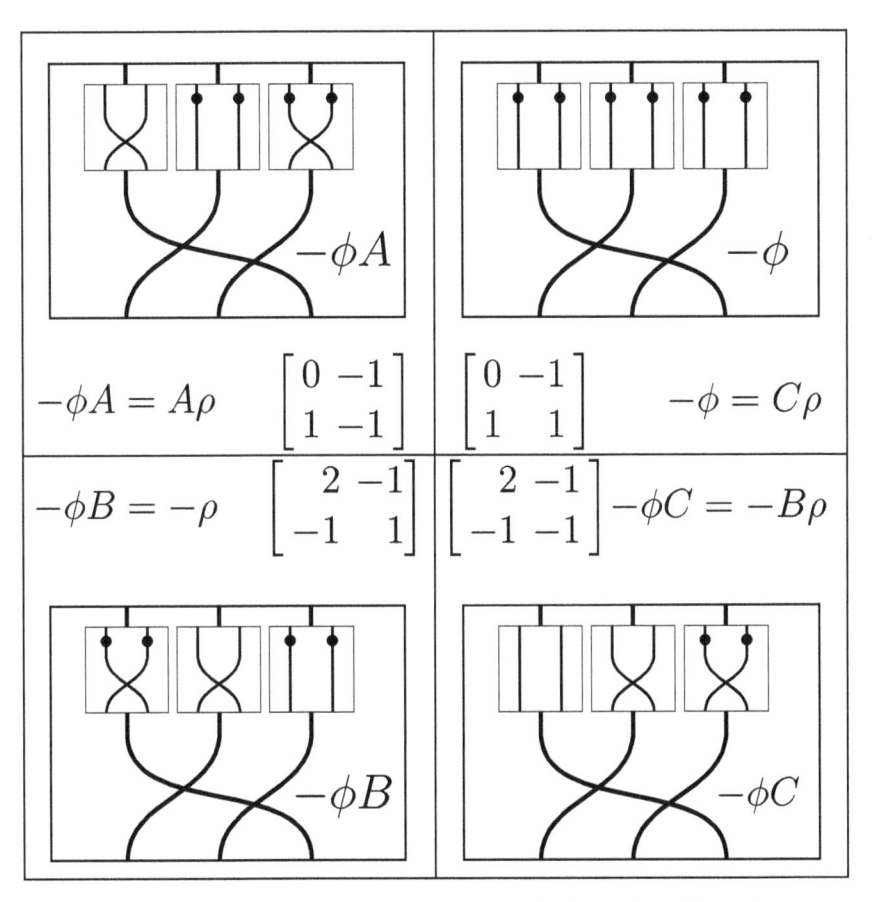

$$-\phi A = A\rho \quad \begin{bmatrix} 0 & -1 \\ 1 & -1 \end{bmatrix} \quad \begin{bmatrix} 0 & -1 \\ 1 & 1 \end{bmatrix} \quad -\phi = C\rho$$

$$-\phi B = -\rho \quad \begin{bmatrix} 2 & -1 \\ -1 & 1 \end{bmatrix} \quad \begin{bmatrix} 2 & -1 \\ -1 & -1 \end{bmatrix} -\phi C = -B\rho$$

Figure 5.57. The elements in the coset $\{-\phi, -\phi A, -\phi B, -\phi C\}$.

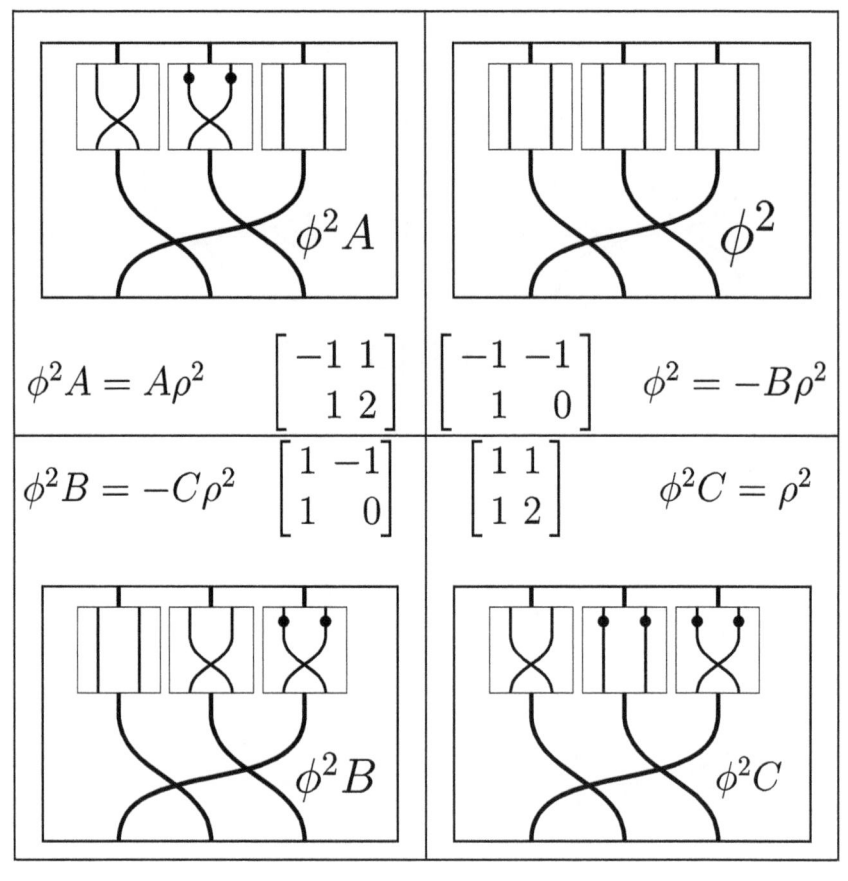

$\phi^2 A = A\rho^2$ $\begin{bmatrix} -1 & 1 \\ 1 & 2 \end{bmatrix}$ $\begin{bmatrix} -1 & -1 \\ 1 & 0 \end{bmatrix}$ $\phi^2 = -B\rho^2$

$\phi^2 B = -C\rho^2$ $\begin{bmatrix} 1 & -1 \\ 1 & 0 \end{bmatrix}$ $\begin{bmatrix} 1 & 1 \\ 1 & 2 \end{bmatrix}$ $\phi^2 C = \rho^2$

Figure 5.58. The elements in the coset $\{\phi^2, \phi^2 A, \phi^2 B, \phi^2 C\}$.

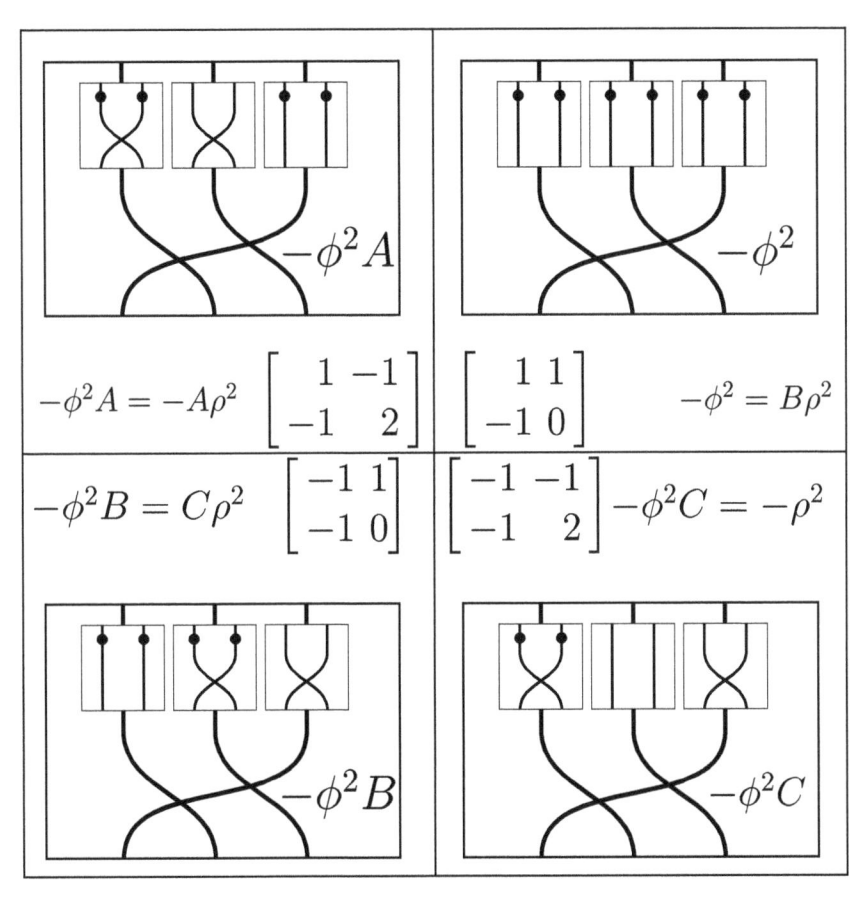

Figure 5.59. The elements in the coset $\{-\phi^2, -\phi^2 A, -\phi^2 B, -\phi^2 C\}$.

$$-\phi^2 A = -A\rho^2 \quad \begin{bmatrix} 1 & -1 \\ -1 & 2 \end{bmatrix} \quad \begin{bmatrix} 1 & 1 \\ -1 & 0 \end{bmatrix} \quad -\phi^2 = B\rho^2$$

$$-\phi^2 B = C\rho^2 \quad \begin{bmatrix} -1 & 1 \\ -1 & 0 \end{bmatrix} \quad \begin{bmatrix} -1 & -1 \\ -1 & 2 \end{bmatrix} \quad -\phi^2 C = -\rho^2$$

The 3-strings-with-K_4-quipu representation of $\mathcal{A} = q_1^{-1}(A_4)$ depicted in Figs. 5.54–5.59 is a permutation representation constructed by means of a methodology (or voodoo) slightly different from the prior, and subsequent, quipu representations made. In particular, the depiction of

$$\mathbb{Z}/2 \times \mathbb{Z}/2 \times \mathbb{Z}/2 = \{\pm 1, \pm A, \pm B, \pm C\}$$

didn't really follow the same rhyme and reason that has been used for other subgroups.

These differences in method become more apparent when we find a representative 3-strings diagram for j which is the representative for the other coset of $\mathcal{A}$. The quipu that adorn the strings in the subsequent representation of j, and also those that adorn any element in the coset $j\mathcal{A}$, are elements in the dihedral group D_8. Yet, D_8 is not a subgroup of $\mathrm{SL}_2(\mathbb{Z}/4)$.

The authors encountered the 3-strings-with-D_8-quipu representation of $\mathrm{SL}_2(\mathbb{Z}/4)$ from an entirely different point of view. Telling the whole story here and now would be too much of a diversion, but we'd like to give a synopsis.

We were looking at a permutation representation of a group that was given by a few generators. At the time, we encountered the group, we didn't recognize it, but we knew that it should map two-to-one onto the permutation group Σ_4. Due to this, we first found a 6-strings-with-(mod 2)-quipu representation of the group. We also were able to bundle strings together into three larger strings, and the quipu became dihedral.

We determined all the elements in that group and also determined their orders. Finally, we compared the group with $\mathrm{SL}_2(\mathbb{Z}/4)$, $\mathrm{GL}_2(\mathbb{Z}/3)$, and the binary octahedral group $\widehat{\Sigma_4}$, and found out that the group we were interested in was $\mathrm{SL}_2(\mathbb{Z}/4)$.

As a side note to this small digression, we point out that both the calculations with the group that seeded our curiosity and the matrix group were done using pad and stylus (the 21st century version of quill and paper). Working in this way has been personally gratifying.

Extensions of the Permutation Group Σ_4 263

The element $j = \begin{bmatrix} 0 & -1 \\ 1 & 0 \end{bmatrix}$ is represented as follows:

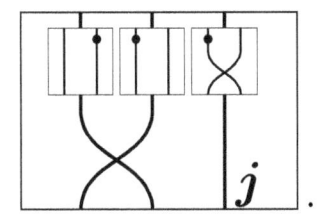

One can check either by using matrices or by means of the 4-strings-with-dicyclic-quipu representations of j, A, and C that

$$jC = Cj = AjA.$$

In Figs. 5.60 and 5.61, we demonstrate these identities and the more mundane equation $j^2 = -1$ in the context of 3-strings-with-D_8-quipu.

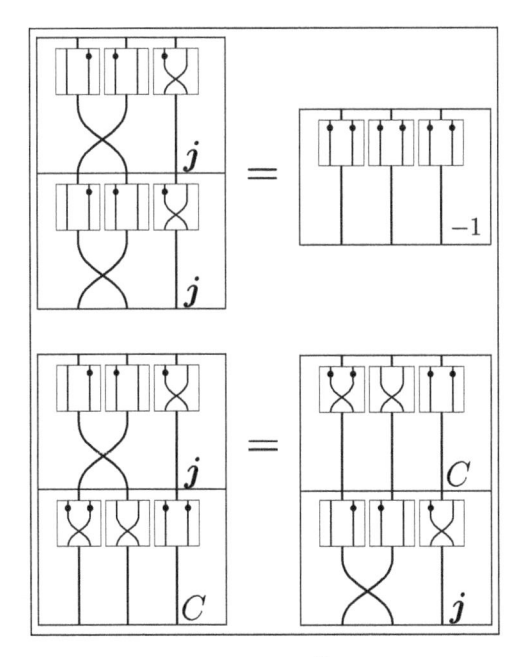

Figure 5.60. The identities $j^2 = -1$ and $Cj = jC$.

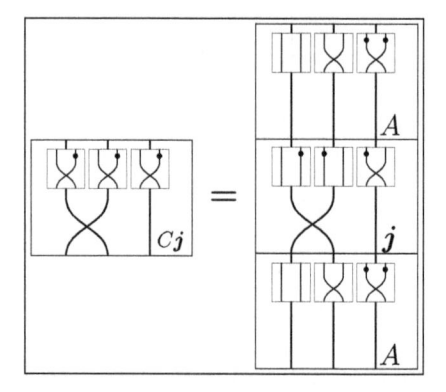

Figure 5.61. The identity $Cj = AjA$.

In Section 5.3.2, these are among the defining relations of a group that is isomorphic to the 2-Sylow subgroup of $\mathrm{SL}_2(\mathbb{Z}/4)$. The relations are listed here to help indicate that 3-strings representation of j seems to be correct.

Figures 5.62–5.67 indicate the elements of $j\mathcal{A}$.

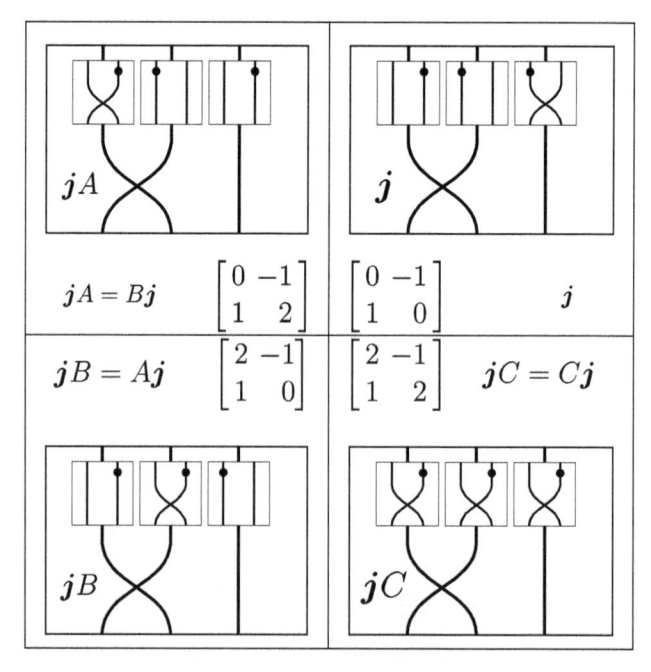

Figure 5.62. The elements in the coset $\{j, jA, jB, jC\}$.

$jA = Bj \quad \begin{bmatrix} 0 & -1 \\ 1 & 2 \end{bmatrix} \quad \begin{bmatrix} 0 & -1 \\ 1 & 0 \end{bmatrix} \quad j$

$jB = Aj \quad \begin{bmatrix} 2 & -1 \\ 1 & 0 \end{bmatrix} \quad \begin{bmatrix} 2 & -1 \\ 1 & 2 \end{bmatrix} \quad jC = Cj$

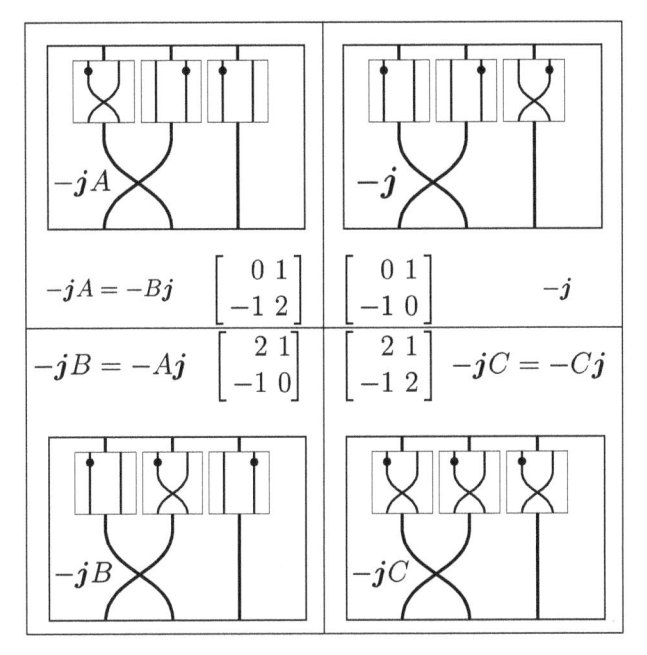

Extensions of the Permutation Group Σ_4 265

$-jA$			$-j$
$-jA = -Bj$	$\begin{bmatrix} 0 & 1 \\ -1 & 2 \end{bmatrix}$	$\begin{bmatrix} 0 & 1 \\ -1 & 0 \end{bmatrix}$	$-j$
$-jB = -Aj$	$\begin{bmatrix} 2 & 1 \\ -1 & 0 \end{bmatrix}$	$\begin{bmatrix} 2 & 1 \\ -1 & 2 \end{bmatrix}$	$-jC = -Cj$
$-jB$			$-jC$

Figure 5.63. The elements in the coset $\{-j, -jA, -jB, -jC\}$.

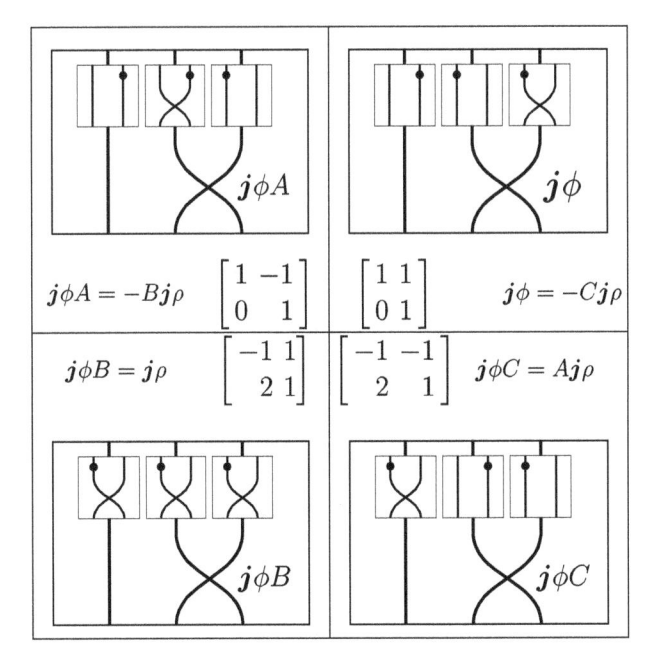

			$j\phi A$			$j\phi$
$j\phi A = -Bj\rho$	$\begin{bmatrix} 1 & -1 \\ 0 & 1 \end{bmatrix}$	$\begin{bmatrix} 1 & 1 \\ 0 & 1 \end{bmatrix}$	$j\phi = -Cj\rho$
$j\phi B = j\rho$	$\begin{bmatrix} -1 & 1 \\ 2 & 1 \end{bmatrix}$	$\begin{bmatrix} -1 & -1 \\ 2 & 1 \end{bmatrix}$	$j\phi C = Aj\rho$
$j\phi B$			$j\phi C$

Figure 5.64. The elements in the coset $\{j\phi, j\phi A, j\phi B, j\phi C\}$.

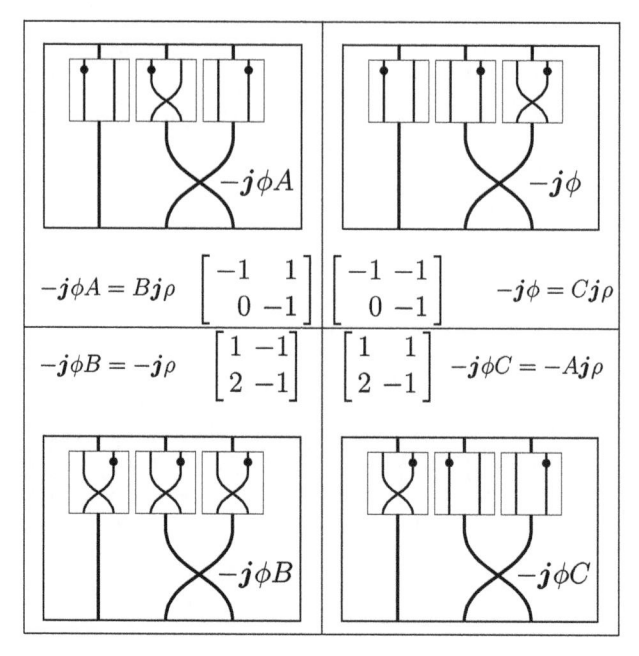

$$-j\phi A = Bj\rho \quad \begin{bmatrix} -1 & 1 \\ 0 & -1 \end{bmatrix} \quad \begin{bmatrix} -1 & -1 \\ 0 & -1 \end{bmatrix} \quad -j\phi = Cj\rho$$

$$-j\phi B = -j\rho \quad \begin{bmatrix} 1 & -1 \\ 2 & -1 \end{bmatrix} \quad \begin{bmatrix} 1 & 1 \\ 2 & -1 \end{bmatrix} \quad -j\phi C = -Aj\rho$$

Figure 5.65. The elements in the coset $\{-j\phi, -j\phi A, -j\phi B, -j\phi C\}$.

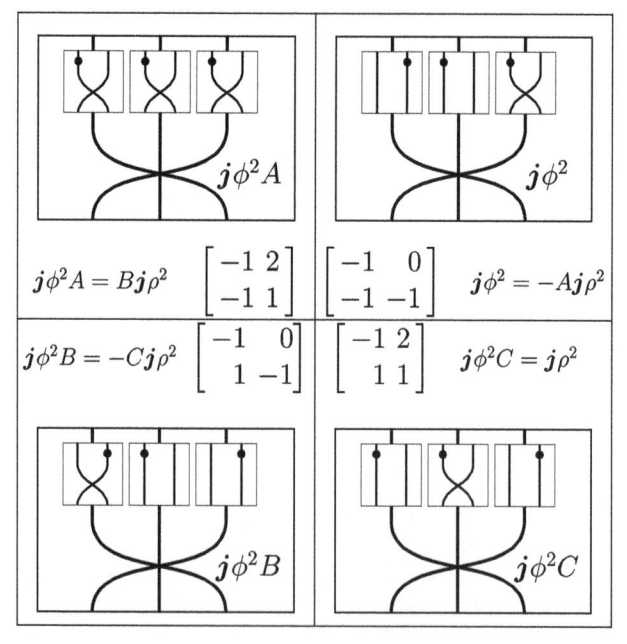

$$j\phi^2 A = Bj\rho^2 \quad \begin{bmatrix} -1 & 2 \\ -1 & 1 \end{bmatrix} \quad \begin{bmatrix} -1 & 0 \\ -1 & -1 \end{bmatrix} \quad j\phi^2 = -Aj\rho^2$$

$$j\phi^2 B = -Cj\rho^2 \quad \begin{bmatrix} -1 & 0 \\ 1 & -1 \end{bmatrix} \quad \begin{bmatrix} -1 & 2 \\ 1 & 1 \end{bmatrix} \quad j\phi^2 C = j\rho^2$$

Figure 5.66. The elements in the coset $\{j\phi^2, j\phi^2 A, j\phi^2 B, j\phi^2 C\}$.

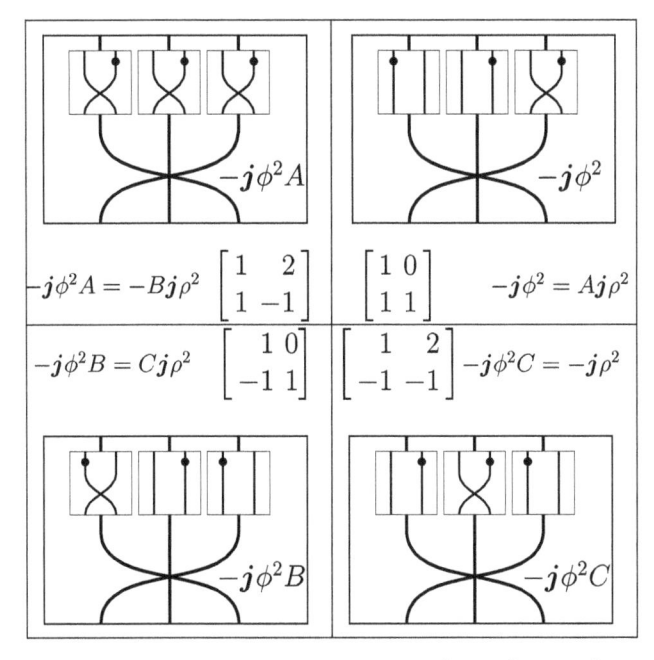

Figure 5.67. The elements in the coset $\{-j\phi^2, -j\phi^2 A, -j\phi^2 B, -j\phi^2 C\}$.

That these quipu upon multiples of j are in the dihedral group D_8 is very curious since D_8 is not a subgroup of $\mathrm{SL}_2(\mathbb{Z}/4)$. In the following section, we will explore some related subgroups. Following that diversion, Section 5.3.3 will give an alternative quotient mapping to Σ_4.

5.3.2. The 2-Sylow subgroup of $\mathrm{SL}_2(\mathbb{Z}/4)$

A sin — for which we are also guilty — is to simply scroll past a long list of illustrations such as those presented in Figs. 5.45–5.52 without actually examining the contents of the figures. In order to rectify any guilt that you may have in this regard, we call your attention to the collection of elements that are illustrated at the horizontally extreme vertices among these eight hexagons. The elements in question are known as ± 1, $\pm j$, $\pm A$, $\pm Aj$, $\pm B$, $\pm Bj$, $\pm C$, and $\pm Cj$. These 16 elements form a subgroup of $SL_2(\mathbb{Z}/4)$ for which we will give an alternative description.

First, let us examine the quipu upon these elements. The elements are characterized by the fact that the four quipu upon any one of them are the same: they are either the quipu that represent ± 1 or $\pm j$ in the dicyclic group Dic_3. Which element has which homogeneous set of quipu is convenient. The negatives are adorned with [figure]. The multiples of j are adorned with [figure]. The multiples of $-j$ are adorned with [figure], and A, B, and C are adorned with identities [figure].

The elements form a group of order 16. That multiplication among the elements is closed follows because the set $\{\pm 1, \pm A, \pm B, \pm C\}$ is also a subgroup that is isomorphic to the product $K_4 \times \mathbb{Z}/2 = \mathbb{Z}/2 \times \mathbb{Z}/2 \times \mathbb{Z}/2$ and because all the quipu on any element are the same and are the multiples of [figure].

Let us call the group

$$H = \{\pm 1, \pm j, \pm A, \pm Aj, \pm B, \pm Bj, \pm C, \pm Cj\}.$$

By skimming through the "groupprops.subwiki" on the groups of order 16, we were able to identify H as smallgroup(16,3) which apparently is its name among the GAP census.

In our notation, the subgroup H has as its presentation

$$H = \left\langle j, C, A : \begin{array}{ll} j^4 = A^2 = C^2 = 1, & CA = AC(= B), \\ jC = Cj, & AjA = jC. \end{array} \right\rangle.$$

There are three different representations of the elements of the subgroup H that are given above. First of all, they are thought of as (2×2)-matrices with entries in $\mathbb{Z}/4$. Next, they were given as 4-strings-with-dicyclic-quipu, and in their most recent incarnation, they are presented as 3-strings-with-D_8-quipu. In Fig. 5.53, the relations $A^2 = C^2 = 1$ and $CA = AC = B$ were demonstrated to hold in that 3-strings representation. That $jC = Cj = AjA$ and $j^4 = 1$ follow from the illustrations in Figs. 5.60 and 5.61. Compare the content of Exercises 15 and 16 for the calculation at the matrix level. The corresponding calculations using 4-strings-with-dicyclic-quipu should also be straightforward for the reader to implement.

Let $\mathscr{J}$ denote the ordered subgroup $\mathscr{J} = (1, j, -1, j)$. As we've pointed out before $\mathscr{J}$ is isomorphic to the multiplicatively cyclic group $\mathbb{Z}/4$ which should not be confused with the ring whose elements populate the matrices in this section. Since $jC = Cj$, then $\mathscr{J}$ and coset $\mathscr{C} = (C, Cj, -C, -Cj)$ form a subgroup $\mathscr{J}\mathscr{C} = \mathscr{J} \cup \mathscr{C}$ that is isomorphic to $\mathbb{Z}/4 \times \mathbb{Z}/2$. The subgroup H can be written as the union of the cosets $\mathscr{J}\mathscr{C} \cup A\mathscr{J}\mathscr{C}$. It is, in fact, a semi-direct product $(\mathbb{Z}/4 \times \mathbb{Z}/2) \rtimes \mathbb{Z}/2$.

Moreover, since the order of H is 16 and the order of $\mathrm{SL}_2(\mathbb{Z}/4)$ is 48, then H is the 2-Sylow subgroup of the matrix group.

Let us comment about Section 5.3.1 in the current context. There is a subgroup of $\mathrm{SL}_2(\mathbb{Z}/4)$, which is generated by the elements $\pm A$, $\pm B$, and ϕ. That subgroup is isomorphic to a semi-direct product $(\mathbb{Z}/2 \times \mathbb{Z}/2 \times \mathbb{Z}/2) \rtimes \mathbb{Z}/3$. The $\mathbb{Z}/3$ factor cyclically permutes the three factors of $\mathbb{Z}/2$. Such a subgroup has 24 elements, and consequently, it is an index 2 subgroup in the larger matrix group. The element j represents the non-trivial coset.

On the other hand, the subgroup H has index 3, and the non-trivial cosets of H are ϕH and $\phi^2 H$.

We think that it is natural to arrange elements in the cyclic group $\mathbb{Z}/4$ around the vertices of a square. In this conception, the elements of $\mathbb{Z}/4 \times \mathbb{Z}/2$ can be thought of as the elements of a cube. Or even more generally, the elements of $H = (\mathbb{Z}/4 \times \mathbb{Z}/2) \rtimes \mathbb{Z}/2$ might easily be thought of as the vertices of the 4-dimensional cube C_4.

Similarly, the powers of ϕ which are $\Delta = (1, \phi, \phi^2)$ might be arranged as the vertices of an equilateral triangle. In this way, the 48 elements of $\mathrm{SL}_2(\mathbb{Z}/4)$ might be conceived as the vertices of a 6-dimensional figure $C_4 \times \Delta$ that is the cartesian product of the 4-dimensional cube and an equilateral triangle.

Various 3-dimensional and 4-dimensional facets can be envisioned. The vertices $\pm 1, \pm \phi, \pm \phi^2$ form a triangular prism $\Delta \times [-1,1]$.[2] The vertices of similar prisms are found in $\Delta \times \{1, A\}$, $\Delta \times \{1, B\}$, and $\Delta \times \{1, C\}$. The powers of j and ϕ form a 4-dimensional hypersolid that is of the form $\Delta \times \square$. Similarly, the elements of Klein 4 group $\{I, A, B, C\}$ are arranged in a square $\square$, and there are hypersolids $\pm\Delta \times \boxed{K_4}$ that are found among the 4-dimensional facets of $C_4 \times \Delta$.

Such arrangements of the group elements are not merely for recreational purposes. As we have seen, the elements in dicyclic groups lie upon Hopf links in the 3-sphere S^3, and in our opinion, arranging the group elements in such configurations help reveal the structure of the group. Furthermore, such configurations give rise to spaces upon which the given group may act as a set of congruences.

We also remind you that the symmetric group Σ_4 can be represented as 3-strings-with-D_8-quipu as in Figs. 5.10–5.12. In these figures, we used a different representation of the elements in the dihedral group, but you might enjoy converting them to the 2-strings-with-(mod 2)-quipu that are employed in this section. Both Σ_4 and $SL_2(\mathbb{Z}/4)$, then, are subgroups of the semi-direct product $(D_8)^3 \rtimes \Sigma_3$, and this is a subgroup of the set of symmetries of the 4-dimensional hypersolid $\square \times \Delta$. It is fun to try and imagine how either group acts. The authors feel as if the action is akin to a Rubik's cube. A twist along one square face results in all three moving.

The subgroups $(1, j, -1, -j)$ and $(1, C)$ form a subgroup of H that is isomorphic to $\mathbb{Z}/4 \times \mathbb{Z}/2$. There are 2-strings representations of these elements with $\mathbb{Z}/4$-quipu. The element j is represented by [diagram], and the representation of the element C is [diagram].

---

[2] While Sir Issac Newton used such a device to scatter light, ours is only metaphorical.

Extensions of the Permutation Group Σ_4 271

We are hopeful that you can construct all the representatives of $\mathbb{Z}/4 \times \mathbb{Z}/2$ using these 2-strings-with-(mod 4)-quipu representations. Meanwhile, H can be decomposed as the union of the subgroup $\mathbb{Z}/4 \times \mathbb{Z}/2$ and the coset $A(\mathbb{Z}/4 \times \mathbb{Z}/2)$. In this way, the elements of H can be imagined as 2-strings-with-$(\mathbb{Z}/4 \times \mathbb{Z}/2)$-quipu.

It is possible to continue this method and represent the elements of $\mathrm{SL}_2(\mathbb{Z}/4)$ in 3-strings pictures for which the quipu are elements of the Sylow subgroup H. But the representations of the elements in H would appear as a 2-strings-with-$(\mathbb{Z}/4 \times \mathbb{Z}/2)$-quipu, and those quipu are also made from a pair of strings. The final construction of the elements of $\mathrm{SL}_2(\mathbb{Z}/4)$ would be an elaborate, baroque entanglement.

The previous paragraph does not preclude you, the reader, from constructing that representation. You might enjoy doing so.

5.3.3. An alternative projection to Σ_4

To begin this section, let us observe that the illustrations in Figs. 5.54–5.59 and 5.63–5.67 suggest a surjective homomorphism $\Sigma_3 \xleftarrow{p} \mathrm{SL}_2(\mathbb{Z}/4)$. The kernel consists of $\pm\{I, A, B, C\}$. So, there is a seseq

$$1 \longleftarrow \Sigma_3 \xleftarrow{p} \mathrm{SL}_2(\mathbb{Z}/4) \xleftarrow{i} \mathbb{Z}/2 \times \mathbb{Z}/2 \times \mathbb{Z}/2 \longleftarrow 1$$

that is obtained by ignoring the quipu upon the 3-strings representatives.

The dihedral quipu upon the 3-strings representations of the elements in $\mathrm{SL}_2(\mathbb{Z}/4)$ consist of rotations

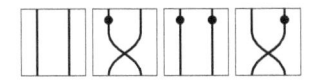

and reflections

,

where we include the identity among the rotations. The reflections will be mapped to (mod 2)-quipu.

In this way, we will obtain a 3-strings-with-(mod 2)-quipu quotient of $SL_2(\mathbb{Z}/4)$. A priori the homomorphic image is just a subgroup of the (3×3) signed permutation group. But we will suppose that the resulting 3-strings-with-(mod 2)-quipu elements correspond to the elements in Σ_4 that are induced by the 4-strings pictures (Figs. 5.46–5.52).

So, a 4-strings-with-dicyclic-quipu diagram projects to an element in Σ_4 by ignoring the quipu. A 3-strings-with-D_8-quipu diagram projects to a 3-strings-with-(mod 2)-quipu diagram by sending the reflections in D_8 to (mod 2)-quipu. See Figs. 5.68–5.72. In both cases, a given element M will project to the same diagram as $-M$ projects. Only the positive elements will be catalogued, and in the catalogs that follow, an alternative seseq

$$1 \longleftarrow \Sigma_3 \xleftarrow{q_3} \Sigma_4 \xleftarrow{i} K_4 \longleftarrow 1$$

is being proposed.

More specifically, $SL_2(\mathbb{Z}/4)$ is projected to Σ_4 by means of ignoring the Dic_3-quipu upon the blue strings. For example, the element Aj projects to the 4-cycle (1243) by means of its blue 4-strings depiction. In teal, on the right of the illustration in Fig. 5.69, that element projects to the transposition (12) in Σ_3, but that element has two (mod 2)-quipu upon it. The one upon the left string causes that 3-strings-with-(mod 2)-quipu to represent an element of order 4. So, the teal representative is labeled by means of the 4-cycle (1243), and the projection q_3 in the seseq above takes (1243) to the transposition (12).

The kernel of q_3 in Σ_4 consists of the elements in K_4 — I, A, B, and C. The kernel of the projection from $SL_2(\mathbb{Z}/4)$ to Σ_3 consists of the set $\pm\{I, A, B, C\}$.

Extensions of the Permutation Group Σ_4 273

$$I = \begin{bmatrix} 1 & 0 \\ 0 & 1 \end{bmatrix}$$

1

$$A = \begin{bmatrix} 1 & 2 \\ 0 & 1 \end{bmatrix}$$

12.34

$$B = \begin{bmatrix} 1 & 0 \\ 2 & 1 \end{bmatrix}$$

13.24

$$C = \begin{bmatrix} 1 & 2 \\ 2 & 1 \end{bmatrix}$$

14.23

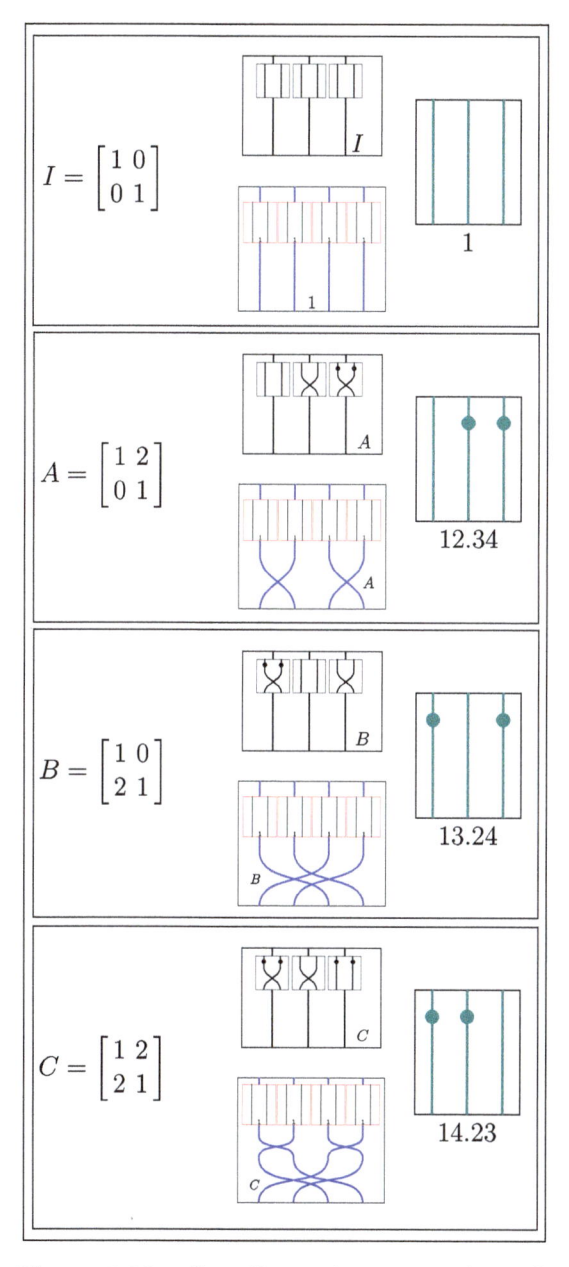

Figure 5.68. Compiling and projecting (part 1).

$$j = \begin{bmatrix} 0 & -1 \\ 1 & 0 \end{bmatrix}$$

23

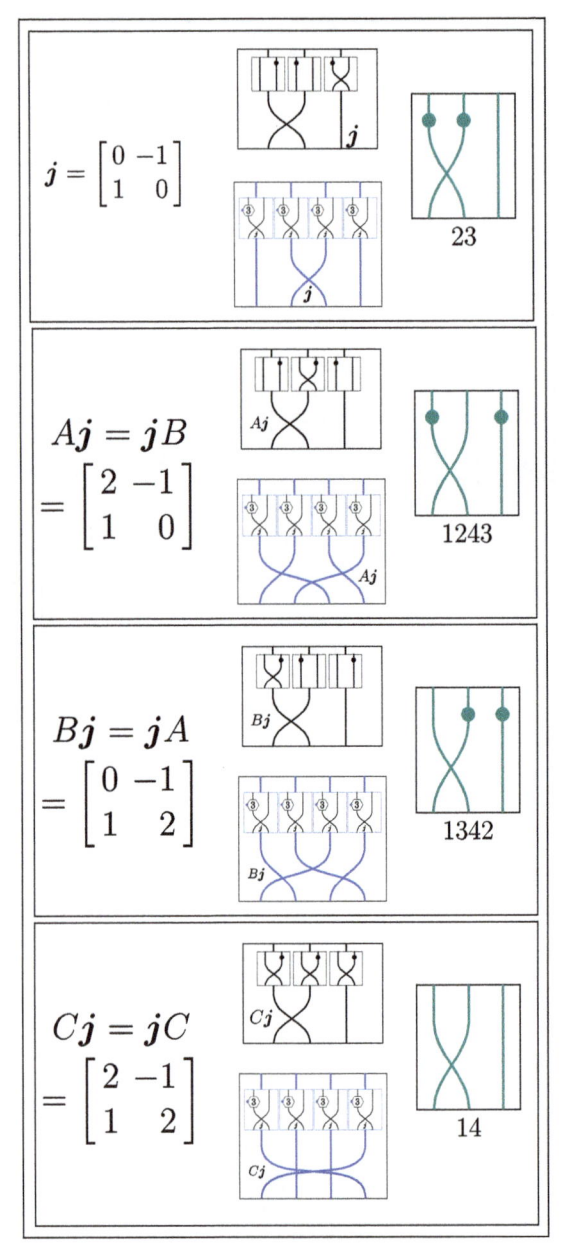

$$Aj = jB = \begin{bmatrix} 2 & -1 \\ 1 & 0 \end{bmatrix}$$

1243

$$Bj = jA = \begin{bmatrix} 0 & -1 \\ 1 & 2 \end{bmatrix}$$

1342

$$Cj = jC = \begin{bmatrix} 2 & -1 \\ 1 & 2 \end{bmatrix}$$

14

Figure 5.69. Compiling and projecting (part 2).

$$j\rho = j\phi B = \begin{bmatrix} -1 & 1 \\ 2 & 1 \end{bmatrix}$$

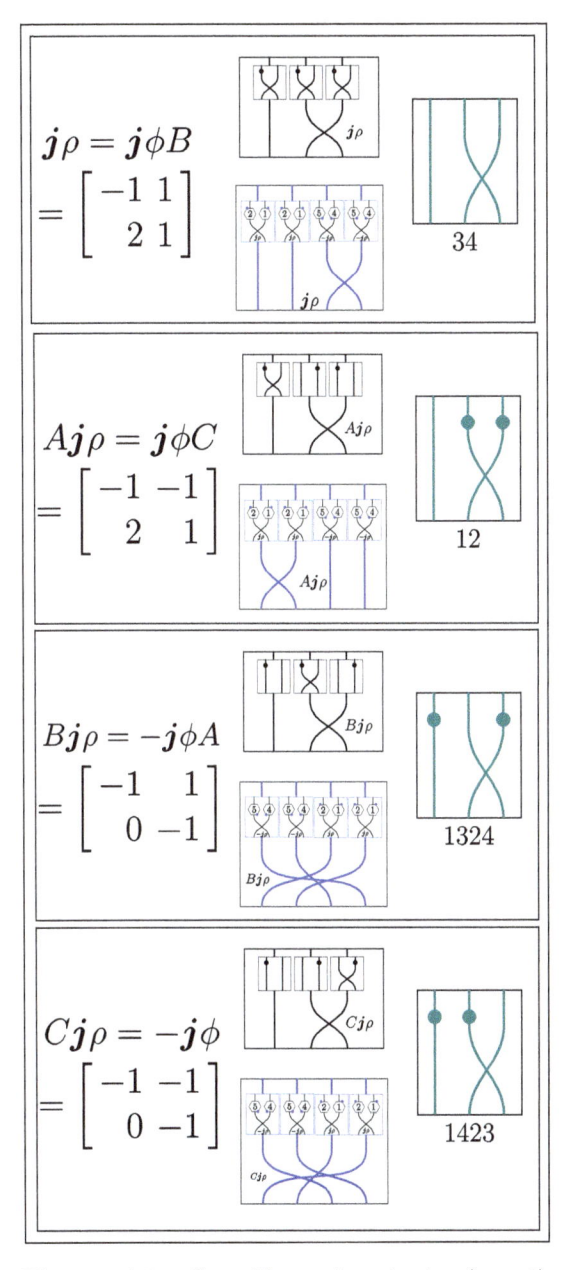

$$Aj\rho = j\phi C = \begin{bmatrix} -1 & -1 \\ 2 & 1 \end{bmatrix}$$

$$Bj\rho = -j\phi A = \begin{bmatrix} -1 & 1 \\ 0 & -1 \end{bmatrix}$$

$$Cj\rho = -j\phi = \begin{bmatrix} -1 & -1 \\ 0 & -1 \end{bmatrix}$$

Figure 5.70. Compiling and projecting (part 3).

$$j\rho^2 = j\phi^2 B$$
$$= \begin{bmatrix} -1 & 2 \\ 1 & 1 \end{bmatrix}$$

24

$$Aj\rho^2 = -j\phi^2$$
$$= \begin{bmatrix} 1 & 0 \\ 1 & 1 \end{bmatrix}$$

1234

$$Bj\rho^2 = -j\phi^2 A$$
$$= \begin{bmatrix} -1 & 2 \\ -1 & 1 \end{bmatrix}$$

13

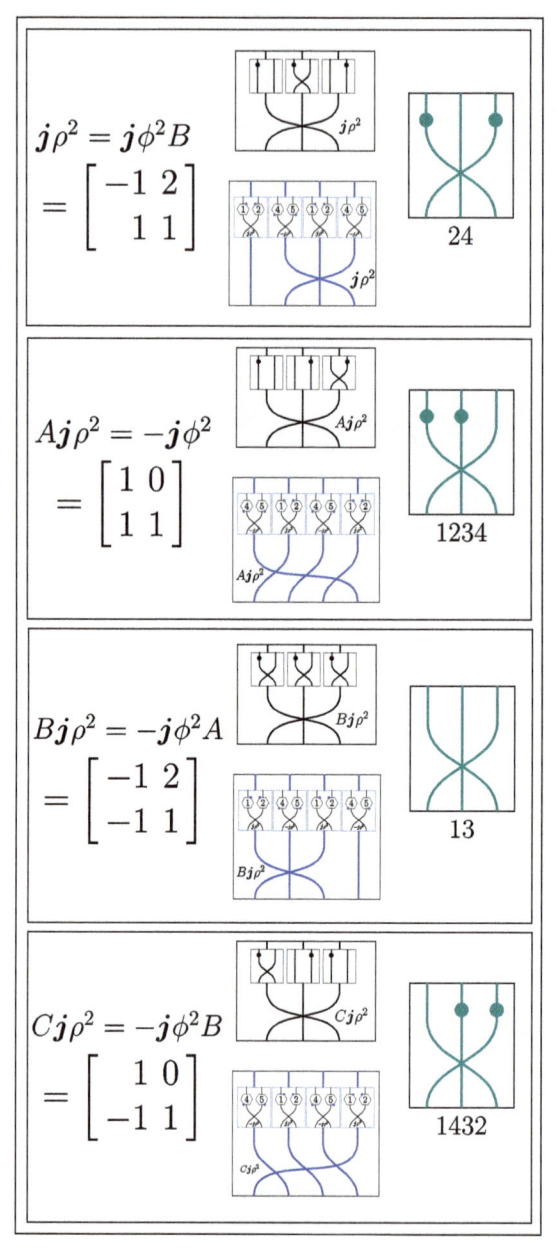

$$Cj\rho^2 = -j\phi^2 B$$
$$= \begin{bmatrix} 1 & 0 \\ -1 & 1 \end{bmatrix}$$

1432

Figure 5.71. Compiling and projecting (part 4).

$\rho = \phi B$
$= \begin{bmatrix} 2 & 1 \\ 1 & -1 \end{bmatrix}$

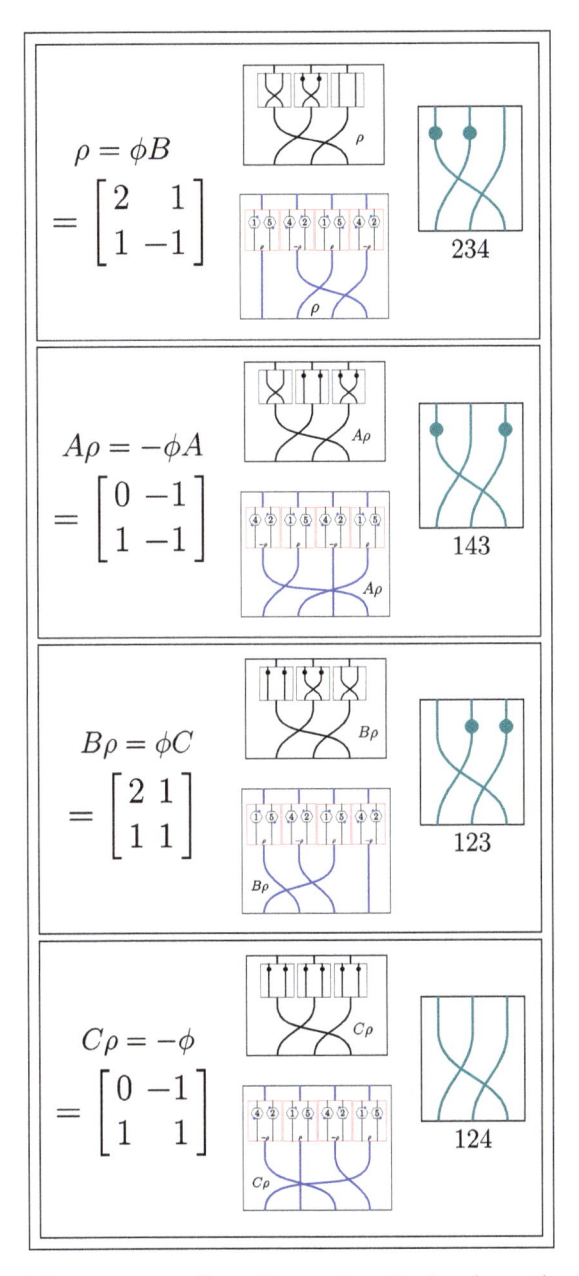

234

$A\rho = -\phi A$
$= \begin{bmatrix} 0 & -1 \\ 1 & -1 \end{bmatrix}$

143

$B\rho = \phi C$
$= \begin{bmatrix} 2 & 1 \\ 1 & 1 \end{bmatrix}$

123

$C\rho = -\phi$
$= \begin{bmatrix} 0 & -1 \\ 1 & 1 \end{bmatrix}$

124

Figure 5.72. Compiling and projecting (part 5).

$\rho^2 = \phi^2 C$
$= \begin{bmatrix} 1 & 1 \\ 1 & 2 \end{bmatrix}$

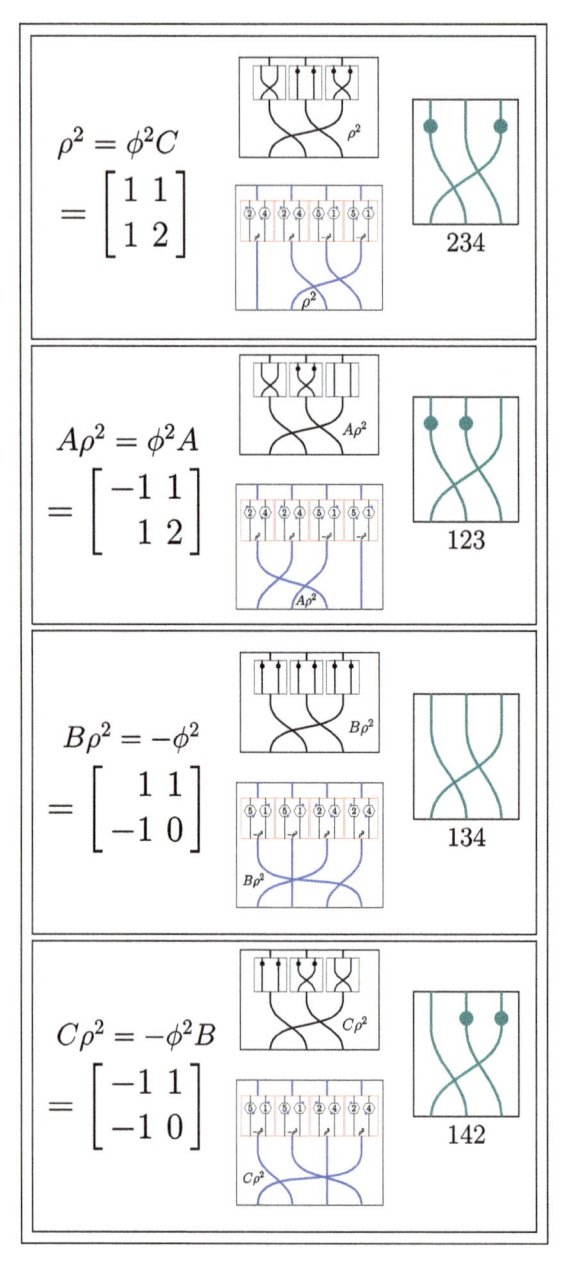

234

$A\rho^2 = \phi^2 A$
$= \begin{bmatrix} -1 & 1 \\ 1 & 2 \end{bmatrix}$

123

$B\rho^2 = -\phi^2$
$= \begin{bmatrix} 1 & 1 \\ -1 & 0 \end{bmatrix}$

134

$C\rho^2 = -\phi^2 B$
$= \begin{bmatrix} -1 & 1 \\ -1 & 0 \end{bmatrix}$

142

Figure 5.73. Compiling and projecting (part 6).

The projection from $SL_2(\mathbb{Z}/4)$ to Σ_3 passes through Σ_4. It is obtained by means of assigning non-trivial (mod 2)-quipu to the reflections that are the dihedral quipu upon the 3-strings picture. Finally, ignore the (mod 2)-quipu to project to Σ_3.

It remains to be shown that the newer, teal-colored, 3-strings-with-(mod 2)-quipu represent the elements in Σ_4 that are used to label them. Various identities need to be checked. In a notational abuse, we have to verify, for example, that $(12)(23)(12) = (23)(12)(23) = (13)$. The similar defining relations among the generating transpositions also need to be checked. In addition, the teal-colored representatives of three and four cycles can be written in standard form by means of the adjacent transpositions. This paragraph is an indication that the authors actually did these verifications using the teal-colored diagrams, but we would rather not illustrate them here.

Or perhaps, more politely, we invite you to perform these verifications.

5.4. The binary octahedral group

The binary octahedral group, $\widetilde{\Sigma_4}$, is the last of the 2-fold extensions of the permutation group Σ_4 that will be presented in this chapter. To be a 2-fold extension means that there is a seseq

$$1 \longleftarrow \Sigma_4 \stackrel{q}{\longleftarrow} \widetilde{\Sigma_4} \stackrel{i}{\longleftarrow} \mathbb{Z}/2 \longleftarrow 1.$$

The group is given via the presentation

$$\widetilde{\Sigma_4} = \langle a, f : a^3 = f^4 = (af)^2 \rangle.$$

The generators are identified with the elements $a = 1/2(1+\boldsymbol{i}+\boldsymbol{j}+\boldsymbol{k})$ and $f = (1+\boldsymbol{i})/\sqrt{2}$ in S^3. Two strings-with-quipu representations will be presented. One of these will involve 4-strings and the other will involve 3-strings.

Interestingly, the binary octahedral group and the matrix group $GL_2(\mathbb{Z}/3)$ both contain subgroups that are isomorphic to the binary tetrahedral group. In the case of $GL_2(\mathbb{Z}/3)$, the determinant 1 matrices, $SL_2(\mathbb{Z}/3)$, form a subgroup that is isomorphic to the binary tetrahedral group, $\widetilde{A_4}$, which is also the 2-fold extension of the alternating group. So, the relationship between $GL_2(\mathbb{Z}/3)$ and $SL_2(\mathbb{Z}/3)$ is analogous to the relationship between the symmetric group Σ_4 and the alternating subgroup A_4. Another analogy occurs between the binary octahedral group $\widetilde{\Sigma_4}$ and the binary tetrahedral group $\widetilde{A_4}$ that, perhaps, is apparent via their names and their notations.

The 4-strings depiction of the binary octahedral group $\widetilde{\Sigma_4}$ will involve quipu that are elements of the 12-element dicyclic group Dic_3. In this sense, the 4-strings picture of $\widetilde{\Sigma_4}$ will be organized analogously to that of $SL_2(\mathbb{Z}/4)$ since these two groups have subgroups that are isomorphic to Dic_3. On the other hand, the 3-strings presentation of $\widetilde{\Sigma_4}$ involves quipu that are elements of the 16 element dicyclic group Dic_4. This subgroup intersects the semi-dihedral group whose elements adorn the 3-strings picture of $GL_2(\mathbb{Z}/3)$. In these 3-strings depictions, you will see — at least we hope that you will see — that the groups $GL_2(\mathbb{Z}/3)$ and $\widetilde{\Sigma_4}$ shared some commonality.

We recall that in the 3-sphere, S^3, the inverse of an element $(w + x\boldsymbol{i} + y\boldsymbol{j} + z\boldsymbol{k})$ is given by the expression $(w - x\boldsymbol{i} - y\boldsymbol{j} - z\boldsymbol{k})$ because $w^2 + x^2 + y^2 + z^2 = 1$. Furthermore, the element $a = 1/2(1+\boldsymbol{i}+\boldsymbol{j}+\boldsymbol{k})$ is of order 6. For example, its square is $a^2 = 1/2(-1+\boldsymbol{i}+\boldsymbol{j}+\boldsymbol{k})$, and so, $-a^{-1} = a^2$, and $a^3 = -1$.

The (ordered) set of powers of a, $A = (1, a, a^2, -1, -a, -a^2)$, and its coset $\frac{i-j}{\sqrt{2}}A$ form a subgroup, $D = A \cup \frac{i-j}{\sqrt{2}}A$ that is isomorphic to the dicyclic group Dic_3 of order 12. There are four cosets: D, $\boldsymbol{i}D$, $\boldsymbol{j}D$, and $\boldsymbol{k}D$. So, there is a 4-strings-with-Dic_3-quipu representation.

We encourage the reader to please verify that D is closed and that the inverse of any element therein is also an element therein.

The subgroup $P = (1, f, \boldsymbol{i}, f^3, -1, -f, -\boldsymbol{i}, -f^3)$ that consists of powers of f and its coset $\boldsymbol{j}P$ form the dicyclic group Dic_4 that has 16 elements. This description of the dicyclic group is the same as that given in Figs. 4.22 and 4.23 of Chapter 4. The generators a and f are depicted as 3-strings-with-Dic_4-quipu in Fig. 5.87. The elements of the dicyclic subgroup and those of the cosets $a(P \cup \boldsymbol{j}P)$

and $b(P \cup jP)$ are represented in Figs. 5.91–5.96. Since the quipu are elements in Dic$_4$, the elements a, b, and f can be presented as (3×3)-matrices in which there is a unique non-zero entry in any row or column, and the non-zero entries are elements of Dic$_4$. See Fig. 3.19 and the text nearby to reacquaint yourself with these ideas.

The matrix representations for a and b are

$$a \leftrightharpoons \begin{bmatrix} 0 & 0 & j \\ 1 & 0 & 0 \\ 0 & j & 0 \end{bmatrix} \quad \text{and} \quad b \leftrightharpoons \begin{bmatrix} 0 & i & 0 \\ 0 & 0 & i \\ 1 & 0 & 0 \end{bmatrix}.$$

The representation for f is

$$f \leftrightharpoons \begin{bmatrix} \frac{(1+i)}{\sqrt{2}} & 0 & 0 \\ 0 & 0 & \frac{(1+i)}{\sqrt{2}} \\ 0 & \frac{(j+k)}{\sqrt{2}} & 0 \end{bmatrix}.$$

Thus, the binary octahedral group $\widetilde{\Sigma_4}$ is a subgroup of $(\text{Dic}_4)^3 \rtimes \Sigma_3$. See also Figs. 5.87 and 5.95.

The 4-strings-with-Dic$_3$-quipu representation of $\widetilde{\Sigma_4}$ are constructed in Section 5.4.1. The 3-strings-with-Dic$_4$-quipu representation of $\widetilde{\Sigma_4}$ are constructed in Section 5.4.2. These two constructions follow the usual procedure. A cyclic subgroup is identified and ordered, then this subgroup and one of its cosets are conjoined to form a larger subgroup. The larger subgroup in either case is dicycle. The cases differ in the order of the dicyclic group. The group is then considered as a union of cosets of the dicyclic subgroup. Subgroups, cosets, and collections of cosets are ordered somewhat arbitrarily. The actions of a and f upon the cosets and their orderings are computed to give strings-with-quipu representations of the generators.

As we feel it is necessary, or beneficial to you, some calculations that use quipu representations will be summarized in the interim. Let us proceed.

5.4.1. Cosets of $A = (1, a, a^2, -1, -a, -a^2)$

The subgroup A and its translate xA, where $x = \frac{(i-j)}{\sqrt{2}}$, form a subgroup of the binary octahedral group $\widetilde{\Sigma_4}$ that is isomorphic to a dicyclic group Dic_3 of order 12. Again, we implore you to verify this. Write $D = A \cup xA$ with the usual understanding that both A and xA are cyclically ordered as indicated above.

We take a moment to remind you that the values $b = 1/2(1 + i + j - k)$, $c = 1/2(1 + i - j - k)$, and $d = 1/2(1 + i - j + k)$ are depicted in Fig. 4.12. In this figure and also within Figs. 5.19–5.22, the powers of these elements are also listed.

If you have not done so already, it is an opportune time to compute these powers directly. You may use the method of your choice. We suggest either multiplying matrices in $SL_2(\mathbb{Z}/3)$, using either of the quipu representations within that group, substituting values for a_ℓ, b_ℓ etc. in Table 4.2, or adding appropriate signs to Table 4.1.

Since there is not yet a suitable quipu representative for the element x, we display Table 5.1 that computes $\sqrt{2}x \cdot 2a$. Consequently, $xa = \frac{k-j}{\sqrt{2}}$. It is not too difficult to use this table to also determine that $xa^2 = \frac{k-i}{\sqrt{2}}$. Since $a^3 = -1$, the remaining products xa^ℓ can also be determined.

Table 5.1. A tabular computation showing $(i - j) \cdot a = 2(k - j)$.

$\cdot$	1	i	j	k
i	i	-1	k	$-j$
$-j$	$-j$	k	1	$-i$

The rows in Table 4.1 determine that

$$ia = 1/2(-1 + i - j + k) = d^2,$$
$$ja = 1/2(-1 + i + j - k) = b^2,$$

and

$$ka = 1/2(-1 - i + j + k) = -c.$$

Extensions of the Permutation Group Σ_4

These last three products can also be computed in several different ways. Similarly,

$$ia^2 = 1/2(-1 - i - j + k) = -b,$$
$$ja^2 = 1/2(-1 + i - j - k) = c^2,$$
$$ka^2 = 1/2(-1 - i + j - k) = -d.$$

These values determine the next four cosets of A.

$$A = (1, a, a^2, -1, -a, -a^2), \quad iA = (i, d^2, -b, -i, -d^2, b),$$
$$jA = (j, b^2, c^2, -j, -b^2, -c^2), \quad kA = (k, -c, -d, -k, c, d).$$

Only those values are needed since $a^3 = -1$, $a^4 = -a$, and $a^5 = -a^2$. Don't loose track of the signs!

We compute directly that

$$ix = -\frac{k+1}{\sqrt{2}}, \quad jx = \frac{1-k}{\sqrt{2}}, \quad \text{and} \quad kx = \frac{i+j}{\sqrt{2}}.$$

To prevent typographical and transcription errors, we present tabulations (Tables 5.2–5.7) that allow the determination of ixa, ixa^2, jxa, jxa^2, kxa, and kxa^2.

Table 5.2. A tabular computation showing $-(1+k) \cdot a = -2(j+k)$.

$\cdot$	1	i	j	k
-1	-1	$-i$	$-j$	$-k$
$-k$	$-k$	$-j$	i	1

Table 5.3. A tabular computation showing $-(1+k) \cdot a^2 = 2(1-j)$.

$\cdot$	-1	i	j	k
-1	1	$-i$	$-j$	$-k$
$-k$	k	$-j$	i	1

Table 5.4. A tabular computation showing $(1-k)\cdot a = 2(1+i)$.

$\cdot$		1	i	j	k
1		1	i	j	k
$-k$		$-k$	$-j$	i	1

Table 5.5. A tabular computation showing $(1-k)\cdot a^2 = 2(i+k)$.

$\cdot$		-1	i	j	k
1		-1	i	j	k
$-k$		k	$-j$	i	1

Table 5.6. A tabular computation showing $(i+j)\cdot a = 2(i-1)$.

$\cdot$		1	i	j	k
i		i	-1	k	$-j$
j		j	$-k$	-1	i

Table 5.7. A tabular computation showing $(i+j)\cdot a^2 = -2(1+j)$.

$\cdot$		-1	i	j	k
i		$-i$	-1	k	$-j$
j		$-j$	$-k$	-1	i

Now, we recompile and reorder the results that were tabulated above to determine the elements in the remaining cosets of A:

$$xA = \frac{1}{\sqrt{2}}(i-j, k-j, k-i, j-i, j-k, i-k),$$

$$ixA = \frac{1}{\sqrt{2}}(-(1+k), -(j+k), 1-j, 1+k, j+k, j-1),$$

$$jxA = \frac{1}{\sqrt{2}}(1-k, 1+i, i+k, k-1, -(1+i), -(i+k)),$$

and

$$kxA = \frac{1}{\sqrt{2}}(i+j, i-1, -(1+j), -(i+j), 1-i, 1+j).$$

To help justify the results in the following, we refer again to Table 4.1. By reading the columns therein, we get

$$ai = 1/2(-1+i+j-k) = b^2,$$
$$aj = 1/2(-1-i+j+k) = -c,$$

and

$$ak = 1/2(-1+i-j+k) = d^2.$$

So, to determine aiA, locate b^2 as the second element of jA. This gives $aiA = [jA, 1]$. Similarly, $-c$ is the second element of kA. Thus, $ajA = [kA, 1]$. Since d^2 is the second element of iA, $akA = [iA, 1]$. As before, we write $aA = [A, 1]$ to indicate that the cyclic group A has been rotated by $\pi/3$.

To continue the analysis of the actions of the elements a upon each of x, xi, xj, and xk are computed are computed in Table 5.8.

Table 5.8. $ax = \frac{i-k}{\sqrt{2}}$, $aix = -\frac{i+k}{\sqrt{2}}$, $ajx = \frac{1+j}{\sqrt{2}}$, $akx = \frac{j-1}{\sqrt{2}}$.

·	i	$-j$
1	i	$-j$
i	-1	$-k$
j	$-k$	1
k	j	i

·	-1	$-k$
1	-1	$-k$
i	$-i$	j
j	$-j$	$-i$
k	$-k$	1

·	1	$-k$
1	1	$-k$
i	i	j
j	j	$-i$
k	k	1

·	i	j
1	i	j
i	-1	k
j	$-k$	-1
k	j	$-i$

Now, $ax = \frac{i-k}{\sqrt{2}}$ is the last element of xA. So, $a(xA) = [xA, 5]$. Similarly, $aix = -\frac{i+k}{\sqrt{2}}$ is the last element of jxA. So, $a(ixA) = [xjA, 5]$. Continuing, $ajx = \frac{1+j}{\sqrt{2}}$ is the last element of kxA. So, $a(jxA) = [xkA, 5]$. Finally, $akx = \frac{-1+j}{\sqrt{2}}$ is the last element of ixA. So, $a(kxA) = [xiA, 5]$.

We summarize as follows:

$$aA = [A, 1], \quad axA = [xA, 5],$$
$$aiA = [jA, 1], \quad aixA = [jxA, 5],$$
$$ajA = [kA, 1], \quad ajxA = [kxA, 5],$$

and

$$akA = [iA, 1], \quad akxA = [ixA, 5].$$

To compute the actions of $f = \frac{1+i}{\sqrt{2}}$, first note that f is the second element of jxA. So, $fA = [jxA, 1]$. Next, $fi = \frac{i-1}{\sqrt{2}}$ is the second element of kxA. So, $fiA = [kxA, 1]$. Thirdly, $fj = \frac{j+k}{\sqrt{2}}$ is the fifth element of ixA. So, $fjA = [ixA, 4]$. Finally, $fk = \frac{k-j}{\sqrt{2}}$ is the second element of xA. So, $fkA = [xA, 1]$.

Then compute the products of these four elements with $x = \frac{i-j}{\sqrt{2}}$. While it may seem excessively prudent to do so, we provide Table 5.9 that facilitates these computations. We do so for your convenience and for our own. Otherwise, either you or we will be flipping through pages, scribbling down partial calculations, and generally feeling foolish upon making errors with signs.

Table 5.9. $fx = c^2$, $fix = -d$, $fjx = b$, $fkx = a^2$.

·	i	$-j$	·	i	$-j$	·	i	$-j$	·	i	$-j$
1	i	$-j$	-1	$-i$	j	j	$-k$	1	$-j$	k	-1
i	-1	$-k$	i	-1	$-k$	k	j	i	k	j	i

We get that fx is the third element of jA, fix is the third element of kA, fjx is the sixth element of iA, and fkx is the third element

Extensions of the Permutation Group Σ_4

of A. In summary, the following eight equations hold:

$$fA = [jxA, 1], \quad f(xA) = [jA, 2],$$
$$fiA = [kxA, 1], \quad f(xiA) = [kA, 2],$$
$$fjA = [ixA, 4], \quad f(xjA) = [iA, 5],$$
$$fkA = [xA, 1], \quad f(xkA) = [A, 2].$$

The representations of these elements as strings-with-Dic$_3$-quipu are presented in Fig. 5.74.

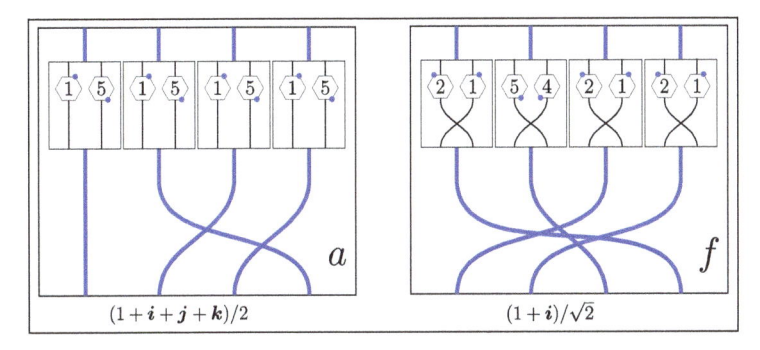

Figure 5.74. The representations of a and f using the subgroup $A \cup xA$.

Since the binary octahedral group, $\widetilde{\Sigma_4}$, is generated by these two elements, they may be multiplied in various ways to obtain representatives for the remaining elements in $\widetilde{\Sigma_4}$. On the other hand, we need to know which products represent which elements, and the goal for this section is to arrange the elements around the vertices of eight hexagons, so that the analogy between this group and $\mathrm{SL}_2(\mathbb{Z}/4)$ is visually apparent.

We point out that tabulations that are similar to those given in Tables 5.1–5.9 can be used to determine that $f^2 = i$ and $a \cdot ia^{-1} = j$. Meanwhile,

$$ia^{-1} = i(1 - i - j - k)/2 = (1 + i + j - k)/2 = b.$$

Furthermore, $x = \frac{i-j}{\sqrt{2}} = -jf^{-1}b$ (compare with the caption in Table 5.9). In Figs. 5.75–5.78, these calculations are replicated to get the strings-with-quipu presentations of these elements.

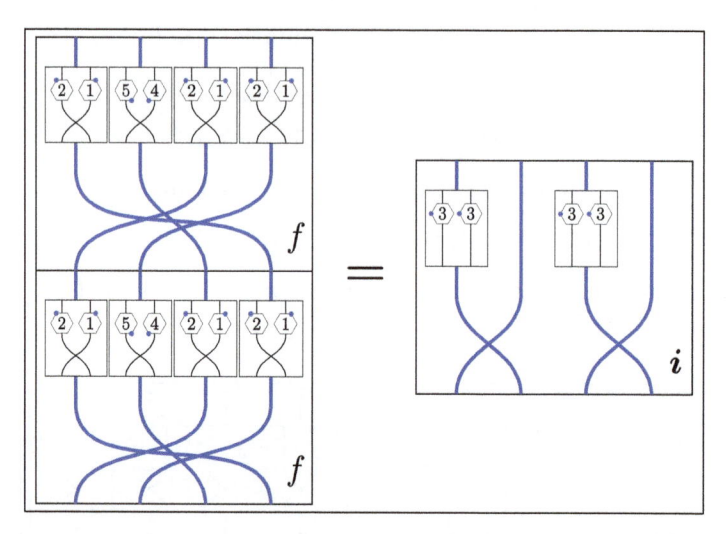

Figure 5.75. The strings-with-Dic$_3$-quipu description that $f^2 = i$.

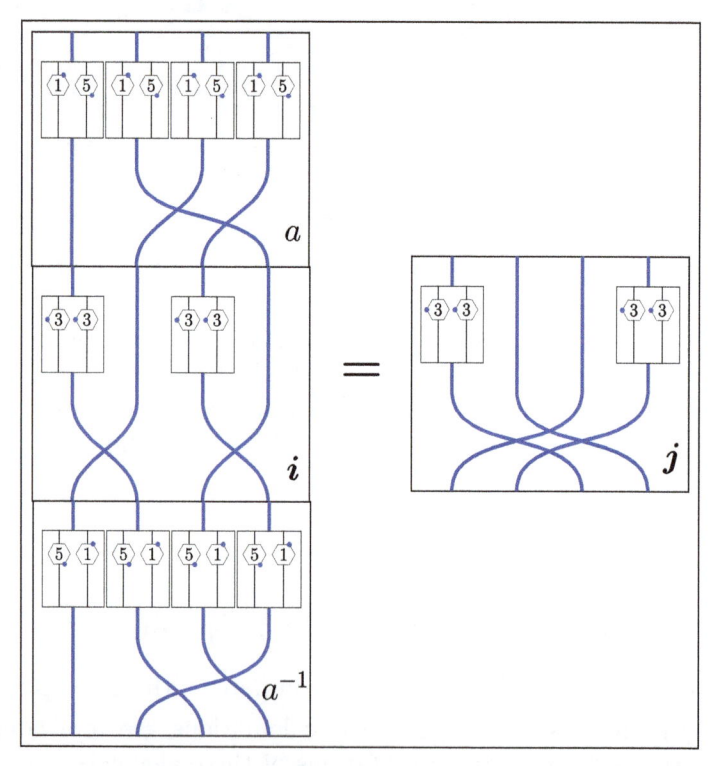

Figure 5.76. The strings-with-Dic$_3$-quipu description that $aia^{-1} = j$.

Before illustrating $x = \frac{i-j}{\sqrt{2}}$ in terms of the quantities illustrated thus far, it might be good to clean out a few negative signs from the expression $-\boldsymbol{j}f^{-1}b$. Specifically, since $f^2 = \boldsymbol{i}$, $f^4 = -1$ and $f^{-1} = f^7 = -f^3 = -f\boldsymbol{i}$. Furthermore, $b = \boldsymbol{i}a^{-1} = -\boldsymbol{i}a^2$. Now, all the minus signs commute with everything. So,

$$x = -\boldsymbol{j}f^{-1}b = (-\boldsymbol{j})(-f\boldsymbol{i})(-\boldsymbol{i}a^2) = -\boldsymbol{j}f\boldsymbol{i}^2 a^2 = \boldsymbol{j}fa^2,$$

and we compute this last expression in Fig. 5.78.

The previous paragraph might cause some frustration upon the few few readings thereof. The element f acts as a generator of a cyclic group of order 8, and the element a acts as a generator of a cyclic group of order 6. Furthermore, $a^3 = f^4 = \boldsymbol{i}^2 = -1$. So, we get to play fast and free, but if you are not used to these type of calculations, please take them slowly.

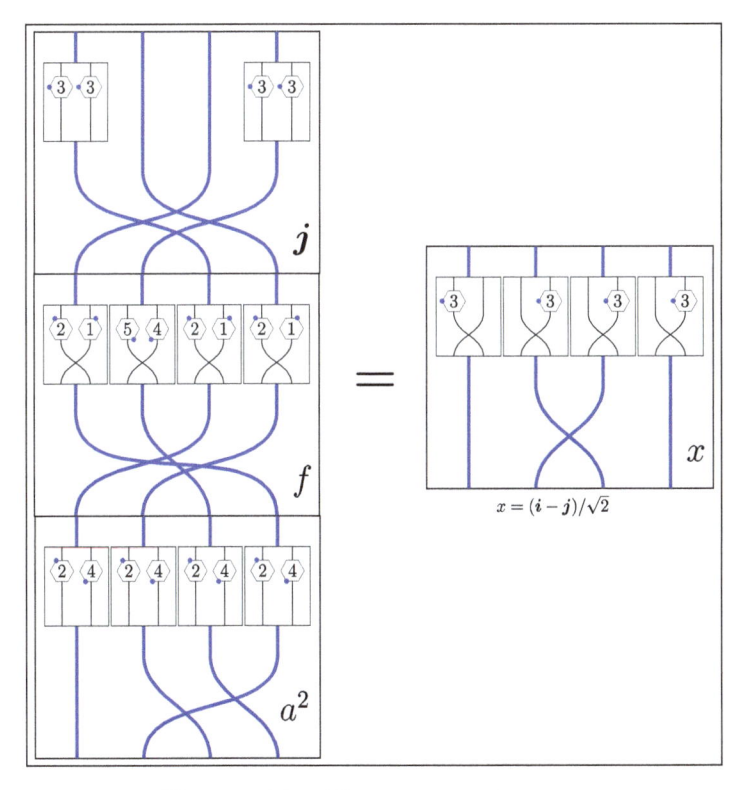

Figure 5.77. The representative for x.

We point out that the non-trivial quipu that adorn the strings of i, j, and, consequently, k are representative of (-1) as it appears in the dicyclic group Dic_3. The element (-1) in $\widetilde{\Sigma}_4$ is an identity permutation on 4-strings with a "(3,3)-straight"-quipu upon all 4-strings, and as always, the projection from $\widetilde{\Sigma}_4$ to Σ_4 is given by ignoring the quipu.

In the representations of a^2 and $a^{-1} = -a^2$ that were given in Fig. 5.77, we simply squared the diagram for a (which projects to (234)) and multiplied by the representative of (-1). But we did so within our own notes. At this stage, we implore you to verify that the remaining elements in $\widetilde{\Sigma}_4$ are as depicted in Figs. 5.78–5.85.

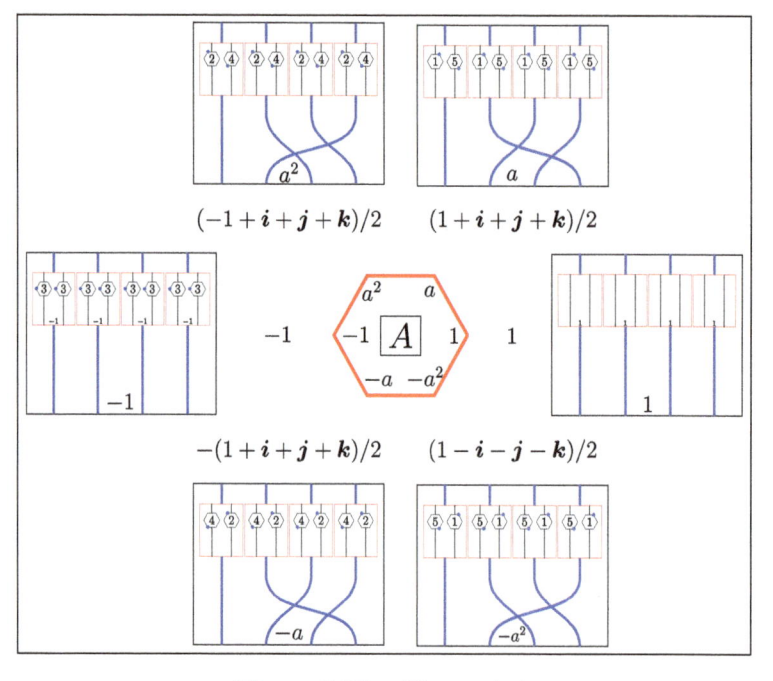

Figure 5.78. The coset A.

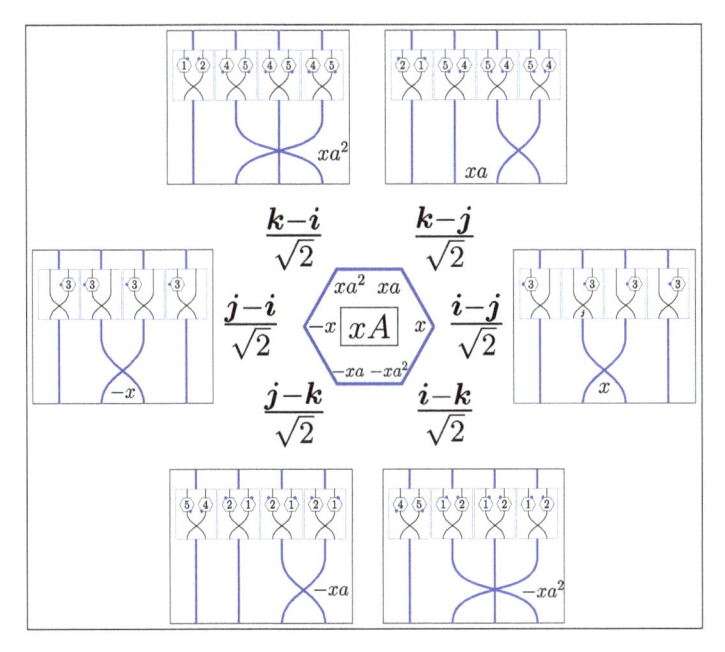

Figure 5.79. The coset xA.

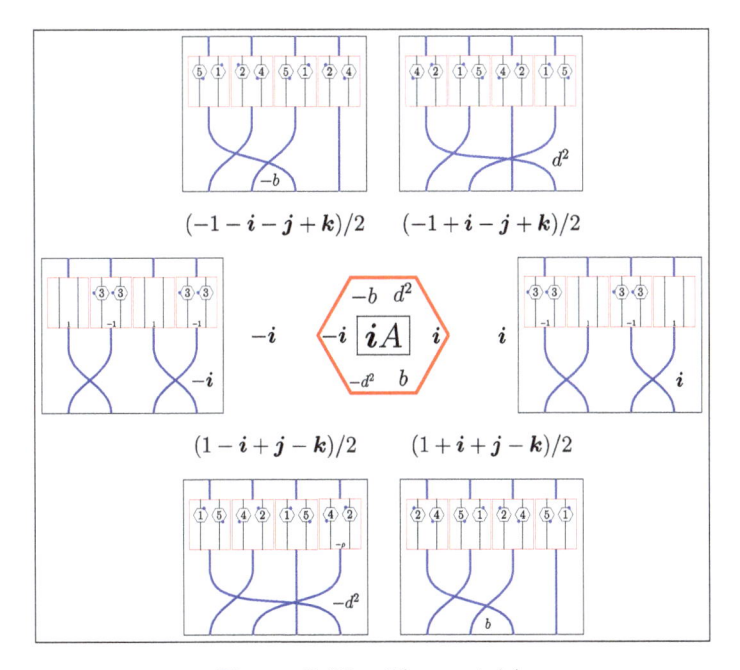

Figure 5.80. The coset iA.

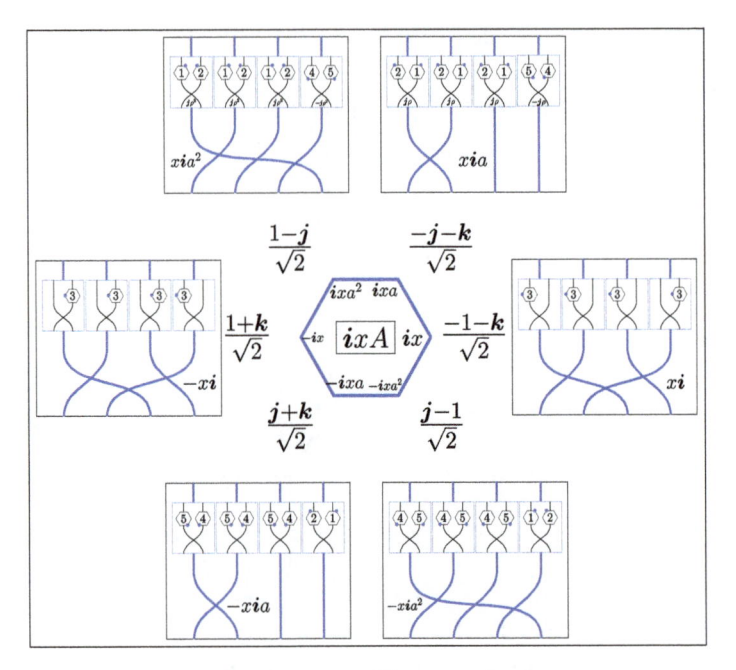

Figure 5.81. The coset ixA.

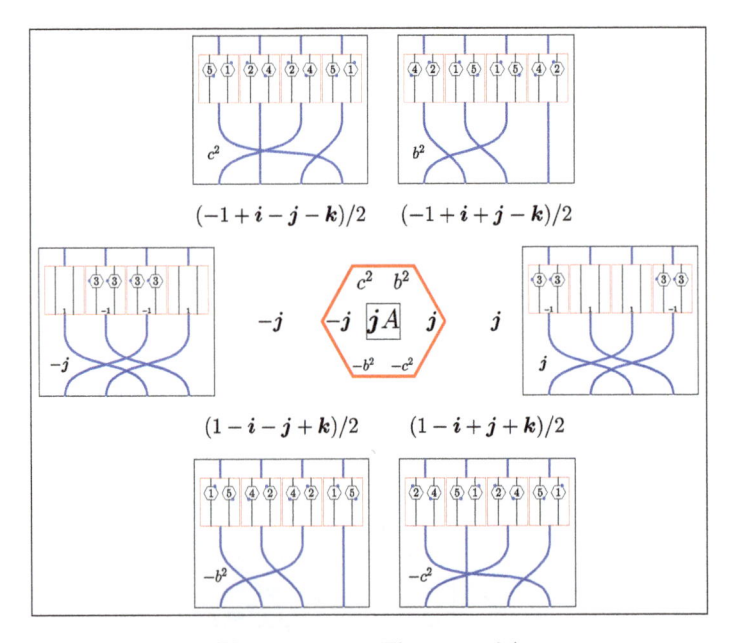

Figure 5.82. The coset jA.

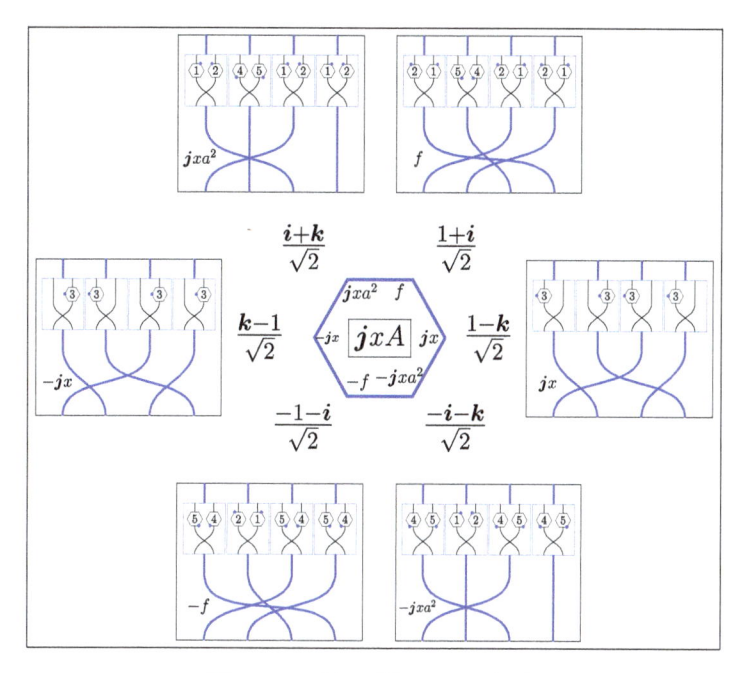

Figure 5.83. The coset jxA.

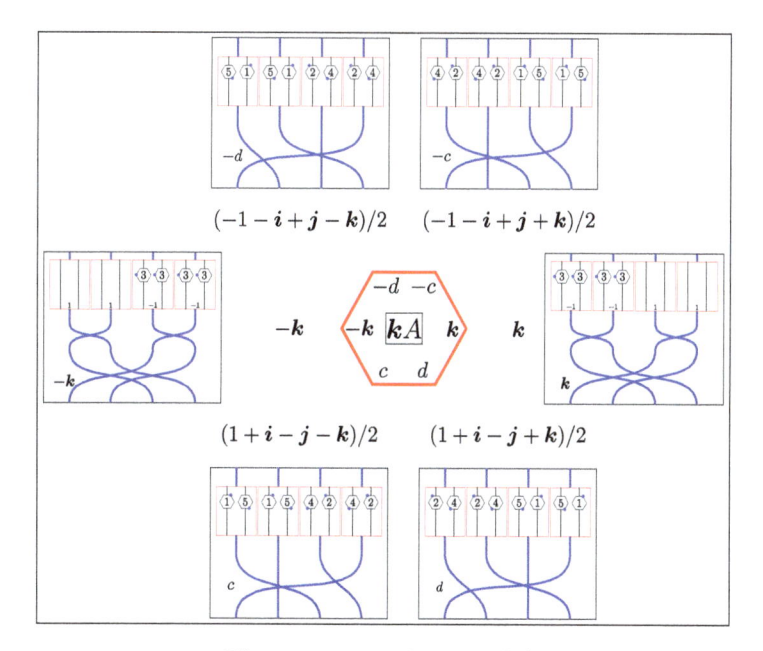

Figure 5.84. The coset kA.

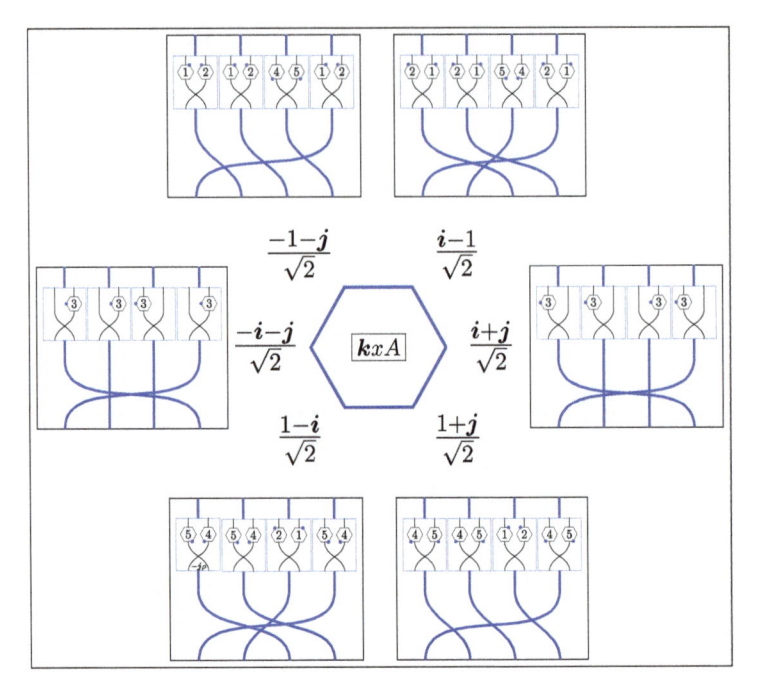

Figure 5.85. The coset kxA.

5.4.2. Cosets of the dicyclic group Dic_4 of order 16

Let

$$P = \langle f \rangle = (f^\ell : \ell = 0, \ldots, 7) = (1, f, i, f^3, -1, -f, -i, -f^3)$$

denote the (ordered) subgroup of the binary octahedral group, $\widetilde{\Sigma_4}$ that consists of the powers of the element $f = (1+i)/\sqrt{2}$. On the $(1, i)$ circle in the 3-sphere, S^3, the element f subtends an angle of $45°$. Consequently, the subgroup P is isomorphic to $\mathbb{Z}/8$.

The union $P \cup jP$ is the dicyclic group Dic_4 that is of order 16. This is the standard description of the dicyclic group that was given in Figs. 4.22 and 4.23. We recall from Fig. 4.23 (or you may recompute) that the ordering of jP that is induced from that of P is

$$jP = \frac{1}{\sqrt{2}}(j, j-k, -k, -(j+k), -j, k-j, k, j+k).$$

Extensions of the Permutation Group Σ_4

The binary octahedral group is decomposed as the union of three cosets of $P \cup jP$, $a(P \cup jP)$, and $b(P \cup jP)$. We recall that $a = (1 + i + j + k)/2$ and $b = (1 + i + j - k)/2$.

We imagine that it is straightforward to compute jP. Some additional calculations from above are both recalled and modified to suit our current needs:

$$ai = 1/2(-1 + i + j - k) = b^2,$$
$$aj = 1/2(-1 - i + j + k) = -c,$$

and

$$a(-k) = 1/2(1 - i + j - k) = -d^2.$$

So,

$$a(1 + i)/\sqrt{2} = (i + j)/\sqrt{2}$$

and

$$a(-1 + i)/\sqrt{2} = -(1 + k)/\sqrt{2}.$$

There is enough information given here to verify that aP is as listed in the following. To determine the second entry of ajP, compute

$$a(j - k)/\sqrt{2} = -(j - i)/\sqrt{2}.$$

For the fourth entry,

$$a(-j - k)/\sqrt{2} = (1 - k)/\sqrt{2}.$$

The remaining elements in the first four cosets are determined as negatives of the priors:

$$P = \left(1, \frac{1+i}{\sqrt{2}}, i, \frac{-1+i}{\sqrt{2}}, -1, \frac{-1-i}{\sqrt{2}}, -i, \frac{1-i}{\sqrt{2}}\right),$$

$$jP = \left(j, \frac{j-k}{\sqrt{2}}, -k, \frac{-j-k}{\sqrt{2}}, -j, \frac{-j+k}{\sqrt{2}}, k, \frac{j+k}{\sqrt{2}}\right),$$

$$aP = \left(a, \frac{i+j}{\sqrt{2}}, b^2, \frac{-1-k}{\sqrt{2}}, -a, \frac{-i-j}{\sqrt{2}}, -b^2, \frac{1+k}{\sqrt{2}}\right),$$

$$ajP = \left(-c, \frac{-i+j}{\sqrt{2}}, -d^2, \frac{1-k}{\sqrt{2}}, c, \frac{i-j}{\sqrt{2}}, d^2, \frac{-1+k}{\sqrt{2}}\right).$$

Let us compute that

$$bi = 1/2(1+i+j-k)i = 1/2(-1+i-j-k) = c^2,$$
$$bj = 1/2(1+i+j-k)j = 1/2(-1+i+j+k) = a^2,$$

and

$$b(-k) = 1/2(-1-i-j+k)k = 1/2(-1-i+j-k) = -d.$$

Furthermore,

$$b(1+i)/\sqrt{2} = (i-k)/\sqrt{2}$$

and

$$b(-1+i)/\sqrt{2} = -(1+j)/\sqrt{2}.$$

There is enough information given here to verify that bP is as listed in the following. To determine the second entry of bjP, compute

$$b(j-k)/\sqrt{2} = (-1+j)/\sqrt{2}.$$

For the fourth entry,

$$b(-j-k)/\sqrt{2} = -(i+k)/\sqrt{2}.$$

Therefore, the cosets bP and bjP are as follows:

$$bP = \left(b, \frac{i-k}{\sqrt{2}}, c^2, \frac{-1-j}{\sqrt{2}}, -b, \frac{-i+k}{\sqrt{2}}, -c^2, \frac{1+j}{\sqrt{2}}\right),$$

$$bjP = \left(a^2, \frac{-1+j}{\sqrt{2}}, -d, \frac{-i-k}{\sqrt{2}}, -a^2, \frac{1-j}{\sqrt{2}}, d, \frac{i+k}{\sqrt{2}}\right).$$

The actions of a and f upon the ordered cosets $[(P \cup jP), a(P \cup jP), b(P \cup jP)]$ will now be considered.

Since $(P \cup jP)$ is the dicyclic group Dic_4, the generator f moves elements in P forward one space. So, $f(P) = [P, 1]$, and $(1+i)j = j+k$. So, $f(jP) = [jP, 7]$. To compute the action of f upon aP, ajP, bP, and bjP, we first write that

$$ia = (-1+i-j+k)/2 = d^2.$$

So, $fa = (i+k)/\sqrt{2}$. This is the last element of bjP:
$$iaj = i(-1-i+j+k)/2 = (1-i-j+k)/2 = -b^2.$$
So, $faj = (-i+k)/\sqrt{2}$. This is the sixth element of bP:
$$ib = i(1+i+j-k)/2 = (-1+i+j+k)/2 = a^2.$$
So, $fb = (i+j)/\sqrt{2}$. This is the second element of aP, and
$$ibj = i(-1+i+j+k)/2 = (-1-i-j+k)/2 = -b.$$
So, $fbj = (-1+k)/\sqrt{2}$. This is the last element of ajP.

We summarize the computations above:
$$\begin{aligned} f(P) &= [P,1], & f(jP) &= [jP,7], \\ f(aP) &= [bjP,7], & f(ajP) &= [bP,5], \\ f(bP) &= [aP,1], & f(bjP) &= [ajP,7]. \end{aligned}$$

To compute the action of a, multiply the coset representatives on the left by a. It should be clear that P and jP move to aP and ajP upon multiplication by a. Furthermore, $bj = a^2$, so $a(aP) = bjP$ and no rotation occurs. We compute
$$a^2 j = (-1+i+j+k)j/2 = (-1-i-j+k)/2 = -b.$$

We also have $ab = j$. Take a few minutes to puzzle about Fig. 5.76.

In summary,
$$\begin{aligned} a(P) &= [aP,0], & a(jP) &= [ajP,0], \\ a(aP) &= [bjP,0], & a(ajP) &= [bP,4], \\ a(bP) &= [jP,0], & a(bjP) &= [P,4]. \end{aligned}$$

In examining the representations depicted in Fig. 5.86 of a and f in comparison with a and g of Fig. 5.28, please note that the representation of a is identical and that f shares some analogous properties with g, namely, both project to the transposition $(23) \in \Sigma_3$. As far as the depiction of a goes, the binary tetrahedral group, $\widetilde{A}_4$, is isomorphic to $SL_2(\mathbb{Z}/3) \subset GL_2(\mathbb{Z}/3)$ and it is also an index 2 subgroup of the binary octahedral group $\widetilde{\Sigma}_4$. So, the 3-strings representations of the elements in $\widetilde{A}_4$ should be closely related, if not identical to, the elements in the previous group.

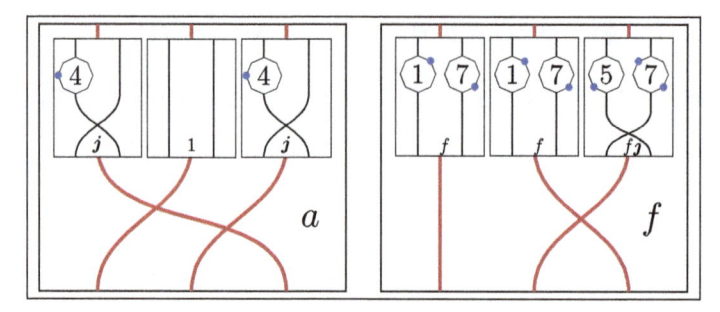

Figure 5.86. The 3-strings-with-Dic$_4$-quipu representatives of a and f in $\widetilde{\Sigma}_4$.

Some figures that are analogous to Figs. 5.75 and 5.76 will be helpful to complete the catalog of the elements in $\widetilde{\Sigma}_4$. In particular, in Fig. 5.87, we compute the 3-strings-with-Dic$_4$-quipu representatives of i and f^3. In Fig. 5.88, we compute a^2 and a^3. In Fig. 5.89, we use the previous diagrams (or their negatives), to compute $a(-i)a^2 = aia^{-1} = j$ as 3-strings-with-Dic$_4$-quipu.

In Figs. 5.90–5.95, the elements of the binary octahedral group, $\widetilde{\Sigma}_4$ are compiled in the 3-strings-with-Dic$_4$-quipu. This compilation

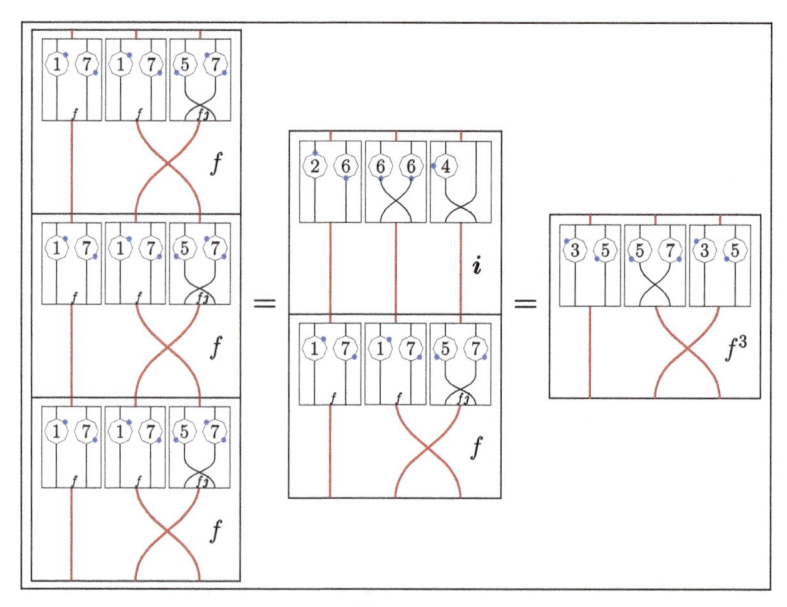

Figure 5.87. Computing i and f^3 representations with 3-strings-with-Dic$_4$-quipu.

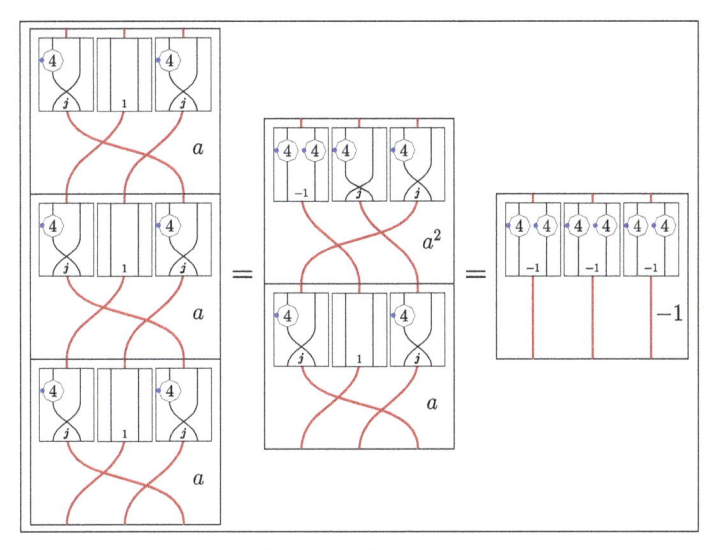

Figure 5.88. Computing a^2 and a^3 using the 3-strings-with-Dic$_4$-quipu representatives.

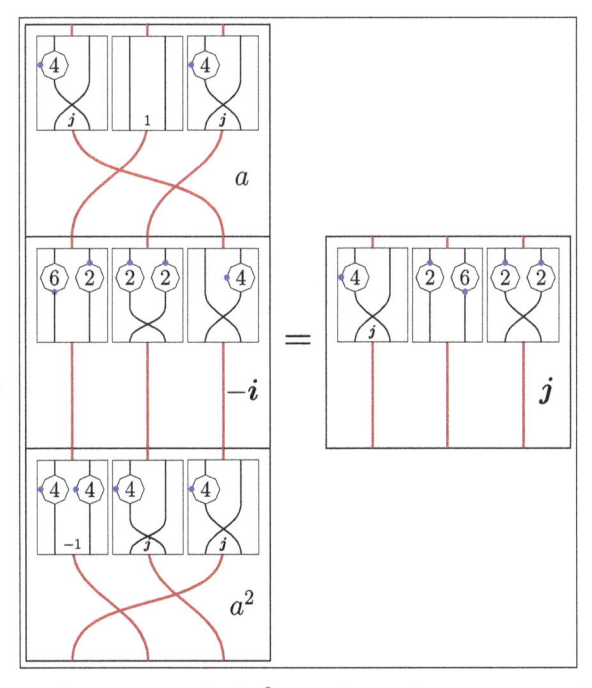

Figure 5.89. Computing $a(-i)a^2 = j$ in the 3-strings-with-Dic$_4$-quipu representation.

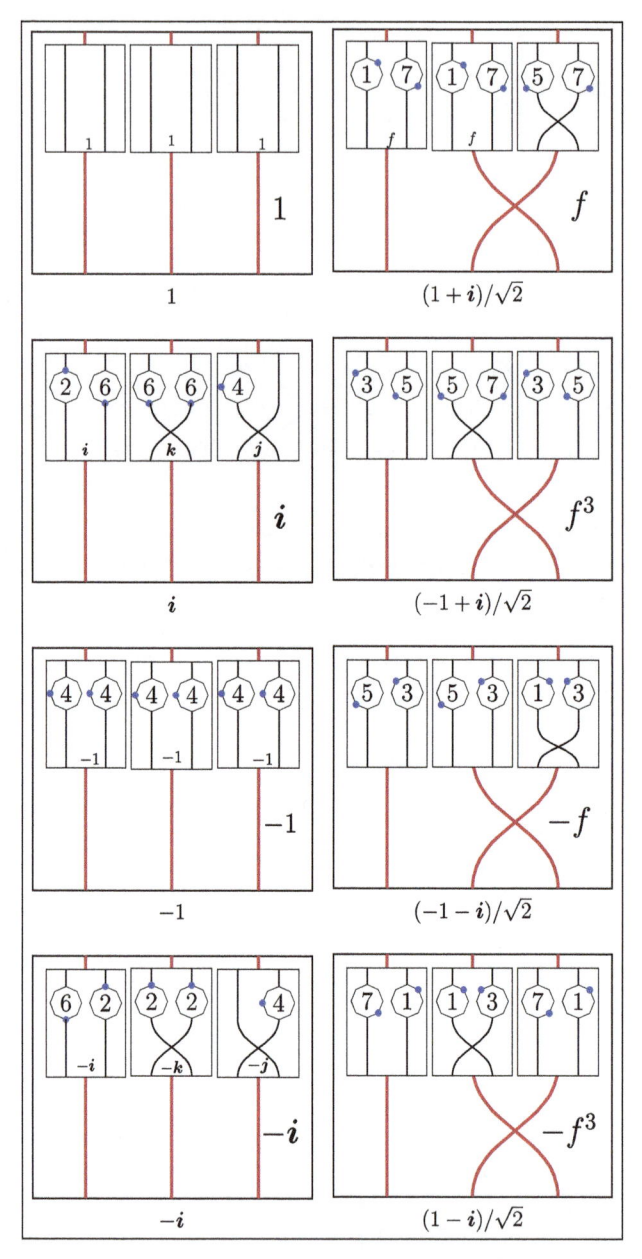

Figure 5.90. 3-strings-with-Dic$_4$-quipu for the subgroup $P \subset \widetilde{\Sigma_4}$.

Extensions of the Permutation Group Σ_4

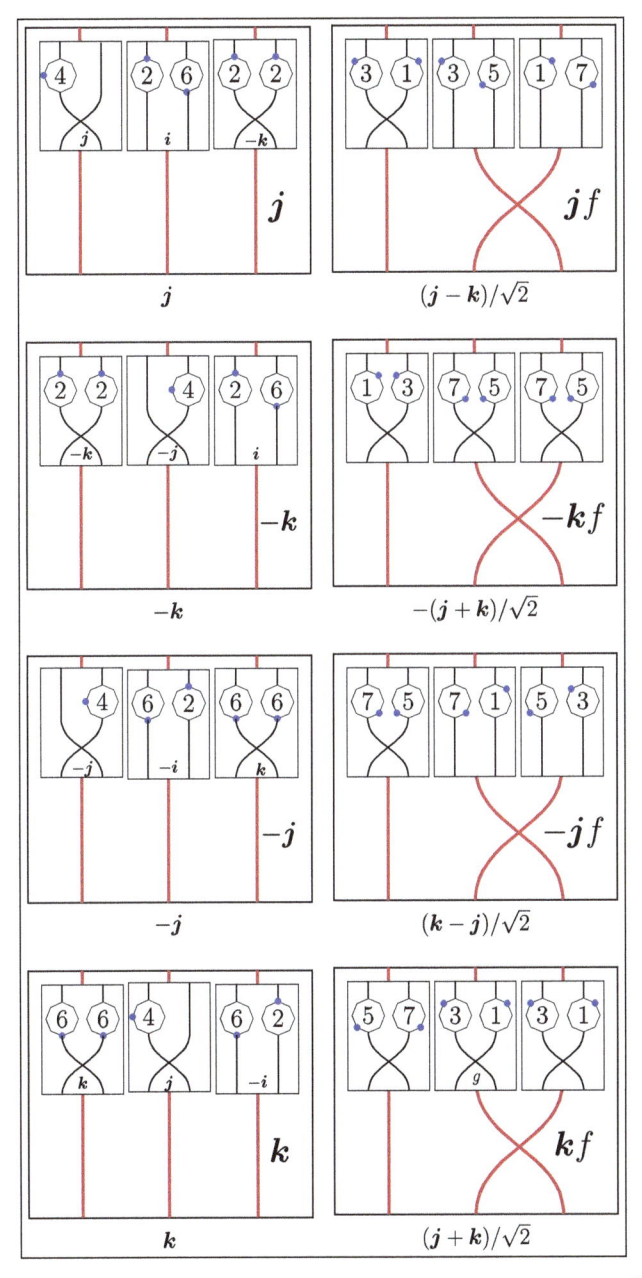

Figure 5.91. 3-strings-with-Dic$_4$-quipu for the coset $jP \subset \widetilde{\Sigma_4}$.

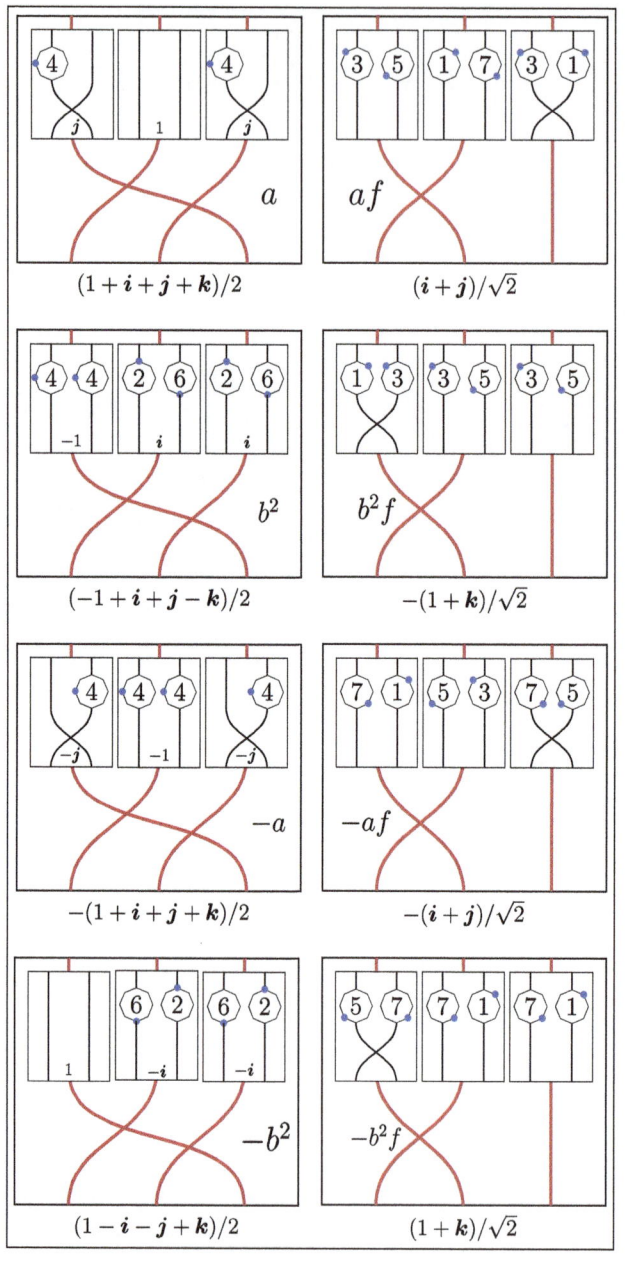

Figure 5.92. 3-strings-with-Dic$_4$-quipu for the coset $aP \subset \widetilde{\Sigma_4}$.

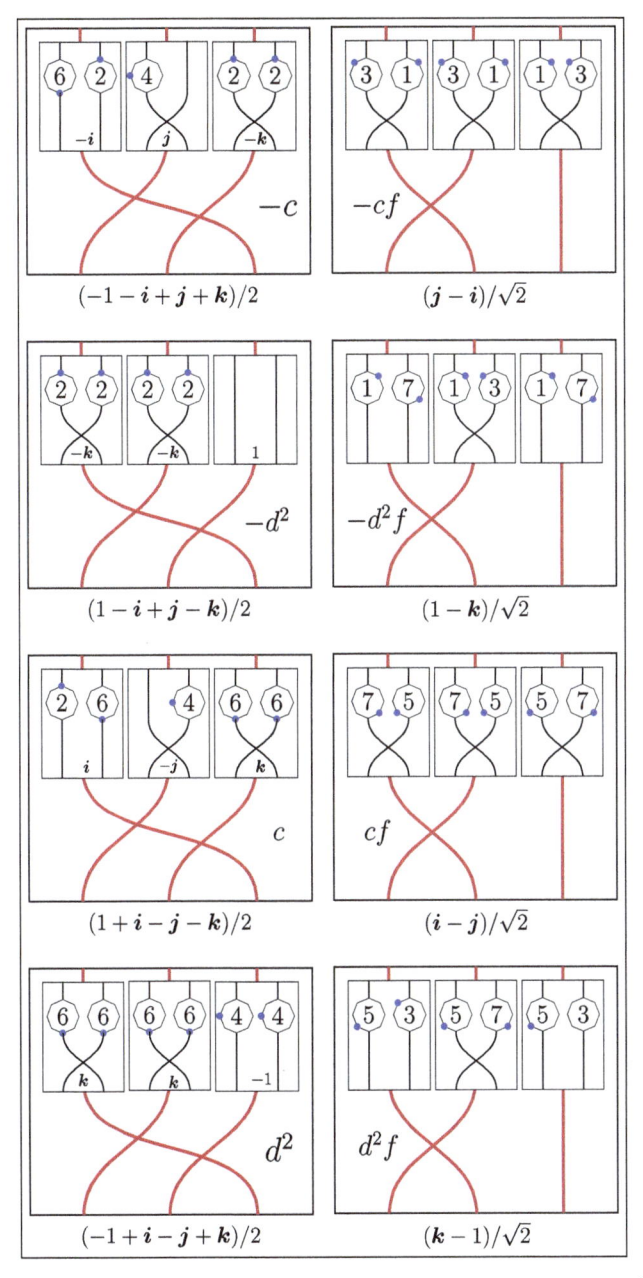

Figure 5.93. 3-strings-with-Dic$_4$-quipu for the coset $ajP \subset \widetilde{\Sigma_4}$.

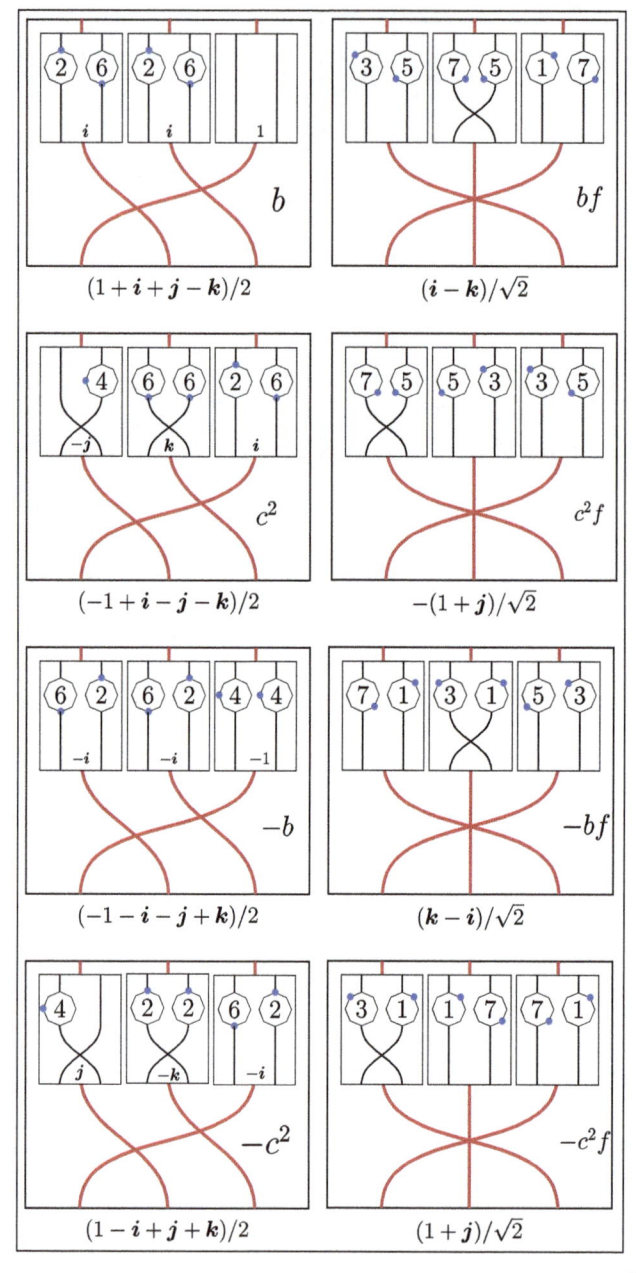

Figure 5.94. 3-strings-with-Dic$_4$-quipu for the coset $bP \subset \widetilde{\Sigma_4}$.

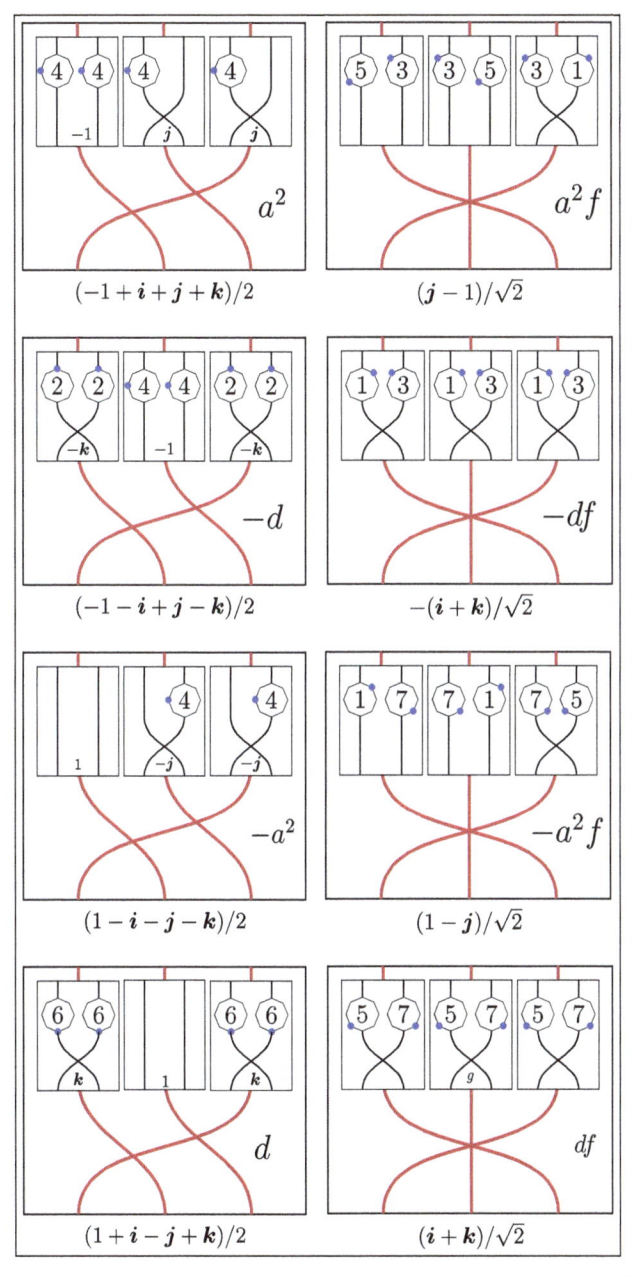

Figure 5.95. 3-strings-with-Dic$_4$-quipu for the coset $bjP \subset \widetilde{\Sigma_4}$.

is the result of using the representations that have been determined thus far and systematically computing the elements in the cosets. The details of the calculations are, perhaps, time-consuming.

For example, once you know the powers $1, f, f^2$, and f^3 as well as the representative of -1, the elements of P are apparent. We have found the representation of j in Fig. 5.89. So, juxtapose the diagram of j above the diagrams that represent the elements of P. The elements of aP can be obtained by vertically juxtaposing the representative of a above the elements of P. Among those elements is $ai = b^2 = -b^{-1}$. A diagram that represents b, for example, can be found since $b^4 = -b$.

Proceed in that fashion. Determine a coset representative (in order: $1, j, a, aj = -c, b = ai, bj = c^2$) and juxtapose each above the representatives of the powers of f.

After juxtaposing the elements, imagine sliding the quipu upon the lower strings up the strings, and let the quipu interact as elements in the dicyclic group. Each quipu involves a pair of strings that are either crossed or uncrossed, and each of these strings has a mod-8-quipu upon it. Simplify the mod-8-quipu and eliminate pairs of crossings between the strings in the dicyclic quipu as needed. At the last stage, the three thick purple strings are simplified as elements in Σ_3.

In this way, we (the authors) hope that you (the reader) have gained enough experience with permutations-with–quipu to verify the diagrams enumerated are correct.

Figures 5.90–5.95 should be compared, respectively, to each of Figs. 5.29–5.34. The elements in the first columns of either are identical. The elements in the second columns share some properties.

5.4.3. Section summary

The binary octahedral group $\widetilde{\Sigma_4}$ shares features of both matrix groups $\mathrm{GL}_2(\mathbb{Z}/3)$ and $\mathrm{SL}_2(\mathbb{Z}/4)$. All three groups are 2-fold extensions of the symmetric group Σ_4 of permutations upon four elements. The projection onto Σ_4 is obtained by ignoring the dicyclic (Dic$_3$) quipu in Figs. 5.78–5.85.

It is possible to compile and project the 3-strings and the 4-strings representations of the elements in $\widetilde{\Sigma_4}$ in a manner analogous to the illustrations in Figs. 5.68–5.73. In such a compilation, the Dic$_4$-quipu

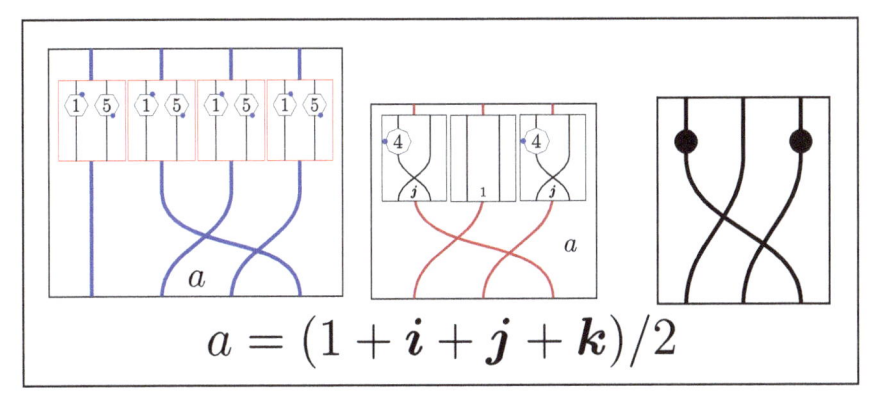

Figure 5.96. The two representatives of a and the projection to 3-strings-with-(mod 2)-quipu.

$$a = (1 + i + j + k)/2$$

in the 3-strings pictures would project to (mod-2)-quipu in a way that the elements in jP project to the non-trivial element of $\mathbb{Z}/2$ and the elements in P project to the identity. For example, Fig. 5.96 indicates the two representations of the element $a = (1+i+j+k)/2$. The element a projects to the 3-cycle (234) in Σ_4, and in the group of signed permutations, it projects to 123[13] since the element j projects to a reflection and j adorns the top of the first and third strings. In full generality, the elements in jP are the dicyclic elements that project to reflections in the dihedral group D_8.

Let us reintroduce and revise the notation in which an element in Σ_3 is adorned by (mod 2)-quipu. The element in Σ_3 is specified by its cycle structure, and the location of the (mod 2)-quipu as a dot (from left to right) is specified in brackets. For example, in Fig. 5.71, the teal-colored 3-strings-with-(mod 2)-quipu reading from top to bottom are 13[13], 13[12], 13[0], and 13[23]. The [0] indicates that none of the strings has a quipu.

Rather than compiling the correspondences in an analogous fashion to Figs. 5.68–5.73, the elements in $\widetilde{\Sigma_4}$, their projection to Σ_4, and the corresponding projections to 3-strings-with-(mod 2)-quipu are presented in Table 5.10. In the third column of the table only, the positive elements are rendered projected to 3-strings-with-(mod 2)-quipu. Thus, $+a$ projects to 123[13], while $-a$ projects to the negative of this representative, 123[2]. In general, the negative of a

3-strings-with-(mod 2)-quipu has quipu on the complement of its corresponding positive counterpart.

Table 5.10. The projections of the elements in the binary octahedral group $\widetilde{\Sigma}_4$.

± 1	1.2.3.4	1.2.3[0]
$\pm i$	12.34	1.2.3[23]
$\pm j$	13.24	1.2.3[13]
$\pm k$	14.23	1.2.3[12]
$\pm(1+i)/\sqrt{2}$	1324	23[3]
$\pm(1-i)/\sqrt{2}$	1423	23[2]
$\pm(j-k)/\sqrt{2}$	34	23[1]
$\pm(j+k)/\sqrt{2}$	12	23[123]
$\pm(1+i+j+k)/2$	234	123[13]
$\pm(-1+i+j-k)/2$	132	123[0]
$\pm(1+i-j-k)/2$	143	123[23]
$\pm(1-i+j-k)/2$	124	123[12]
$\pm(i+j)/\sqrt{2}$	14	12[3]
$\pm(1+k)/\sqrt{2}$	1243	12[1]
$\pm(i-j)/\sqrt{2}$	23	12[123]
$\pm(1-k)/\sqrt{2}$	1342	12[2]
$\pm(1+i+j-k)/2$	123	132[0]
$\pm(1-i+j+k)/2$	134	132[12]
$\pm(-1+i+j+k)/2$	243	132[23]
$\pm(1+i-j+k)/2$	142	132[13]
$\pm(i-k)/\sqrt{2}$	24	13[2]
$\pm(1+j)/\sqrt{2}$	1432	13[1]
$\pm(1-j)/\sqrt{2}$	1234	13[3]
$\pm(i+k)/\sqrt{2}$	13	13[123]

This concludes the current chapter.

Chapter 6

The Binary Tetrahedral Group

The binary tetrahedral group $\widetilde{A_4}$ is isomorphic to the group $\mathrm{SL}_2(\mathbb{Z}/3)$ of (2×2) matrices with entries in $\mathbb{Z}/3$ whose determinant is 1. It is an index 2 subgroup of $\mathrm{GL}_2(\mathbb{Z}/3)$. It is also an index 2 subgroup of the binary octahedral group $\widetilde{\Sigma_4}$. We recall that a subgroup H of a group G is of *index* 2 if there is an element $n \notin H$ such that the larger group is written as the (disjoint) union of two cosets $G = H \cup nH$. In this case, H is normal and G/H is isomorphic to $\mathbb{Z}/2$.

In addition, the binary tetrahedral group is a 2-fold extension of the alternating group A_4 of even permutations on four letters — permutations that can be written as an even number adjacent transpositions $t_\ell = (\ell, \ell+1)$. That is to say there is a seseq

$$1 \longleftarrow A_4 \xleftarrow{q} \widetilde{A_4} \xleftarrow{i} \mathbb{Z}/2 \longleftarrow 1.$$

Consequently, the inclusion relationship $\widetilde{A_4} \subset \widetilde{\Sigma_4}$ is analogous to the inclusion $A_4 \subset \Sigma_4$.

In Chapter 5, $\widetilde{A_4}$ was encountered as a subgroup of both of the larger two groups. But it is worthwhile to study the group in and of itself. In this way, some anomalies between quipu in the 4-strings representations can be rectified. This chapter will be far more brief than its predecessor, and it will also serve as a segue to the sequel.

The chapter begins with a brief discussion of the elements in $\widetilde{A_4}$ as elements in S^3. Section 6.1 provides a summary of some little things that the authors have noticed in our workings with these elements. Section 6.2 discusses the correspondences among elements of S^3, matrices with entries in $\mathbb{Z}/3$, and permutations upon 8-strings, and the section compiles the 4-strings-with-quipu representations in which the quipu are either elements of $\mathbb{Z}/2$ or elements of $\mathbb{Z}/6$, the 3-strings-with-Q_8-quipu, and the matrix representations that have been discussed.

Much of this chapter, then, will also be a review of aspects of both Chapters 4 and 5.

6.1. The binary tetrahedral group as a subgroup of S^3

The binary tetrahedral group, $\widetilde{A_4}$, can be described succinctly via the presentation

$$\widetilde{A_4} = \langle a, b : (ab)^2 = a^3 = b^3 \rangle.$$

That is, every element can be written as a product of a's and b's, and those expressions are subject to the relations that $(ab)^2 = a^3 = b^3$. This presentation is known as the group's *abstract presentation*. Many authors prefer to define groups in such an abstract manner and then find an alternative description, say via a matrix representation, as a subgroup of a more well-known larger group or as a set of symmetries of a geometric object.

A geometric object that corresponds to $\widetilde{A_4}$ is known as the 24-cell. It resides in 4-dimensional space. It has 24 octahedral faces that are of dimension 3, it has 96 triangular faces, 96 edges, and has 24-vertices. The vertices correspond to the set of points in the 3-dimensional sphere that have coordinates

$$\pm 1, \pm i, \pm j, \pm k,$$

and

$$(\pm 1 \pm i \pm j \pm k)/2,$$

where all of the possible 16 signs in the latter expression occur. In this way, we are very quickly thinking of $\widetilde{A_4}$ as a subgroup of the group S^3. The chapter closes with Fig. 6.6 that is a very rough depiction of the set of edges and vertices of the 24-cell.

The Binary Tetrahedral Group

It is not too difficult to check that the elements $a = (1 + i + j + k)/2$ and $b = (1 + i + j - k)/2$ satisfy the relations $(ab)^2 = a^3 = b^3$. Indeed, Table 4.1 can be used to compute that $a^2 = (-1+i+j+k)/2$. Table 6.3 is used to compute b^2. Moreover, in Section 5.2.1, we encountered the relationships that $ab = j$ and $ba = i$. The tabular calculations that appeared in that section are summarized in Tables 6.1–6.5 wherein the factors of $1/2$ are neglected. See also Fig. 5.12. Since $ab = j$, we have that $(ab)^2 = a^3 = b^3 = -1$.

Note further that $ba^2b = k$. In a few moments, the elements of $\widetilde{A_4}$ will be partitioned as cosets A, iA, jA, and kA where $A = (1, a, a^2, -1, -a, -a^2)$ is the subgroup generated by a. In this way, all the elements of the binary tetrahedral group can be written in terms of a and b.

Table 6.1. $ab = j$.

·	1	i	j	$-k$
1	1	i	j	$-k$
i	i	-1	k	j
j	j	$-k$	-1	$-i$
k	k	j	$-i$	1

Table 6.2. $ba = i$.

·	1	i	j	k
1	1	i	j	k
i	i	-1	k	$-j$
j	j	$-k$	-1	i
$-k$	$-k$	$-j$	i	1

For your convenience, we also tabulate computations of b^2, c^2, and d^2 again by way of ignoring the factors of $1/2$.

Table 6.3. $b^2 = (-1 + i + j - k)/2$.

·	1	i	j	$-k$
1	1	i	j	$-k$
i	i	-1	k	j
j	j	$-k$	-1	$-i$
$-k$	$-k$	$-j$	i	-1

Table 6.4. $c^2 = (-1 + i - j - k)/2$.

·	1	i	$-j$	$-k$
1	1	i	$-j$	$-k$
i	i	-1	$-k$	j
$-j$	$-j$	k	-1	i
$-k$	$-k$	$-j$	$-i$	-1

Table 6.5. $d^2 = (-1 + i - j + k)/2$.

·	1	i	$-j$	k
1	1	i	$-j$	k
i	i	-1	$-k$	$-j$
$-j$	$-j$	k	-1	$-i$
k	k	j	i	-1

In Section 4.2, we demonstrated that

$$(w + x\boldsymbol{i} + y\boldsymbol{j} + z\boldsymbol{k})^{-1} = w - x\boldsymbol{i} - y\boldsymbol{j} - z\boldsymbol{k}.$$

So, in the case of $a = (1 + \boldsymbol{i} + \boldsymbol{j} + \boldsymbol{k})/2$, we obtain that $a^{-1} = (1 - \boldsymbol{i} - \boldsymbol{j} - \boldsymbol{k})/2$. From Table 4.1, $a^2 = -a^{-1}$. So, $a^3 = -1$, and consequently, $a^6 = 1$. This yields that $a^5 = -a^2 = a^{-1}$.

The choices

$$a = (1 + \boldsymbol{i} + \boldsymbol{j} + \boldsymbol{k})/2 \qquad d = (1 + \boldsymbol{i} - \boldsymbol{j} + \boldsymbol{k})/2$$
$$b = (1 + \boldsymbol{i} + \boldsymbol{j} - \boldsymbol{k})/2 \qquad c = (1 + \boldsymbol{i} - \boldsymbol{j} - \boldsymbol{k})/2$$

were made so that these four elements are the vertices of a square. Similar arguments give that each of b, c, d are also of order 6. Thus,

$$b^{-1} = b^5 = -b^2 = (1 - \boldsymbol{i} - \boldsymbol{j} + \boldsymbol{k})/2,$$

and so,

$$b^2 = (-1 + \boldsymbol{i} + \boldsymbol{j} - \boldsymbol{k})/2.$$

Similarly,

$$c^{-1} = c^5 = -c^2 = (1 - \boldsymbol{i} + \boldsymbol{j} + \boldsymbol{k})/2,$$

and so,

$$c^2 = (-1 + \boldsymbol{i} - \boldsymbol{j} - \boldsymbol{k})/2.$$

Finally,

$$d^{-1} = d^5 = -d^2 = (1 - \boldsymbol{i} + \boldsymbol{j} - \boldsymbol{k})/2,$$

so that

$$d^2 = (-1 + \boldsymbol{i} - \boldsymbol{j} + \boldsymbol{k})/2.$$

The negatives of a, b, and c are easily written.

It is a good time to review Fig. 4.12 in which the powers of each of the elements a–d are color-coded. In the diagram, the set of powers of each element form a line through ∞ which corresponds to 1, and these lines also intersect at -1. In another projection (Fig. 6.6), each set of powers forms a hexagon.

At times, it may be difficult to recall which sequence of ±s in the expressions $(\pm 1 \pm i \pm j \pm k)/2$ corresponds to the power of which element. But the key facts to put together are (1) a, b, c, and d are the vertices of a square, (2) the fifth power of any of these is the inverse, and (3) the inverse is given by the "conjugate" formula:

$$(w + xi + yj + zk)^{-1} = w - xi - yj - zk,$$

and (4) the second power is the negative of the fifth power. Finally, Fig. 4.12 explicitly locates all the elements in the binary tetrahedral group with the exceptions of $-i$, $-j$, and $-k$ and 1. The former three are determined by tracing the undrawn coordinate axes, while 1 corresponds to the point at ∞ in the stereographic projection from S^3 to $\mathbb{R}^3$.

Often, one can use four fingers to help one compute the powers of any of the elements a–d. Think of your fingers as binary digits. After all, they are digits, and they can either be up or down. By convention, a finger pointing up indicates a negative coefficient. Let's say that 1 is indicated by your index finger, i is indicated by the middle finger, j is indicated by the annular finger, and k is indicated by the pinky. Remember that c corresponds to the last two fingers being up. Then $-c$ might be confused with a gesture that is deemed to be obscene in many cultures. Be careful! Nonetheless, c^2 is indicated as the negative in which the three other fingers point upward. Come to think of it, you may want to compute these powers in privacy.

The chapter and its predecessor present not only traditional calculations but also novel methods of computing products in the group $\widetilde{A_4}$ using strings-with-quipu.

6.2. Correspondences among representations

Figure 5.10 indicates a 4-dimensional hyper solid that is of the form $\Delta \times \Delta$. It represents the vector space $\mathbb{Z}/3 \times \mathbb{Z}/3$ upon which the group $SL_2(\mathbb{Z}/3)$ acts. The vertices are labeled so that antipodal vertices have labels that are congruent modulo 4. The following array was also given in Section 5.2. It indicates the labels upon the column vectors in $\mathbb{Z}/3 \times \mathbb{Z}/3$. In the same section, matrices that correspond to the elements a and b in the binary tetrahedral group act upon the vectors

in $\mathbb{Z}/3 \times \mathbb{Z}/3$ to give a permutation action on $\{0,1,2,3,4,5,6,7\}$:
$$\begin{bmatrix} 0 & 1 & 2 & 3 & 4 & 5 & 6 & 7 \\ 1 & 0 & 1 & -1 & -1 & 0 & -1 & 1 \\ -1 & -1 & 1 & 0 & 1 & 1 & -1 & 0 \end{bmatrix}$$

In particular,
$$a \leftrightarrow \begin{bmatrix} 0 & 1 \\ -1 & 1 \end{bmatrix} \leftrightarrow (04)(163527)$$

and
$$b \leftrightarrow \begin{bmatrix} -1 & -1 \\ 0 & -1 \end{bmatrix} \leftrightarrow (37)(056412).$$

Figure 5.11 contains illustrations and correspondences among the matrix representations, the permutation representation, the elements in S^3, and the 4-strings-with-(mod 2)-quipu depiction of a and b. It also includes the element g which is not an element of the binary tetrahedral group. In Fig. 5.12, the products $ab = j$ and $ba = i$ are illustrated as products of 4-strings-with-(mod 2)-quipu. Finally, in Figs. 5.14–5.19, the red-colored elements that appear at the cardinal directions (NESW) along the octagons are all of the elements in the binary tetrahedral group as presented in 4-strings-with-(mod 2)-quipu.

In this section, we will revisit those illustrations as well as the 3-strings-with-Q_8-quipu. Moreover, we will analyze $\widetilde{A_4}$ from the point of view of the cyclic subgroup that consists of powers of A. However, only half the elements will be illustrated. The identity always consists of straight strings with identity quipu upon it, and the negatives of the elements are obtained by juxtaposing the strings-with-quipu illustration of -1.

6.2.1. Cosets of $A = \langle a : a^6 = 1 \rangle$

From Section 5.4.1, we recall that $ia = d^2$, $ia^2 = -b$, $ja = b^2$, $ja^2 = c^2$, $ka = -c$, and $ka^2 = -d$. Also, in that section, these cosets
$$A = (1, a, a^2, -1, -a, -a^2); \quad iA = (i, d^2, -b, -i, -d^2, b);$$
$$jA = (j, b^2, c^2, -j, -b^2, -c^2); \quad kA = (k, -c, -d, -k, c, d)$$

exhaust $\widetilde{A_4}$. The consequences of several calculations in that section are as follows:

$$aA = [A, 1]; \quad aiA = [jA, 1];$$

$$ajA = [kA, 1]; \quad akA = [iA, 1].$$

Since the binary tetrahedral group is generated by a and b, a similar set of calculations to compute the action of b upon these cosets will be helpful:

$$bi = (1 + i + j - k)i/2 = (-1 + i - j - k)/2 = c^2;$$

$$bj = (1 + i + j - k)j/2 = (-1 + i + j + k)/2 = a^2;$$

$$bk = (1 + i + j - k)k/2 = (1 + i - j + k)/2 = d.$$

So, we get the actions of b upon the cosets as follows:

$$bA = [iA, 5]; \quad biA = [jA, 2];$$

$$bjA = [A, 2]; \quad bkA = [kA, 5].$$

As a consequence of these calculations, the 4-strings-with-(mod 6)-quipu representations of the generators a and b are depicted in Fig. 6.1. Earlier, we wrote of "anomalies" between the 4-strings-with-(mod 2)-quipu representations and the 4-strings-with-Dic$_3$-quipu representations of the elements in $\widetilde{A_4}$. But by comparing the representations of a and b that are given in Fig. 6.1 with those given in Figs. 5.74 and 5.76, you will note that the (mod 6)-quipu are the first components of the dicyclic quipu. Furthermore, the parity of the (mod 6)-quipu agrees with the existence or non-existence of dots in the 4-strings-with-(mod 2)-quipu. So, probably, "anomaly" was not the correct word. Things appear congruent when examined with the correct lenses.

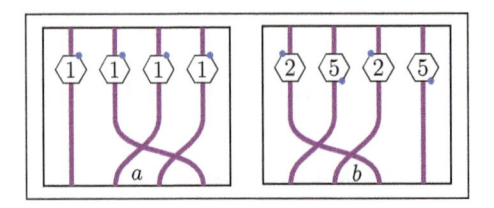

Figure 6.1. 4-strings-with-(mod 6)-quipu representations of a and b.

Our primary goal in this section is to compile the various descriptions of the elements in $\widetilde{A_4}$, but before developing that catalog, it seems like a good idea to illustrate the computations of i, j, and k using 4-strings-with-(mod 6)-quipu. These computations are given in Fig. 6.2.

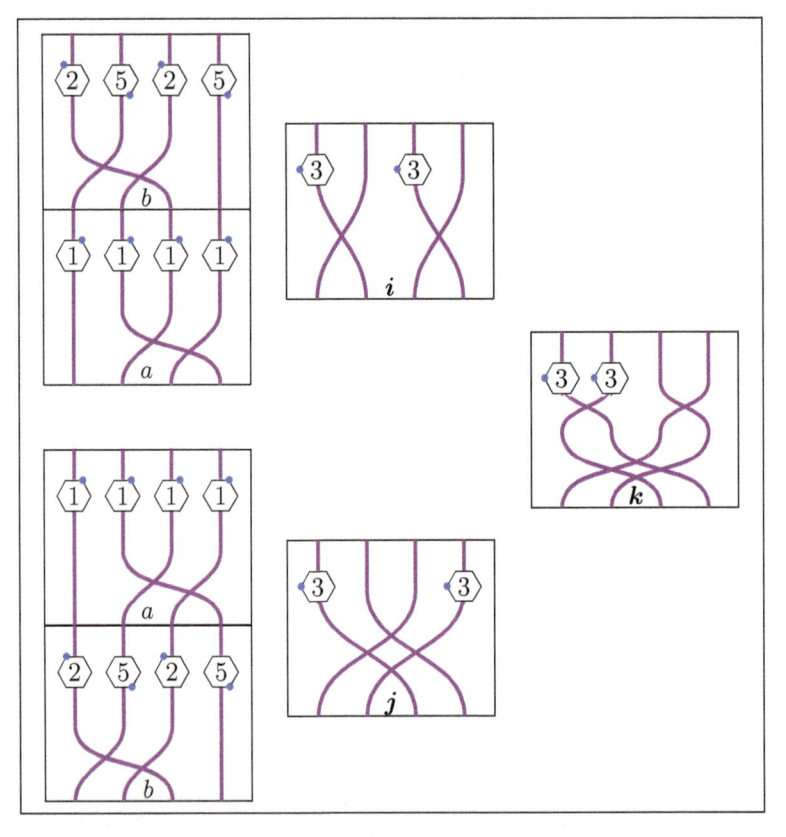

Figure 6.2. 4-strings-with-(mod 6)-quipu calculations of i, j, and k.

6.2.2. The subgroup Q_8 of $\widetilde{A_4}$

The structure of the 3-strings-with-Q_8-quipu representations of the elements in the binary tetrahedral group $\widetilde{A_4}$ is in form similar to the representation in the larger binary octahedral group. The difference is that the smaller quipu are elements of $\mathbb{Z}/4$ rather than of $\mathbb{Z}/8$; also, the indices of rotations are halved. Most of the calculations that are needed for this section can be found in Section 5.4.2.

More specifically, write Q_8 as a union

$$Q_8 = (1, i, -1, -i) \cup (j, -k, -j, k).$$

Let $\mathcal{I} = (1, i, -1, -i)$. Then $j\mathcal{I} = (j, -k, -j, k)$, and $\widetilde{A_4}$ can be written as the disjoint union of six cosets

$$\widetilde{A_4} = \mathcal{I} \cup j\mathcal{I} \cup a\mathcal{I} \cup aj\mathcal{I} \cup b\mathcal{I} \cup bj\mathcal{I}.$$

The cosets are combined into three groups of 2, and Q_8 is also described as the dicyclic group Dic_2.

Note that the cosets here consist of every other element of the cosets of $-P$ that were presented in Section 5.4.2. Thus,

$$a\mathcal{I} = (a, b^2, -a, -b^2), \quad aj\mathcal{I} = (-c, -d^2, c, d^2),$$
$$b\mathcal{I} = (b, c^2, -b, -c^2), \quad bj\mathcal{I} = (a^2, -d, -a^2, d).$$

Since $\mathcal{I}$ is a cyclic group of order 4, the actions of a and b can be computed by determining the products $a \cdot 1 = a$, $a \cdot j = -c$, $a \cdot a = a^2$, $a \cdot c = -b$, $a \cdot b = j$, $a \cdot a^2 = -1$, $b \cdot 1 = b$, $b \cdot j = a^2$, $b \cdot a = i$, $b \cdot (-c) = k$, $b \cdot b = b^2$, and $b \cdot a^2 = d^2$. Some products are obvious; most of the others are found in Section 5.4.2. The remaining products are exercises.

Then

$$a\mathcal{I} = [a\mathcal{I}, 0]; \quad aj\mathcal{I} = [aj\mathcal{I}, 0];$$
$$a(a\mathcal{I}) = [bj\mathcal{I}, 0]; \quad a(aj\mathcal{I}) = [b\mathcal{I}, 2];$$
$$a(b\mathcal{I}) = [j\mathcal{I}, 0]; \quad a(bj\mathcal{I}) = [\mathcal{I}, 2].$$

Furthermore,

$$b\mathcal{I} = [b\mathcal{I}, 0]; \quad bj\mathcal{I} = [bj\mathcal{I}, 0];$$
$$b(a\mathcal{I}) = [\mathcal{I}, 1]; \quad b(aj\mathcal{I}) = [j\mathcal{I}, 3];$$
$$b(b\mathcal{I}) = [a\mathcal{I}, 1]; \quad b(bj\mathcal{I}) = [aj\mathcal{I}, 3].$$

Compare this to the representation of b that was given in Fig. 5.90.

In Figs. 6.3–6.5, the various descriptions of a, b, c, d, a^2, b^2, c^2, d^2, -1, i, j, and k are compiled. In the images on the left of the illustrations, the 4-strings-with-(mod 6)-quipu representation is demonstrated. The parity projection of $\mathbb{Z}/6$ to $\mathbb{Z}/2$ (for which $(0,2,6) \mapsto +1$ and $(1,3,5) \mapsto -1$ in multiplicative notation for the latter group) and the 4-strings-with-(mod 2)-quipu are illustrated in the second column. Therein the elements are also expressed as elements in S^3. The next column gives the matrix representation in $\mathrm{SL}_2(\mathbb{Z}/3)$. The numeric expression, e.g., $b \leftrightarrow 123[24]$, is a rewriting of the 4-strings-with-(mod 2)-quipu. In this case, [24] indicates that the quipu appears on the top of the second and fourth strings. The permutation representation as the non-zero elements in $\mathbb{Z}/3 \times \mathbb{Z}/3$ are also indicated in this column. The fourth column illustrates the 3-strings-with-Q_8-quipu. The fifth column indicates the corresponding matrix representation as a (3×3) "permutation" matrix in which the non-zero entries are elements of Q_8, i.e., as an element of $(Q_8)^3 \rtimes \mathbb{Z}/3$.

We expect that you, the reader, will find a favorite method to do routine calculations in the binary tetrahedral group $\widetilde{A_4}$. While the entries in the first columns of the table include the actions upon the cosets of the cyclic group $A = (1, a, a^2, -1, -a, -a^2)$, it is not as convenient as the 4-strings-with-(mod 2)-quipu representation. If you remain skeptical of the quipu methodology, the (3×3) permutation-like matrices with entries in Q_8 may suit your fancy. It is not up to us to choose your favorite method. We find one representation or another more convenient as the context varies.

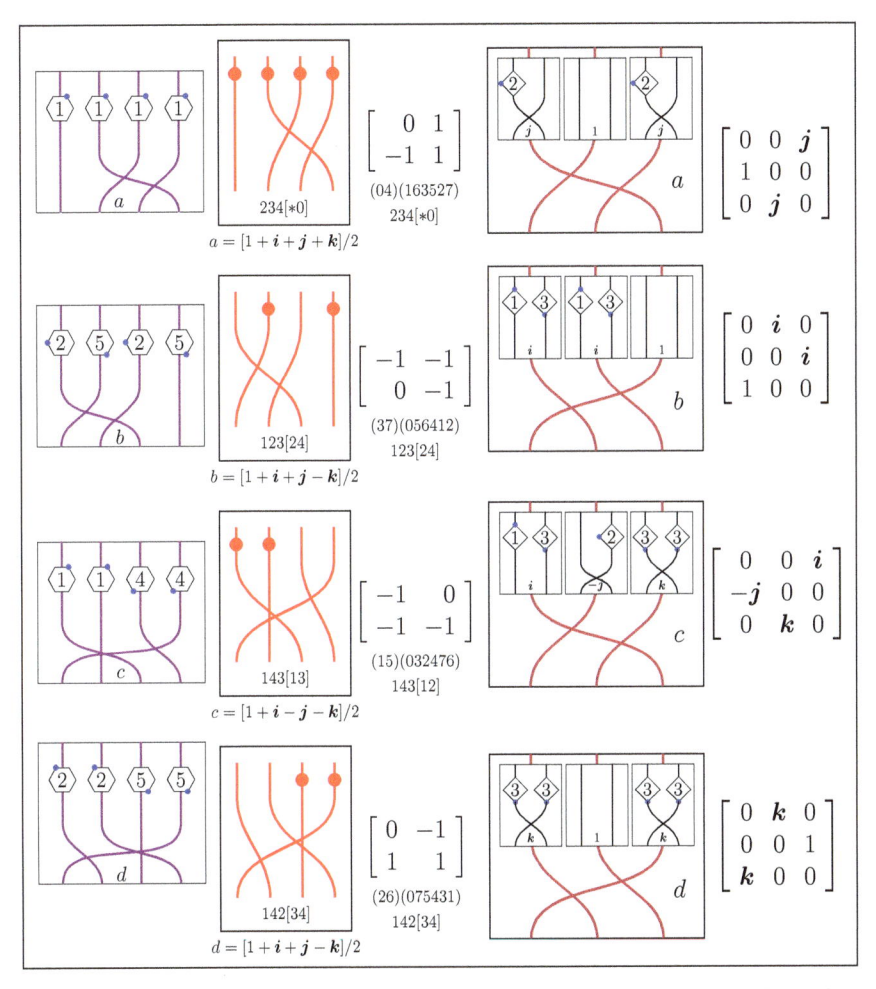

Figure 6.3. Different matrix, permutation, and quipu expressions for $a, b, c, d \in \widetilde{A}_4$.

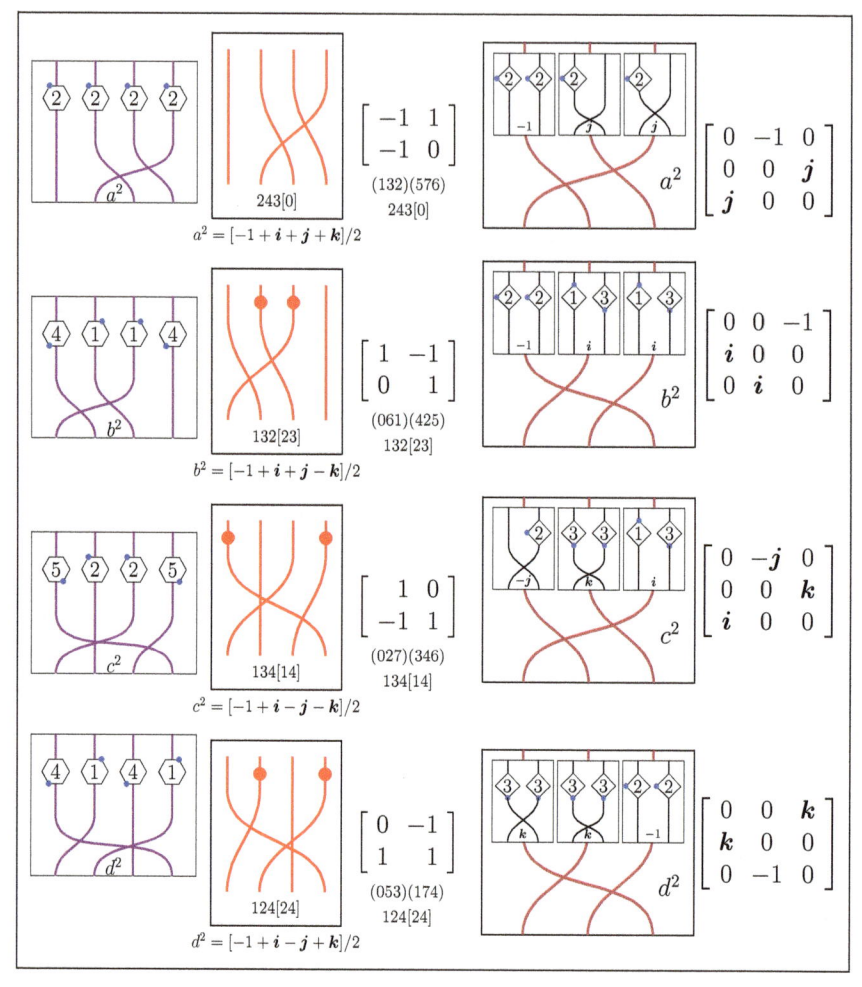

Figure 6.4. Different matrix, permutation, and quipu expressions for $a^2, b^2, c^2, d^2 \in \widetilde{A_4}$.

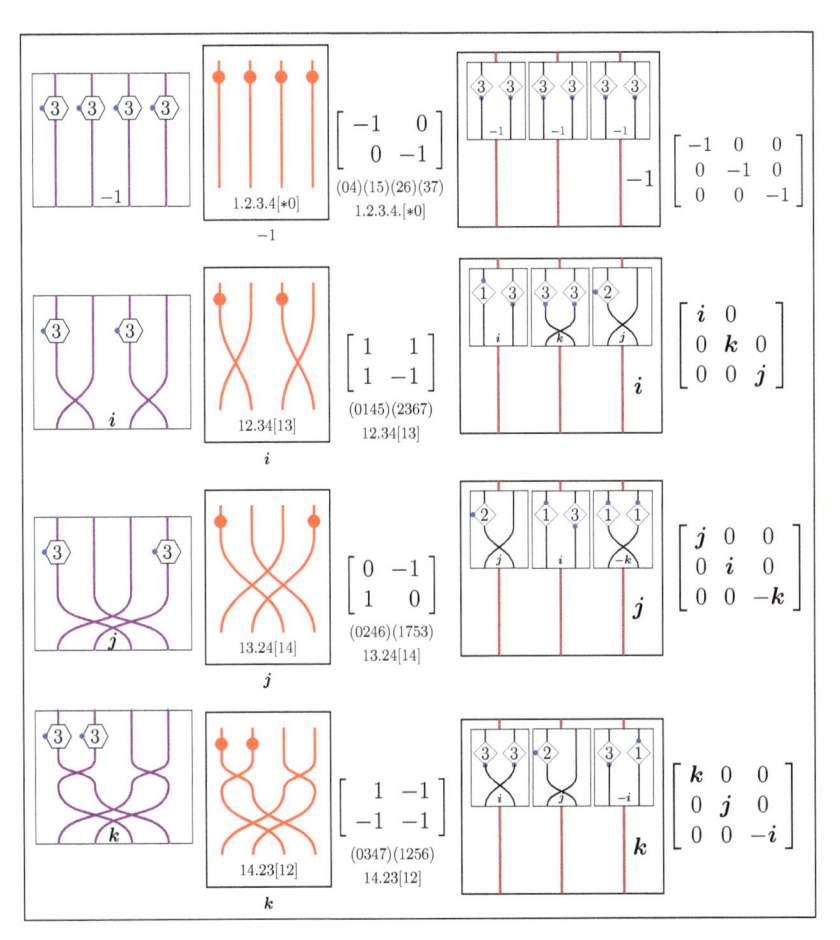

Figure 6.5. Different matrix, permutation, and quipu expressions for $-1, i, j, k \in \widetilde{A_4}$.

Please also note that the projections to the alternating group A_4 is obtained by ignoring the (mod 2)-quipu upon the 4-strings representations, e.g., a projects to the 3-cycle (234).

The 24-cell is a 4-dimensional hyper solid whose vertices are the elements of the binary tetrahedral group. There are many excellent illustrations of it that are freely available on the web. In Fig. 6.6, we have attempted to draw its vertices and the edges by means of a linear projection from $\mathbb{R}^4$ to $\mathbb{R}^2$.

The square that is formed by a, b, c, and d appears on the upper right of the figure as a parallelogram, and the powers of these elements form a hyper cube. We invite you to find more features within the figure.

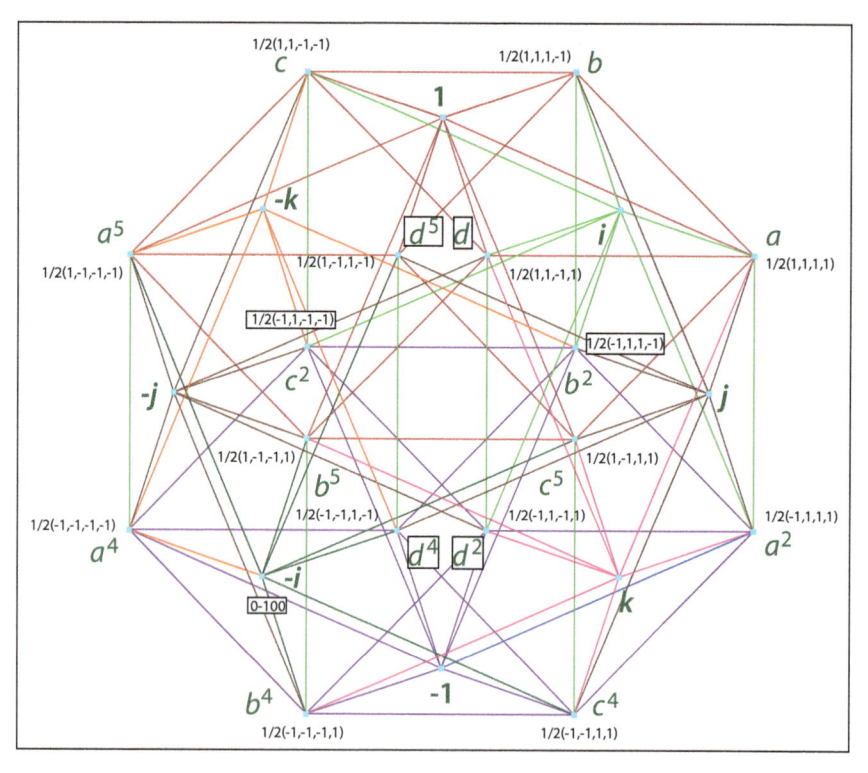

Figure 6.6. A drawing of the 1-skeleton of the 24-cell.

Chapter 7

The Binary Icosahedral Group

This chapter gives two representations of the binary icosahedral group. The first involves quipu that are elements of the cyclic group of order 10. In this representation, there are 12 strings. The cyclic subgroup is not too difficult to understand, but the 12-strings diagrams get fairly messy.

Meanwhile, the binary tetrahedral group $\widetilde{A_4}$ is an index 5 subgroup of the binary icosahedral group $\widetilde{A_5}$. So, there is a 5-strings-with-$\widetilde{A_4}$-quipu depiction of $\widetilde{A_5}$. We have choices about which quipu description of $\widetilde{A_4}$ that we use.

From an accounting point of view, the 4-strings-with-(mod 6)-quipu representation of $\widetilde{A_4}$ keeps track of all of the elements of the larger group $\widetilde{A_5}$. On the other hand, the 4-strings-with-(mod 2)-quipu representation involves less book-keeping at the final stages. So, the 5-strings that correspond to the cosets of the binary tetrahedral subgroup will be adorned with 4-strings-with-(mod 2)-quipu representatives of $\widetilde{A_4}$. Ultimately, this is a 20-strings picture, but it is a cleaner representation than the 12-strings picture.

A dodecahedron, as illustrated in Fig. 7.1, has 20 vertices, 12 pentagonal faces, and 30 edges. Each vertex is incident to three pentagons and three edges. The group of symmetries is isomorphic to the alternating group of permutations on five symbols, A_5. This group can be presented as

$$A_5 = \langle \alpha, \tau : (\alpha\tau)^2 = \alpha^3 = \tau^5 = 1 \rangle.$$

Here, α represents a rotation through an angle of $2\pi/3$ around an axis that connects antipodally opposite vertices. Such a transformation acts as a cyclic permutation among three pentagons that are incident to either end of the rotational axis, and τ represents a rotation of $\frac{2\pi}{5}$ about an axis that pierces antipodal pentagonal faces. To be sure, the vertices for the "α"-axis are also antipodal vertices of the pentagonal faces that define the "τ"-axis.

It is, perhaps, a little difficult to see, in our description of the actions of α and τ, that $(\alpha\tau)^2 = 1$. Moreover, it takes some additional work to demonstrate that the presentation above is indeed A_5. However, one can show that the 3-cycle (253) and the 5-cycle (12345) generate A_5, and $[(253)(12345)]^2 = 1$. So, let $\alpha \mapsto (253)$ and $\tau \mapsto (12345)$ to achieve the isomorphism between the group as it has been presented and the alternating group A_5.

Imagine the two rotations α and τ by looking at Fig. 7.1. The regular pentagon that is entirely visible has one vertex at the planar point $(\cos(2\pi/5), \sin(2\pi/5))$ (in the standard coordinate system in which the center of the regular pentagon is at $(0,0)$). Imagine α to be an anti-clockwise rotation in the axis defined by that vertex and its antipode on the dodecahedron. Imagine τ to be an anti-clockwise rotation of this main pentagon. It takes a few minutes of concentrated thought to imagine the kinematics and to convince yourself that $(\alpha\tau)^2 = 1$. You might imagine image of the edge from the point $(1, 0)$ to $(\cos(2\pi/5), \sin(2\pi/5))$ under the successive transformations.

A dodecahedron is the platonic dual to the icosahedron. These two 3-dimensional solids are analogous to the pair (cube, octahedron). In particular, the icosahedron has the same symmetry group, A_5, as the dodecahedron does.

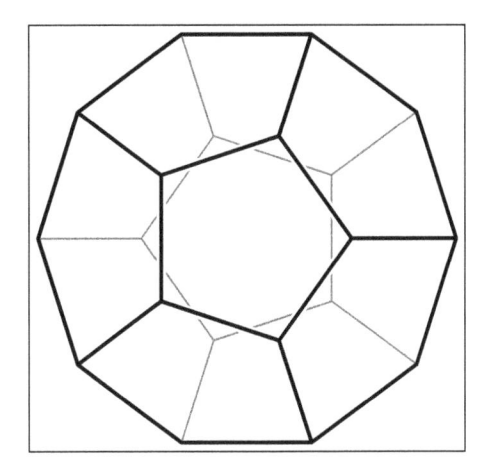

Figure 7.1. A dodecahedron.

The binary icosahedral group, $\widetilde{A_5}$, is a 2-fold extension of the alternating group on 5-elements, A_5, which, in turn is the group of symmetries of the icosahedron or equivalently of the dodecahedron. That is to say, there is a seseq

$$1 \leftarrow A_5 \xleftarrow{q} \widetilde{A_5} \xleftarrow{i} \mathbb{Z}/2 \leftarrow 1,$$

in which the kernel $\mathbb{Z}/2$ is written multiplicatively as ± 1. The extension is given by the presentation:

$$\widetilde{A_5} = \langle a, t : (at)^2 = a^3 = t^5 \rangle.$$

The elements $(at)^2$, t^5, and a^3 are not equal to 1, but, when $\widetilde{A_5}$ is considered as a subgroup of S^3, they are equal to -1, and the quotient map $A_5 \xleftarrow{q} \widetilde{A_5}$ takes $\pm a$ to α and $\pm t$ to τ.

The generators correspond to the elements $a = (1 + i + j + k)/2$ and $t = (\phi + (\phi - 1)i + j)/2 \in S^3$, where $\phi = \frac{1+\sqrt{5}}{2}$ denotes the golden ratio.

In the following, we'll demonstrate that $t^{10} = 0$. The group $\widetilde{A_5}$ has 120 elements and the subgroup $T = \langle t \rangle$ is cyclic of order 10. So, there is a 12-strings-with-($\mathbb{Z}/10$)-quipu description of the elements of the group. It would be extreme, even for this treatise, to list all (or half) of the elements of $\widetilde{A_5}$ in this graphical form. So, instead, the generators, a and t, will be illustrated as 12-strings pictures, and the relations $(at)^2 = t^5 = a^3 = -1$ will be demonstrated.

An alternative 5-strings-with-$\widetilde{A_4}$-quipu will be developed. The elements of the binary tetrahedral group, $\widetilde{A_4}$, will be depicted as 4-strings-with-($\mathbb{Z}/2$)-quipu, but to determine this representation, we compute as if the quipu are elements of $\mathbb{Z}/6$.

Section 7.1 describes some of the arithmetic of powers of t. It defines the cosets of the cyclic subgroup $T = \langle t \rangle$, and it describes the actions of a and t upon these (cyclically ordered) cosets. As a result, 12-strings-with-($\mathbb{Z}/10$)-quipu depictions of the generators a and t are presented in Fig. 7.2. The relations that describe the group are presented in subsequent figures. Section 7.2 describes the cosets of the cyclic subgroup $A = \mathbb{Z}/6$ in the binary tetrahedral group. This process is used to give a 5-strings-with-$\widetilde{A_4}$-quipu description of $\widetilde{A_5}$. The generators of the bigger group are depicted, and the relations verified in these depictions.

7.1. Powers of t

Let us start examining the powers of $t = (\phi + (\phi - 1)\boldsymbol{i} + \boldsymbol{j})/2$. We remind the reader that $1/\phi = \phi - 1$; also, $\phi^2 = \phi + 1$ and $(1/\phi)^2 = 2 - \phi$. These relationships among quadratic and linear functions of the golden ratio will be needed often in the computations that give the quipu representations. To compute t^2, we present Table 7.1.

Table 7.1. $t^2 = ((\phi - 1) + \boldsymbol{i} + \phi\boldsymbol{j})/2$.

$r \cdot c$	ϕ	$\phi^{-1}\boldsymbol{i}$	$\boldsymbol{j}$
ϕ	$\phi + 1$	$\boldsymbol{i}$	$\phi\boldsymbol{j}$
$\phi^{-1}\boldsymbol{i}$	$\boldsymbol{i}$	$\phi - 2$	$\phi^{-1}\boldsymbol{k}$
$\boldsymbol{j}$	$\phi\boldsymbol{j}$	$-\phi^{-1}\boldsymbol{k}$	-1

Since $a^3 = -1 = t^5$, the order of the generator t divides 10. Since $t^2 = ((\phi - 1) + i + \phi j)/2$, the order of t is not 2; so, it must be 10. Once we know t and t^2, the other powers follow since the reciprocals in S^3 are computed as $(w + xi + yj + zk)^{-1} = (w - xi - yj - zk)$ and since $t^5 = -1$.

We have $t^9 = t^{-1}$. Thus, $t^4 = -t^{-1}$ and $t^6 = t^5 \cdot t = -t$. Meanwhile, $t^8 = t^{-2}$. Thus, $t^3 = -t^{-2}$ and $t^7 = t^5 \cdot t^2 = -t^2$. These facts are summarized in Table 7.2.

Table 7.2. Powers of $t = [\phi + \phi^{-1}i + j]/2$.

1	t^0	1
t	t	$[\phi + \phi^{-1}i + j]/2$
t^2	t^2	$[\phi^{-1} + i + \phi j]/2$
t^3	$-t^{-2}$	$[-\phi^{-1} + i + \phi j]/2$
t^4	$-t^{-1}$	$[-\phi + \phi^{-1}i + j]/2$
t^5	-1	-1
t^6	$-t$	$[-\phi - \phi^{-1}i - j]/2$
t^7	$-t^2$	$[-\phi^{-1} - i - \phi j]/2$
t^8	t^{-2}	$[\phi^{-1} - i - \phi j]/2$
t^9	t^{-1}	$[\phi - \phi^{-1}i - j]/2$

In general, the elements of the binary icosahedral group $\widetilde{A_5}$ come in three "flavors." These are more easily expressed by writing the expressions of the form $w + xi + yj + zk$ in a coordinate form:

$$[w|x|y|z] := w + xi + yj + zk.$$

The eight elements in $Q_8 \subset \widetilde{A_5}$ are expressed as

$$[\pm 1|0|0|0], \quad [0|\pm 1|0|0],$$
$$[0|0|\pm 1|0], \quad [0|0|0|\pm 1].$$

The remaining elements in $\widetilde{A_4}$ are the 16 elements of the form

$$[\pm 1|\pm 1|\pm 1|\pm 1]/2.$$

There are eight choices for the configuration of signs in the expression

$$\left[\pm\phi\,|\pm\phi^{-1}|\pm 1\,|0\right]/2,$$

and there are 12 even permutations (in A_4) that act upon the positions of the coordinates. Thus, there are 96 elements of this last form. These are the $8 + 12 + 96 = 120$ elements in the binary icosahedral group.

7.1.1. Cosets of T

There are many computations that we used in order to compile and compute the cosets of the cyclic subgroup T. We are hopeful that at this point, so well into the manuscript, you can replicate many of these calculations. To aid you in your pursuit, we compile products of the form ht^ℓ for $h = 1, i, j, k$ and $\ell = 0, \ldots, 4$ in Tables 7.3–7.6. By sprinkling the tables with the appropriate ±s, you should be able to verify that the ordered cosets of T are as indicated in Tables 7.8–7.12.

Table 7.3. Products of the form ht.

$1 \cdot t = [$	ϕ	$\|\phi - 1\|$	1	0	$]/2$
$i \cdot t = [$	$1 - \phi$	ϕ	0	1	$]/2$
$j \cdot t = [$	-1	0	ϕ	$\|1 - \phi\|$	$]/2$
$k \cdot t = [$	0	-1	$\|\phi - 1\|$	ϕ	$]/2$

Table 7.4. Products of the form ht^2.

$1 \cdot t^2 = [$	$\|\phi - 1\|$	1	ϕ	0	$]/2$
$i \cdot t^2 = [$	-1	$\|\phi - 1\|$	0	ϕ	$]/2$
$j \cdot t^2 = [$	$-\phi$	0	$\|\phi - 1\|$	-1	$]/2$
$k \cdot t^2 = [$	0	$-\phi$	1	$\|\phi - 1\|$	$]/2$

Table 7.5. Products of the form ht^3.

$1 \cdot t^3 =$	$[\,1-\phi\,|$	$1\quad\,|$	$\phi\quad\,|$	$0\quad\,]/2$
$i \cdot t^3 =$	$[\quad -1\,|$	$1-\phi\,|$	$0\quad\,|$	$\phi\quad\,]/2$
$j \cdot t^3 =$	$[\quad -\phi\,|$	$0\quad\,|$	$1-\phi\,|$	$-1\,]/2$
$k \cdot t^3 =$	$[\quad 0\,|$	$-\phi\,|$	$1\quad\,|$	$1-\phi\,]/2$

Table 7.6. Products of the form ht^4.

$1 \cdot t^4 =$	$[\quad -\phi\,|$	$\phi-1\,|$	$1\quad\,|$	$0\quad\,]/2$
$i \cdot t^4 =$	$[\,1-\phi\,|$	$-\phi\,|$	$0\quad\,|$	$1\quad\,]/2$
$j \cdot t^4 =$	$[\quad -1\,|$	$0\quad\,|$	$-\phi\,|$	$1-\phi\,]/2$
$k \cdot t^4 =$	$[\quad 0\,|$	$-1\,|$	$\phi-1\,|$	$-\phi\,]/2$

For example, the values at^ℓ are obtained by adding the columns of the above tables. See Table 7.7.

Table 7.7. Products of the form at^ℓ.

$a \cdot t =$	$[\quad 0\,|$	$\phi-1\,|$	$\phi\quad\,|$	$1\quad\,]/2$
$a \cdot t^2 =$	$[\quad -1\,|$	$0\quad\,|$	$\phi\quad\,|$	$\phi-1\,]/2$
$a \cdot t^3 =$	$[\quad -\phi\,|$	$1-\phi\,|$	$1\quad\,|$	$0\quad\,]/2$
$a \cdot t^4 =$	$[\quad -\phi\,|$	$-1\,|$	$0\quad\,|$	$1-\phi\,]/2$

Please observe that as an ordered set the cyclic subgroup T is written as

$$T = (1, t, t^2, t^3, t^4, -1, -t, -t^2, -t^3, -t^4).$$

In a few lines, we will tabulate the cosets of T. In doing so, we will only list the first five elements in each coset since the next five are the respective negatives and they are antipodal in the 3-sphere S^3. The cosets will be ordered (and regrouped) as

$$((T, iT, jT, kT), (aT, d^2T, b^2T, cT), (a^2T, bT, c^2T, dT)).$$

Here, a–d and their powers are the elements in the binary tetrahedral group $\widetilde{A_4}$ as before. The ordering and the grouping are chosen because the element a respects this grouping (see also Table 7.21). Its 12-strings-with-($\mathbb{Z}/10$)-quipu looks quite nice in this ordering. The element t, however, is a bit of a mess.

The top-half cosets are tabulated in Tables 7.8–7.19.

Table 7.8. The top-half of the subgroup T.

$1 = [1\|0\|0\|0]$
$t = [\phi\|\phi - 1\|1\|0]/2$
$t^2 = [\phi - 1\|1\|\phi\|0]/2$
$t^3 = [1 - \phi\|1\|\phi\|0]/2$
$t^4 = [-\phi\|\phi - 1\|1\|0]/2$

Table 7.9. The top-half of the coset iT.

$i = [0\|1\|0\|0]$
$i \cdot t = [1 - \phi\|\phi\|0\|1]/2$
$i \cdot t^2 = [-1\|\phi - 1\|0\|\phi]/2$
$i \cdot t^3 = [-1\|1 - \phi\|0\|\phi]/2$
$i \cdot t^4 = [1 - \phi\|-\phi\|0\|1]/2$

Table 7.10. The top-half of the coset jT.

$j = [\,0\,|\,0\,|\,1\,|\,0\,]$
$j \cdot t = [\,-1\,|\,0\,|\,\phi\,|\,1 - \phi\,]\,/2$
$j \cdot t^2 = [\,-\phi\,|\,0\,|\,\phi - 1\,|\,-1\,]\,/2$
$j \cdot t^3 = [\,-\phi\,|\,0\,|\,1 - \phi\,|\,-1\,]\,/2$
$j \cdot t^4 = [\,-1\,|\,0\,|\,-\phi\,|\,1 - \phi\,]\,/2$

Table 7.11. The top-half of the coset kT.

$k = [\,0\,|\,0\,|\,0\,|\,1\,]$
$k \cdot t = [\,0\,|\,-1\,|\,\phi - 1\,|\,\phi\,]\,/2$
$k \cdot t^2 = [\,0\,|\,-\phi\,|\,1\,|\,\phi - 1\,]\,/2$
$k \cdot t^3 = [\,0\,|\,-\phi\,|\,1\,|\,1 - \phi\,]\,/2$
$k \cdot t^4 = [\,0\,|\,-1\,|\,\phi - 1\,|\,-\phi\,]\,/2$

Table 7.12. The top-half of the coset aT.

$a = [\,1\,|\,1\,|\,1\,|\,1\,]\,/2$
$a \cdot t = [\,0\,|\,\phi - 1\,|\,\phi\,|\,1\,]\,/2$
$a \cdot t^2 = [\,-1\,|\,0\,|\,\phi\,|\,\phi - 1\,]\,/2$
$a \cdot t^3 = [\,-\phi\,|\,1 - \phi\,|\,1\,|\,0\,]\,/2$
$a \cdot t^4 = [\,-\phi\,|\,-1\,|\,0\,|\,1 - \phi\,]\,/2$

Table 7.13. The top-half of the coset d^2T.

$d^2 = \begin{bmatrix} -1 \mid 1 \mid -1 \mid 1 \end{bmatrix}/2$
$d^2 \cdot t = \begin{bmatrix} 1 - \phi \mid 0 \mid -1 \mid \phi \end{bmatrix}/2$
$d^2 \cdot t^2 = \begin{bmatrix} 0 \mid -1 \mid 1 - \phi \mid \phi \end{bmatrix}/2$
$d^2 \cdot t^3 = \begin{bmatrix} \phi - 1 \mid -\phi \mid 0 \mid 1 \end{bmatrix}/2$
$d^2 \cdot t^4 = \begin{bmatrix} 1 \mid -\phi \mid \phi - 1 \mid 0 \end{bmatrix}/2$

Table 7.14. The top-half of the coset b^2T.

$b^2 = \begin{bmatrix} -1 \mid 1 \mid 1 \mid -1 \end{bmatrix}/2$
$b^2 \cdot t = \begin{bmatrix} -\phi \mid 1 \mid 0 \mid 1 - \phi \end{bmatrix}/2$
$b^2 \cdot t^2 = \begin{bmatrix} -\phi \mid \phi - 1 \mid -1 \mid 0 \end{bmatrix}/2$
$b^2 \cdot t^3 = \begin{bmatrix} -1 \mid 0 \mid -\phi \mid \phi - 1 \end{bmatrix}/2$
$b^2 \cdot t^4 = \begin{bmatrix} 0 \mid 1 - \phi \mid -\phi \mid 1 \end{bmatrix}/2$

Table 7.15. The top-half of the coset cT.

$c = \begin{bmatrix} 1 \mid 1 \mid -1 \mid -1 \end{bmatrix}/2$
$c \cdot t = \begin{bmatrix} 1 \mid \phi \mid 1 - \phi \mid 0 \end{bmatrix}/2$
$c \cdot t^2 = \begin{bmatrix} \phi - 1 \mid \phi \mid 0 \mid 1 \end{bmatrix}/2$
$c \cdot t^3 = \begin{bmatrix} 0 \mid 1 \mid \phi - 1 \mid \phi \end{bmatrix}/2$
$c \cdot t^4 = \begin{bmatrix} 1 - \phi \mid 0 \mid 1 \mid \phi \end{bmatrix}/2$

Table 7.16. The top-half of the coset $a^2 T$.

$a^2 = [-1\|1\|1\|1]/2$
$a^2 \cdot t = [-\phi\|0\|\phi-1\|1]/2$
$a^2 \cdot t^2 = [-\phi\|-1\|0\|\phi-1]/2$
$a^2 \cdot t^3 = [-1\|-\phi\|1-\phi\|0]/2$
$a^2 \cdot t^4 = [0\|-\phi\|-1\|1-\phi]/2$

Table 7.17. The top-half of the coset bT.

$b = [1\|1\|1\|-1]/2$
$b \cdot t = [0\|\phi\|1\|1-\phi]/2$
$b \cdot t^2 = [-1\|\phi\|\phi-1\|0]/2$
$b \cdot t^3 = [-\phi\|1\|0\|\phi-1]/2$
$b \cdot t^4 = [-\phi\|0\|1-\phi\|1]/2$

Table 7.18. The top-half of the coset $c^2 T$.

$c^2 = [-1\|1\|-1\|-1]/2$
$c^2 \cdot t = [1-\phi\|1\|-\phi\|0]/2$
$c^2 \cdot t^2 = [0\|\phi-1\|-\phi\|1]/2$
$c^2 \cdot t^3 = [\phi-1\|0\|-1\|\phi]/2$
$c^2 \cdot t^4 = [1\|1-\phi\|0\|\phi]/2$

Table 7.19. The top-half of the coset dT.

$d = [\,1\,\vert\,1\,\vert\,-1\,\vert\,1\,]\,/2$
$d\cdot t = [\,1\,\vert\,\phi-1\,\vert\,0\,\vert\,\phi\,]\,/2$
$d\cdot t^2 = [\,\phi-1\,\vert\,0\,\vert\,1\,\vert\,\phi\,]\,/2$
$d\cdot t^3 = [\,0\,\vert\,1-\phi\,\vert\,\phi\,\vert\,1\,]\,/2$
$d\cdot t^4 = [\,1-\phi\,\vert\,-1\,\vert\,\phi\,\vert\,0\,]\,/2$

Next, we compute the actions of a and t upon the coset representatives. Then we will locate the resulting element within the corresponding coset. From Chapter 5, we recall that $a\boldsymbol{i} = b^2$, $a\boldsymbol{j} = -c$, and $a\boldsymbol{k} = d^2$. We can also determine the following:

- a^2 is the first element of the coset $a^2 T$;
- $ad^2 = c^2$ is the first element of the coset $c^2 T$;
- $ab^2 = -d$ is the fifth element of the coset dT;
- $ac = b$ is the first element of the coset bT;
- $a^3 = -1$ is the fifth element of the subgroup T;
- $ab = \boldsymbol{j}$ is the first element of the coset $\boldsymbol{j}T$;
- $ac^2 = -\boldsymbol{k}$ is the fifth element of the coset $\boldsymbol{k}T$;
- $ad = \boldsymbol{i}$ is the first element of the coset $\boldsymbol{i}T$.

Thus, the 12-strings-with-($\mathbb{Z}/10$)-quipu representation of a is as depicted in Fig. 7.2.

The computations of the actions of t upon the coset representatives is more complicated. To begin, consider Table 7.20.

Table 7.20. Products of the form th.

$t\cdot 1 = [\ \phi\ \vert\,\phi-1\,\vert\ 1\ \vert\ 0\ ]\,/2$
$t\cdot \boldsymbol{i} = [\,1-\phi\,\vert\ \phi\ \vert\ 0\ \vert\ -1\ ]\,/2$
$t\cdot \boldsymbol{j} = [\ -1\ \vert\ 0\ \vert\ \phi\ \vert\,\phi-1\,]\,/2$
$t\cdot \boldsymbol{k} = [\ 0\ \vert\ 1\ \vert\,1-\phi\,\vert\ \phi\ ]\,/2$

The Binary Icosahedral Group

From Table 7.20, one can add the elements in the column to show that $t \cdot a = [0|\phi|1|\phi-1]/2 = -a^2 \cdot t^4$. The last equality follows by comparing the element to the last row of Table 7.16. Thus, $t(aT) = [a^2T, 9]$. In addition, the table was constructed to demonstrate that $ti = -d^2t^3$, $tj = at^2$, and $tk = -kt^4$. As before, these equalities are found by comparing the values in Table 7.20 with the values in the previous tables. In this way, we can directly determine images of the cosets iT, jT, and kT. So, we itemize here these values along with the values of $t \cdot x$ for x being the coset representative $x \cdot 1$ of each of the cosets. The values (e.g., $ti = -d^2t^3$) are obtained by means of multiplying various rows of the appropriate table by -1 and adding the results:

- $t \cdot 1 = t = [\phi|\phi-1|1|0]/2$ (Table 7.8, row 2); $t(T) = [T, 1]$.
- $ti = -d^2t^3 = [1-\phi|\phi|0|-1]/2$ (Table 7.13, row 4); $t(iT) = [d^2T, 8]$.
- $tj = a^2t^2 = [-1|0|\phi|\phi-1]/2$ (Table 7.16, row 3); $t(iT) = [aT, 2]$.
- $tk = -kt^4 = [0|1|1-\phi|\phi]/2$ (Table 7.11, row 5); $t(kT) = [kT, 9]$.

- $t \cdot a = [0|\phi|1|\phi-1]/2 = -a^2 \cdot t^4$ (Table 7.16, row 5); $t(aT) = [a^2T, 9]$.
- $t \cdot d^2 = [1-\phi|1|-\phi|0]/2 = c^2 \cdot t$ (Table 7.18, row 2); $t(d^2T) = [c^2T, 1]$.
- $t \cdot b^2 = [-\phi|0|\phi-1|-1]/2 = j \cdot t^2$ (Table 7.10, row 3); $t(b^2T) = [jT, 2]$.
- $t \cdot c = [1|\phi-1|0|-\phi]/2 = -i \cdot t^3$ (Table 7.9, row 4); $t(cT) = [iT, 8]$.

- $t \cdot a^2 = [-\phi|1|0|\phi-1]/2 = b \cdot t^3$ (Table 7.17, row 4); $t(a^2T) = [bT, 3]$.
- $t \cdot b = [0|\phi-1|\phi|-1]/2 = -b^2 \cdot t^4$ (Table 7.14, row 5); $t(bT) = [b^2T, 9]$.
- $t \cdot c^2 = [1-\phi|0|-1|-\phi]/2 = -d \cdot t^2$ (Table 7.19, row 3); $t(c^2T) = [dT, 7]$.
- $t \cdot d = [1|\phi|1-\phi|0]/2 = c \cdot t$ (Table 7.15, row 2); $t(dT) = [cT, 1]$.

$$t = [\phi|\phi - 1|1|0]/2.$$

$$a = [1|1|1|1]/2.$$

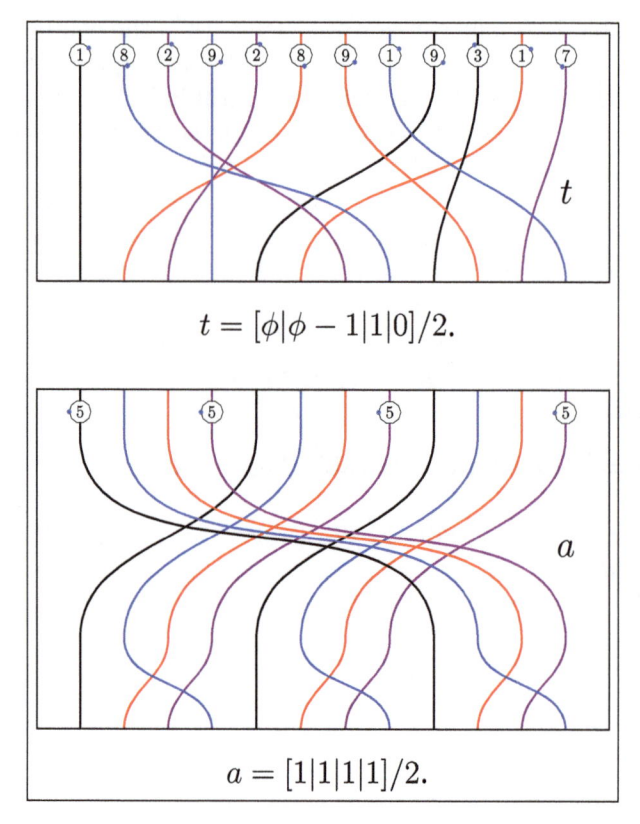

Figure 7.2. Representing the elements a and t using cosets of $\mathcal{T} = \langle t \rangle$.

In Figs. 7.3, 7.4, and 7.6, the identities $a^3 = t^5 = (at)^2 = -1$ are demonstrated using these representations. The remaining figures in this section indicate the addition of the $(\mathbb{Z}/10)$-quipu that appear along the individual strings.

The Binary Icosahedral Group 337

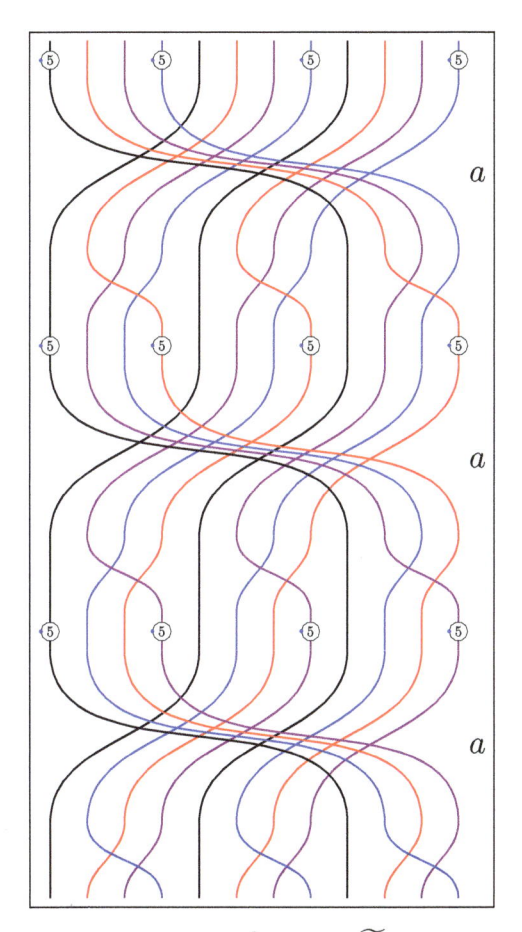

Figure 7.3. The relation $a^3 = -1$ in $\widetilde{A_5}$ using cosets of $\mathcal{T}$.

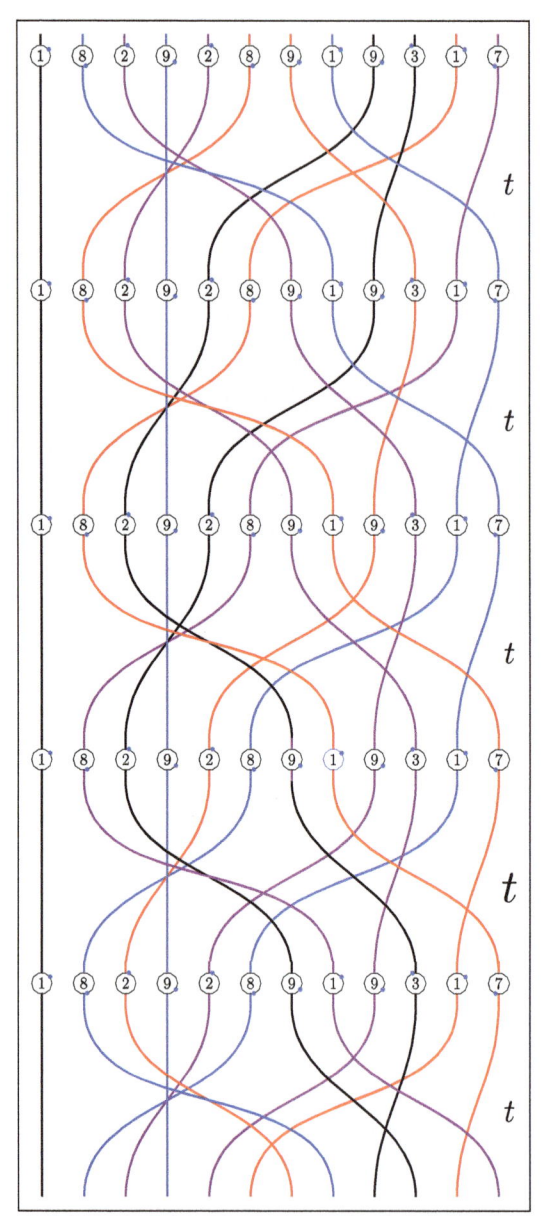

Figure 7.4. The relation $t^5 = -1$ in $\widetilde{A_5}$ group using cosets of $\mathcal{T}$.

The color choices for the strings in these figures are made so that individual strings can be traced through the compositions. At the risk of this being the only understandable computation in the entire book, we compile the five quipu upon each of the 12-strings and indicate their sums as if there are an integral number of twists. Figure 7.5 summarizes the computation of Fig. 7.4 to demonstrate that the 12-string representation of t satisfies $t^5 = -1$.

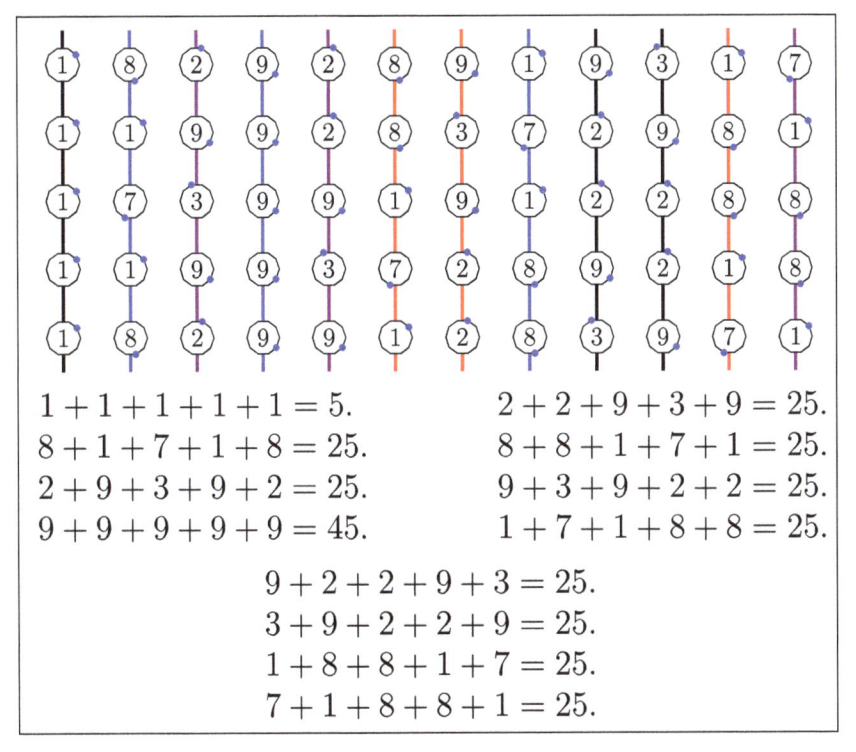

$1+1+1+1+1 = 5.$
$8+1+7+1+8 = 25.$
$2+9+3+9+2 = 25.$
$9+9+9+9+9 = 45.$

$2+2+9+3+9 = 25.$
$8+8+1+7+1 = 25.$
$9+3+9+2+2 = 25.$
$1+7+1+8+8 = 25.$

$9+2+2+9+3 = 25.$
$3+9+2+2+9 = 25.$
$1+8+8+1+7 = 25.$
$7+1+8+8+1 = 25.$

Figure 7.5. Compiling the quipu for the relation $t^5 = -1$.

A similar compilation of the quipu upon the strings for the relation $(at)^2 = -1$ is presented in Fig. 7.7, but the arithmetic computation is left to the reader.

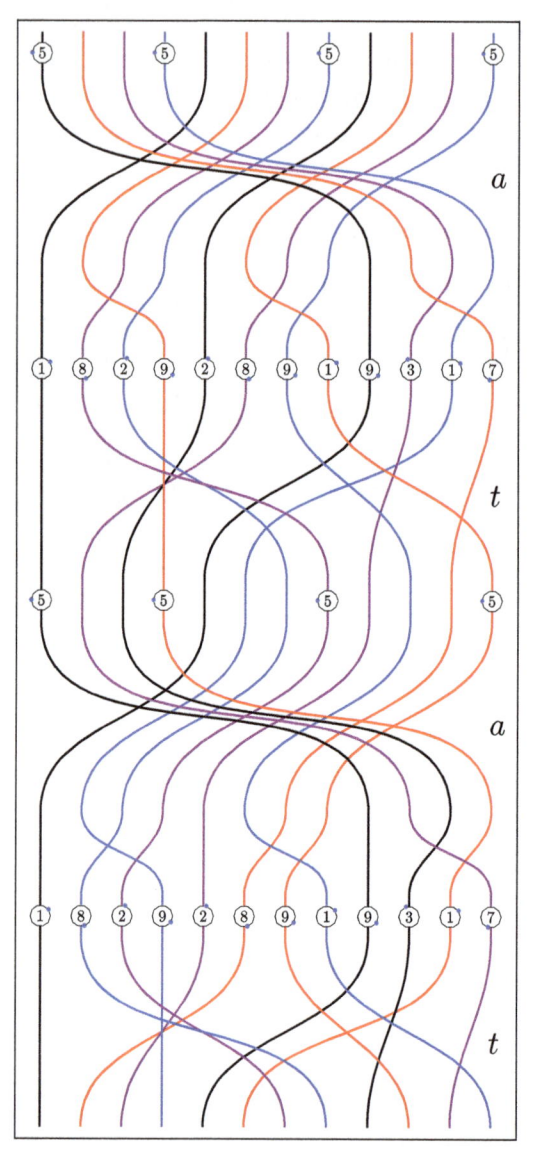

Figure 7.6. The relation $(at)^2 = -1$ in $\widetilde{A_5}$ using cosets of $\mathcal{T}$.

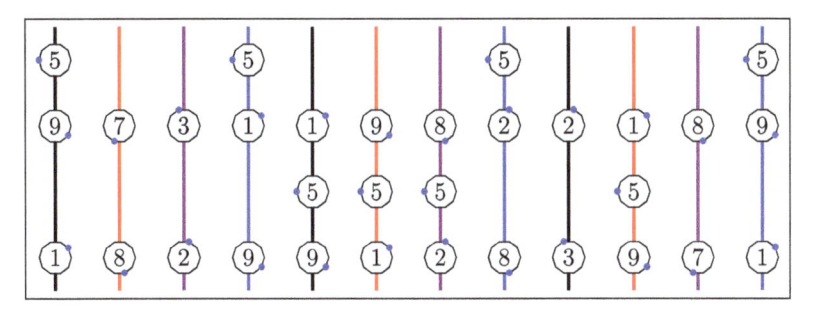

Figure 7.7. The quipu on each string of $(at)^2 = -1$.

In the relation, $a^3 = -1$, we observe that each of the 12 strings that are depicted in Fig. 7.3 has exactly one quipu that is valued at 5 modulo 10. There are, after all, 3×4 quipu depicted.

7.2. Cosets of A

We recall from Sections 5.4.1 and 6.2.1 that the binary tetrahedral group $\widetilde{A_4}$ is written as the union of the four cosets (A, iA, jA, kA), where $A = \langle a \rangle$ is the cyclic group generated by a. We write the following:

$$[0] = [A, iA, jA, kA];$$
$$[1] = [tA, tiA, tjA, tkA];$$
$$[2] = [t^2A, t^2iA, t^2jA, t^2kA];$$
$$[3] = [t^3A, t^3iA, t^3jA, t^3kA];$$
$$[4] = [t^4A, t^4iA, t^4jA, t^4kA]$$

to list the cosets of A in the larger binary icosahedral group. It is not too difficult to observe that $t[0] = [1], t[1] = [2]$, and $t[3] = [4]$. Since $t^5 = -1$, for each $h = 1, i, j, k$, we have that $t^5 hA = hA$, but hA is rotated three steps.

The top three elements of each coset will be listed and sorted according to the outline above. Then we will compute the action of a upon these elements.

From Section 5.4.1, we recall that $ia = d^2$, $ia^2 = -b$, $ja = b^2$, $ja^2 = c^2$, $ka = -c$, and $ka^2 = -d$. So, the "positive" elements of the binary tetrahedral group are listed, as shown in Table 7.21.

Table 7.21. The top-half of the coset [0].

1	i
$a = [1\|1\|1\|1]/2$	$d^2 = [-1\|1\|-1\|1]/2$
$a^2 = [-1\|1\|1\|1]/2$	$-b = [-1\|-1\|-1\|1]/2$
j	k
$b^2 = [-1\|1\|1\|-1]/2$	$-c = [-1\|-1\|1\|1]/2$
$c^2 = [-1\|1\|-1\|-1]/2$	$-d = [-1\|-1\|1\|-1]/2$

Table 7.22. The top-half of the coset [1].

$t = [\phi\|\phi-1\|1\|0]/2$	$ti = [1-\phi\|\phi\|0\|-1]/2$
$ta = [0\|\phi\|1\|\phi-1]/2$	$td^2 = [1-\phi\|1\|-\phi\|0]/2$
$ta^2 = [-\phi\|1\|0\|\phi-1]/2$	$-tb = [0\|1-\phi\|-\phi\|1]/2$
$tj = [-1\|0\|\phi\|\phi-1]/2$	$tk = [0\|1\|1-\phi\|\phi]/2$
$tb^2 = [-\phi\|0\|\phi-1\|-1]/2$	$-tc = [-1\|1-\phi\|0\|\phi]/2$
$tc^2 = [1-\phi\|0\|-1\|-\phi]/2$	$-td = [-1\|-\phi\|\phi-1\|0]/2$

Table 7.23. The top-half of the coset [2].

$t^2 = [\phi-1\|1\|\phi\|0]/2$	$t^2i = [-1\|\phi-1\|0\|-\phi]/2$
$t^2a = [-1\|\phi\|\phi-1\|0]/2$	$t^2d^2 = [0\|\phi-1\|-\phi\|-1]/2$
$t^2a^2 = [-\phi\|\phi-1\|-1\|0]/2$	$-t^2b = [1\|0\|-\phi\|\phi-1]/2$
$t^2j = [-\phi\|0\|\phi-1\|1]/2$	$t^2k = [0\|\phi\|-1\|\phi-1]/2$
$t^2b^2 = [-\phi\|-1\|0\|1-\phi]/2$	$-t^2c = [1-\phi\|0\|-1\|\phi]/2$
$t^2c^2 = [0\|-1\|1-\phi\|-\phi]/2$	$-t^2d = [1-\phi\|-\phi\|0\|1]/2$

Table 7.24. The top-half of the coset [3].

$t^3 = [1-\phi\|1\|\phi\|0]/2$	$t^3i = [-1\|1-\phi\|0\|-\phi]/2$
$t^3a = [-\phi\|1\|0\|1-\phi]/2$	$t^3d^2 = [\phi-1\|0\|-1\|-\phi]/2$
$t^3a^2 = [-1\|0\|-\phi\|1-\phi]/2$	$-t^3b = [\phi\|\phi-1\|-1\|0]/2$
$t^3j = [-\phi\|0\|1-\phi\|1]/2$	$t^3k = [0\|\phi\|-1\|1-\phi]/2$
$t^3b^2 = [-1\|-\phi\|1-\phi\|0]/2$	$-t^3c = [0\|\phi-1\|-\phi\|1]/2$
$t^3c^2 = [\phi-1\|-\phi\|0\|-1]/2$	$-t^3d = [0\|-1\|1-\phi\|\phi]/2$

Table 7.25. The top-half of the coset [4].

$t^4 = [-\phi\|\phi-1\|1\|0]/2$	$t^4i = [1-\phi\|-\phi\|0\|-1]/2$
$t^4a = [-\phi\|0\|1-\phi\|-1]/2$	$t^4d^2 = [1\|1-\phi\|0\|-\phi]/2$
$t^4a^2 = [0\|1-\phi\|-\phi\|-1]/2$	$-t^4b = [\phi\|1\|0\|1-\phi]/2$
$t^4j = [-1\|0\|-\phi\|\phi-1]/2$	$t^4k = [0\|1\|1-\phi\|-\phi]/2$
$t^4b^2 = [0\|-\phi\|-1\|\phi-1]/2$	$-t^4c = [\phi-1\|1\|-\phi\|0]/2$
$t^4c^2 = [1\|-\phi\|\phi-1\|0]/2$	$-t^4d = [\phi-1\|0\|-1\|\phi]/2$

The next step needed to determine the 5-strings-with-$\widetilde{A_4}$-quipu depiction of a is to determine the various products aht^ℓ for $h = 1, i, j, k$ and $\ell = 0, \ldots, 4$. Then we locate these elements among the entries (or their negatives) of Tables 7.21–7.25. Let us itemize as follows:

- $a \cdot 1 = a = [1|1|1|1]/2$; this is the second element of A. So, $a(A) = [A, 1]$.
- $a \cdot i = b^2 = [-1|1|1|-1]/2$; this is the second element of jA. So, $a(iA) = [jA, 1]$.
- $a \cdot j = -c = [-1|-1|1|1]/2$; this is the second element of kA. So, $a(jA) = [kA, 1]$.
- $a \cdot k = d^2 = [-1|1|-1|1]/2$; this is the second element of iA. So, $a(kA) = [iA, 1]$.

- $a \cdot t = [0|\phi - 1|\phi|1]/2$; this is the sixth element of $t^4 A$. So, $a(tA) = [t^4 A, 5]$.
- $a \cdot ti = [1 - \phi|0|1|-\phi]/2$; this is the sixth element of $t^4 kA$. So, $a(tiA) = [t^4 kA, 5]$.
- $a \cdot tj = [-\phi|-1|0|\phi - 1]/2$; this is the sixth element of $t^4 iA$. So, $a(tjA) = [t^4 iA, 5]$.
- $a \cdot tk = [-1|\phi|1 - \phi|0]/2$; this is the sixth element of $t^4 jA$. So, $a(tkA) = [t^4 jA, 5]$.

- $a \cdot t^2 = [-1|0|\phi|\phi - 1]/2$; this is the first element of tjA. So, $a(t^2 A) = [tjA, 0]$.
- $a \cdot t^2 i = [0|-1|\phi - 1|-\phi]/2$; this is the fourth element of tkA. So, $a(t^2 iA) = [tkA, 3]$.
- $a \cdot t^2 j = [-\phi|1 - \phi|-1|0]/2$; this is the fourth element of tA. So, $a(t^2 jA) = [t^2 A, 3]$.

- $a \cdot t^2 k = \left[1 - \phi | \phi | 0 | -1\right]/2$; this is the first element of tiA. So, $a(t^2 kA) = [iA, 0]$.

- $a \cdot t^3 = \left[-\phi | 1 - \phi | 1 | 0\right]/2$; this is the sixth element of $t^3 iA$. So, $a(t^3 A) = [t^3 iA, 5]$.
- $a \cdot t^3 i = \left[\phi - 1 | -\phi | 0 | -1\right]/2$; this is the third element of $t^3 jA$. So, $a(t^3 iA) = [t^3 jA, 2]$.
- $a \cdot t^3 j = \left[-1 | 0 | -\phi | 1 - \phi\right]/2$; this is the third element of $t^3 A$. So, $a(t^3 jA) = [t^3 A, 2]$.
- $a \cdot t^3 k = \left[0 | 1 | \phi - 1 | -\phi\right]/2$; this is the sixth element of $t^3 kA$. So, $a(t^3 kA) = [t^3 kA, 5]$.

- $a \cdot t^4 = \left[-\phi | -1 | 0 | 1 - \phi\right]/2$; this is the second element of $t^2 jA$. So, $a(t^4 A) = [t^2 jA, 1]$.
- $a \cdot t^4 i = \left[1 | -\phi | 1 - \phi | 0\right]/2$; this is the fifth element of $t^2 A$. So, $a(t^4 iA) = [t^2 A, 4]$.
- $a \cdot t^4 j = \left[0 | \phi - 1 | -\phi | -1\right]/2$; this is the second element of $t^2 iA$. So, $a(t^4 jA) = [t^2 iA, 1]$.
- $a \cdot t^4 k = \left[\phi - 1 | 0 | 1 | -\phi\right]/2$; this is the fifth element of $t^2 kA$. So, $a(t^4 kA) = [t^2 kA, 4]$.

Figure 7.8 depicts the 5-strings-with-$\widetilde{A_4}$-quipu representations of a and t. Figure 7.9 demonstrates that $a^3 = -1$. To help with that calculation, Fig. 7.10 compiles the $\widetilde{A_4}$-quipu so that it is easier to see that the resulting compositions are -1 on each of the five strings. Figure 7.11 indicates that $t^5 = -1$. Figure 7.12 demonstrates that $(at)^2 = -1$ in this representation, while Fig. 7.13 compiles the $\widetilde{A_4}$-quipu that appear on these strings.

While we computed the actions of a and t by means of the (mod 6)-quipu, the figures represent the elements of the binary tetrahedral group by using (mod 2)-quipu since they are more economical.

$a = [1|1|1|1]/2.$

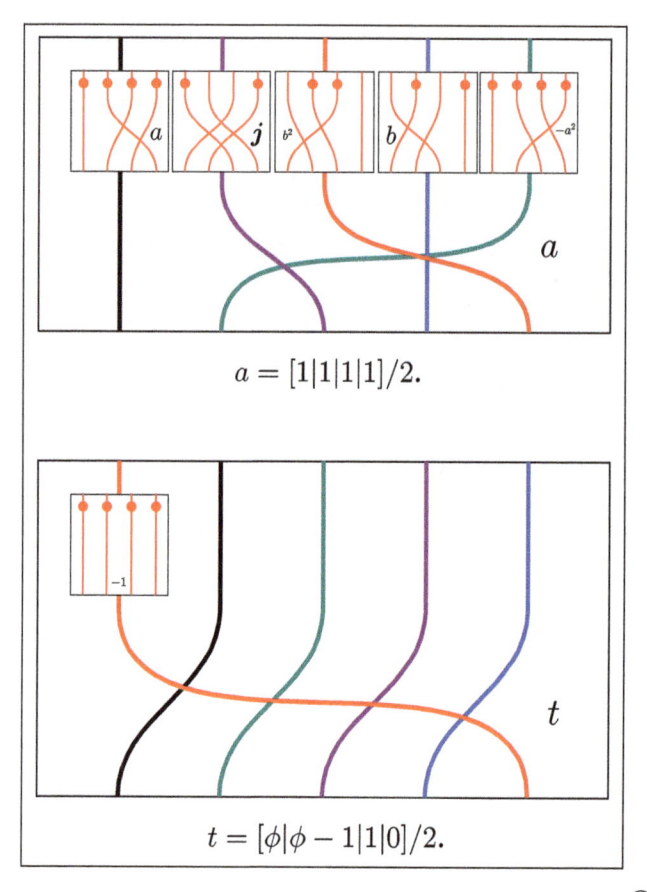

$t = [\phi|\phi - 1|1|0]/2.$

Figure 7.8. Representing the elements a and t using cosets $\widetilde{A_4}$.

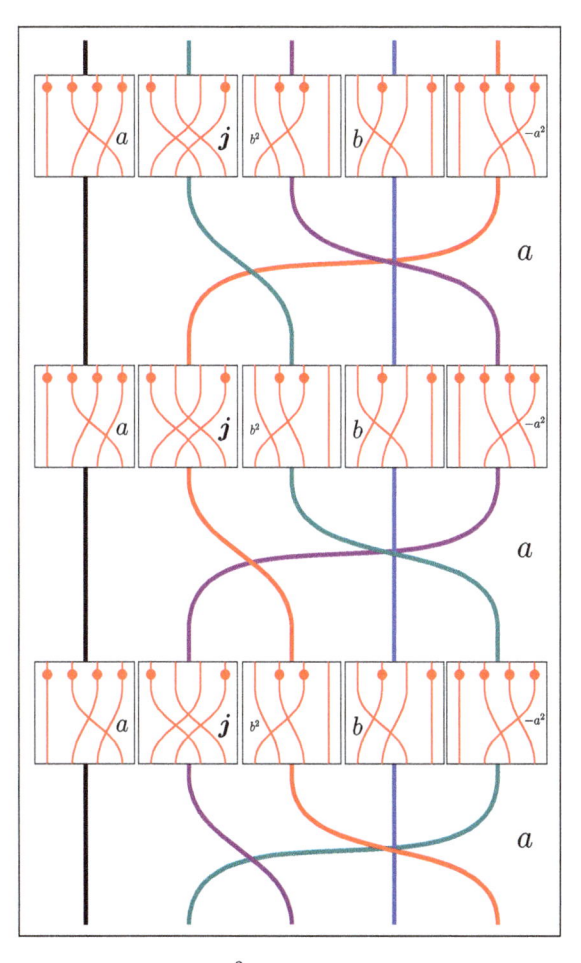

Figure 7.9. The relation $a^3 = -1$ in the binary icosahedral group.

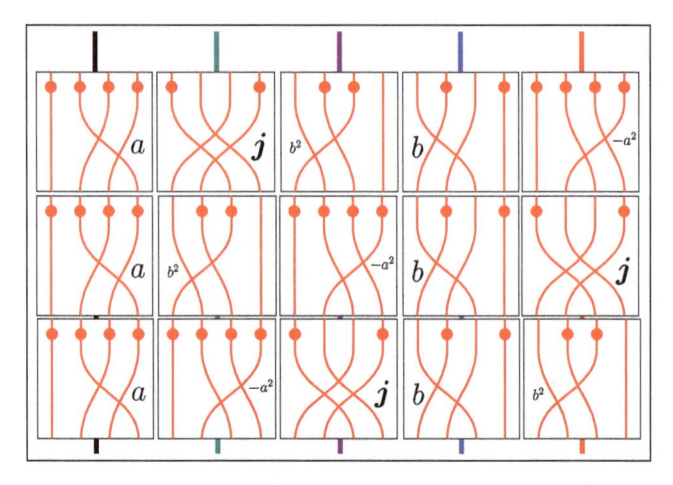

Figure 7.10. Compiling the quipu for the relation $a^3 = -1$.

In Fig. 7.10, the quipu that are the elements of the binary tetrahedral group have been slid to the top of each of the five strings that represent the cosets of $\widetilde{A_4}$ in the binary octahedral group $\widetilde{A_5}$. Upon the first string, the product a^3 appears. As we have observed before, this product is -1, and one can follow each of the red strings upward to verify that an odd number of (mod 2)-quipu appears upon each string. Upon the second string, the product $-jb^2a^2$ appears. Again, it is meant to be visually easy to determine that this product is -1, but of course, you may prefer to demonstrate this algebraically. Upon the third string, a conjugate product $-b^2a^2j$ appears. The fourth string contains the product b^3, and the fifth string contains another conjugate product $-a^2jb^2$. In all cases, slide the (mod 2)-quipu up the strings, and observe that the underlying permutations (in A_4) yield the identity.

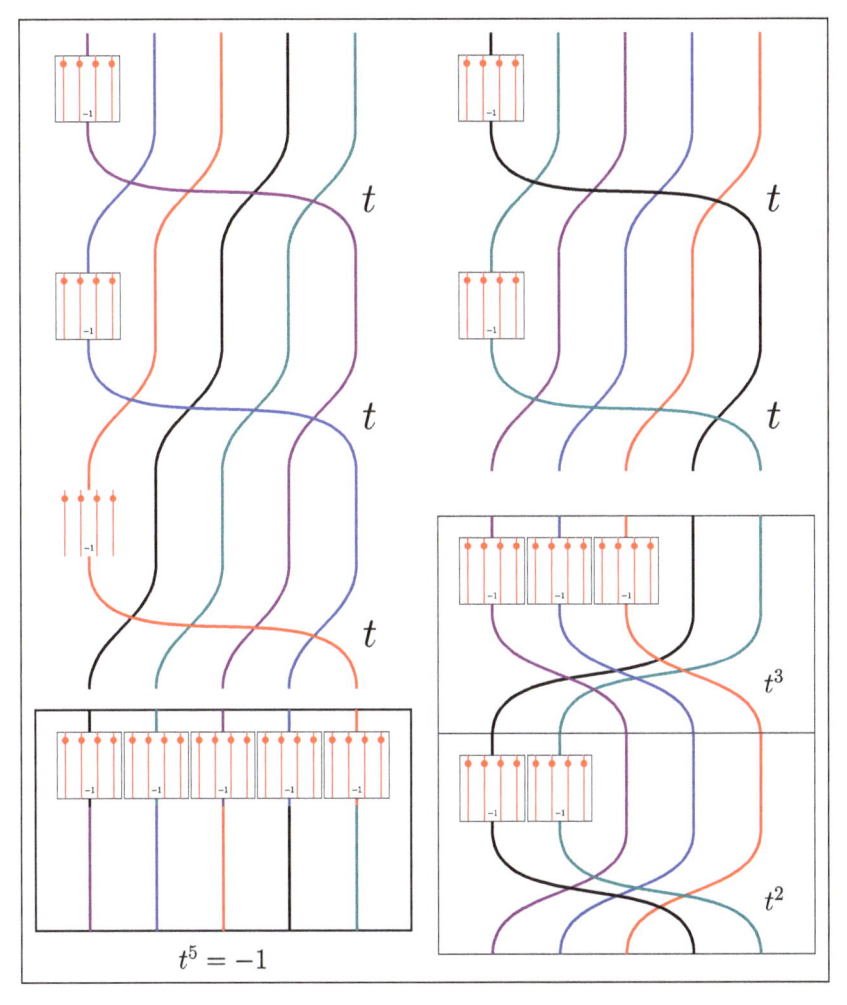

Figure 7.11. The relation $t^5 = -1$ in the binary icosahedral group.

In Fig. 7.11, the relation $t^5 = -1$ is depicted. On the left, the composition t^3 is written, and the result of that computation appears on the bottom right, just above the computation of t^2. Then the bottom left contains the composition $t^3 t^2 = -1$. The arrangement of the various pieces was dictated by the items that could fit upon the page and remain legible.

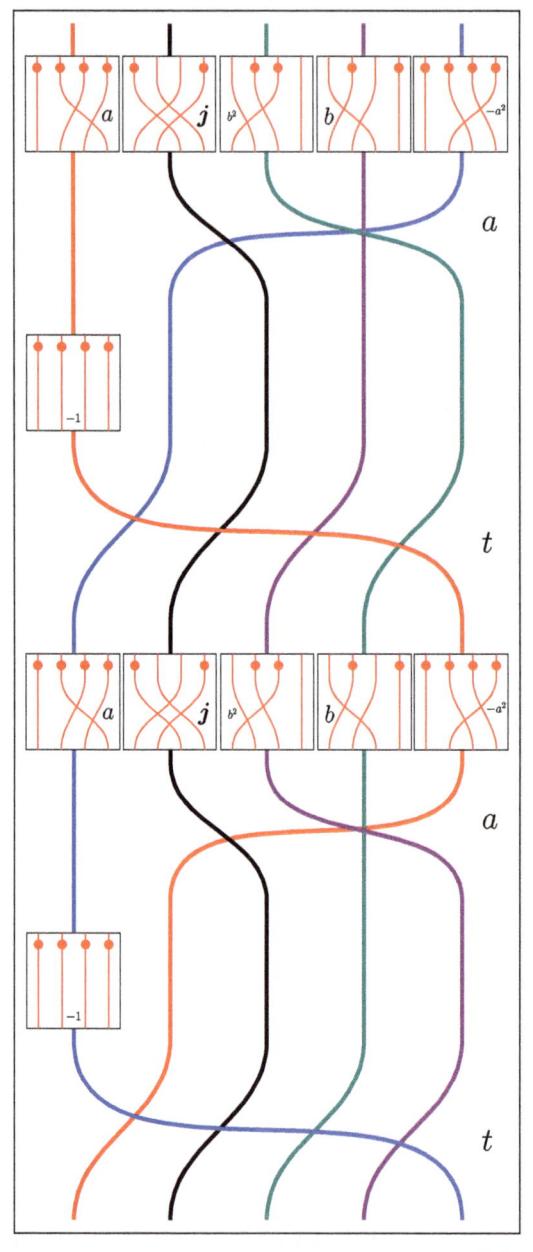

Figure 7.12. The relation $(at)^2 = -1$ in $\widetilde{A_5}$.

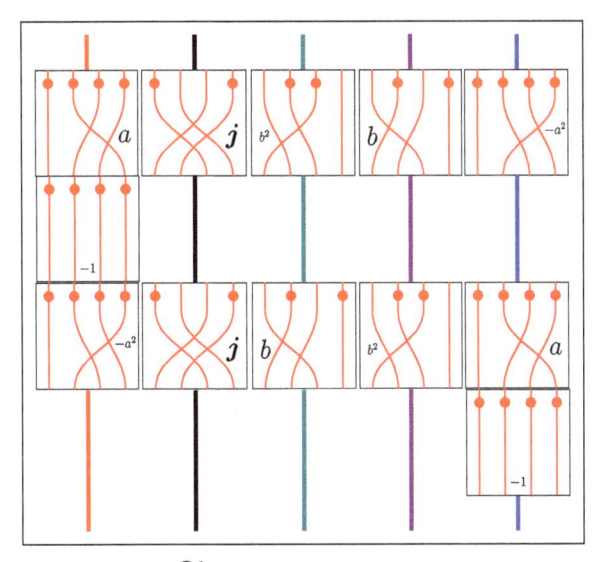

Figure 7.13. Sliding the $\widetilde{A_4}$ quipu upward and straightening the 5-strings in the $(at)^2 = -1$ relation.

In Fig. 7.12, the product $(at)^2$ is depicted. The five colored strings return to their original positions: red in the zeroth position, black in the first, and so forth. In Fig. 7.13, these five strings that represent the cosets of $\widetilde{A_4}$ have each been straightened. The quipu upon the left-most string represent, from top to bottom, the product $a(-1)(-a^2)$. The next string to the right contains the product $(\boldsymbol{j})^2$. The center string contains the product $b^2 b$. To its right, the product $b(b^2)$ is represented. Finally, on the far right, the product $(-a^2)a(-1)$ is represented. Each of these reduces (-1), and we are hopeful that the identities are visually apparent when the red strings are straightened and the (mod 2)-quipu are moved upward and canceled.

Chapter 8

Computing Group 2-cocycles

This is our final chapter. Much of the material here is culled from [1, Chapter IV]. We expand upon the concise descriptions that are given in that excellent text by focusing upon the extensions that were given in our Chapter 5 and some other easy examples. We will demonstrate how to use permutations-with-quipu to compute quantities that are known as "group cocycles." In this way, the differences among the non-trivial extensions of the symmetric group Σ_4 are quantified. You might suppose that this chapter is a collection of exercises that we created for ourselves in order to better understand [1]. You would not be far from the truth in that assumption.

Let us first describe what the term "group cocycle" means.

The noun "group" is used as an adjective to describe the type of cocycle that will be computed. As always in this text, *a group* is a set that has an associative, unitary, and invertible binary operation defined upon it. The term "cocycle" is widely used in mathematics, so we have to state briefly the type of cocycle that will be computed. Hence, we study group cocycles rather than quandle cocycles or the more commonly used algebra cocycles. Even so, there are relationships among all of these concepts that you might want to explore. But the word cocycle is hardly used, if used at all, in common parlance. It is a specialized and technical idea for which this introductory material will give an overview.

First, the prefix "co" is used in the sense of "with." A cocycle is something that goes with a cycle; it is an algebraic dual to the notion of cycle. In general, in mathematics when one is given a theory, its "co-theory" walks side by side with the theory but in an opposite metaphorical direction. Many mathematicians like to make bad puns that involve words which contain the prefix "co." If, for example, a mathematician turns coffee into theorems, then a co-mathematician turns "ffee" into co-theorems.[1] We mention in passing that "sets" and "cosets" are not algebraic duals in this sense.

More to the point, in homology theory, one is interested in cycles that are not boundaries, and in cohomology theory, one is interested in cocycles that are not coboundaries. While the previous sentence is true, it remains as unhelpful as a broken coffee pot without more context.

The notion of "cycle" comes from homological considerations, as first discussed in Section 4.1.3. A *cycle* is a geometric object, X, that is made of simplices (vertices, segments, triangles, tetrahedral, etc.), for which X's boundary is empty. We think of simplices as generating objects, and a union of these is sometimes called a "chain." Or more precisely, a *chain* is a formal sum of geometric simplices. We think of the simplices that have the same dimension as being glued together along some of their faces. The choice of the word "chain" in the current context might be because such a union of line segments appear to be chain-like. A cycle is a chain that has its boundary equal to 0. In homology theory, all of these ideas are made rigorous.

Roughly speaking, a cochain is an assignment of numbers to the simplices in a chain. In a few paragraphs, the ideas of cocycles and coboundaries will also be defined. For now, you may think that a "cocycle" is a numeric quantity which vanishes when it is assigned to the boundary of a chain. The geometric context of a group cocycle is not yet apparent.

The etymological discussion above is still not enough to describe how group cocycles will be computed. So, first, a couple of elementary example will describe the mechanics of computing 2-dimensional cocycles, henceforth 2-*cocycles*. After these examples, the general procedure will be given. Next, we present the textbook-like information that indicates why group cocycles are related to seseqs. Then in a pair of examples — $\mathbb{Z}/9$ versus $\mathbb{Z}/3 \times \mathbb{Z}/3$ — we compute the associated 2-cocycles. Finally, the show will hit the road, and the 2-cocycles for the extensions of Σ_4 will be computed.

---

[1] We warned you that the puns were bad!

First, let us state the general philosophy of this chapter. Equivalence classes of group 2-cocycles correspond to equivalence classes of seseqs:

$$1 \longleftarrow H = G/K \overset{q}{\longleftarrow} G \overset{i}{\longleftarrow} K \longleftarrow 0.$$

The cocycles take values in K and they are parameterized by elements in $H \times H$. More specificity will be given as the text progresses.

8.1. Set-theoretic sections

Let us consider how you or I might compute the sums in a modular arithmetic context. Specifically, consider the addition table of $\mathbb{Z}/3$:

+	0	1	2
0	0	1	2
1	1	2	0
2	2	0	1

To compile this table, you might think to yourself, "$1 + 2 = 2 + 1 = 3$ and $3 = 0 \mod 3$, $2+2 = 4$ and $4 = 1 \mod 3$." In that thought process, the seseq

$$0 \longleftarrow \mathbb{Z}/3 \overset{q}{\longleftarrow} \mathbb{Z} \overset{3 \cdot}{\longleftarrow} \mathbb{Z} \longleftarrow 0$$

was tacitly used, and you momentarily considered the elements in $\mathbb{Z}/3$ to be the integers. More specifically, there is a function $s : \mathbb{Z}/3 \to \mathbb{Z}$ that is not a homomorphism[2] and that has the properties $s(0) = 0$ and $q(s(x)) = x$ for all $x \in \mathbb{Z}/3$. The function in mind is the "ambiguous name" function $s([x]) = x$ that takes the equivalence class to its usual representative. Such a function is called a *set-theoretic section*.

It is set-theoretic since its not a homomorphism, and it is a "section" in that it takes a sample of each equivalence class.

From this section, a function of 2-variables $\mathbb{Z} \overset{\psi}{\longleftarrow} \mathbb{Z}/3 \times \mathbb{Z}/3$ will be defined. That function will satisfy an algebraic condition called the 2-cocycle condition which will be articulated later.

---

[2] Please recall that a function, f, is a *group homomorphism* if $f(a+b) = f(a) + f(b)$. Since the groups are abelian, the group operations are additive.

Here is how to define ψ. Consider a pair of elements x and y that are in $\mathbb{Z}/3$. Pretend that they are integers by means of their ambiguous names. Add their values as if they were integers. That is, compute $s(x)+s(y)$. Now, do this process in reverse. That is, $1+2 = 0$ because these are integers modulo 3. Take the section of the sum and compare it to the sum of the sections: compute $s(x)+s(y)-s(x+y)$. This is an expression in the integers $\mathbb{Z}$. The quotient map q **is** a homomorphism, and s is a set-theoretic section. So, $q(s(x) + s(y) - s(x+y)) = q(s(x)) + q(s(y)) - q(s(x+y)) = x + y - (x+y) = 0$ in $\mathbb{Z}/3$. This is to say that the differences between the sum of the sections and the section of the sum are always divisible by 3. Or back in the integers $1 + 2 - 0 = 2 + 1 - 0 = 3$, and $2 + 2 = 4$ which is compared to $s(2+2) = s(1)$. So, $s(2) + s(2) - s(1) = 4 - 1 = 3$.

The sequence is exact! So, the kernel of q is the image of the function $3 \cdot \_$. The value of ψ is computed as

$$\psi(x,y) = \frac{s(x) + s(y) - s(x+y)}{3}.$$

The function ψ is a 2-cocycle that takes values in the integers $\mathbb{Z}$. It is associated to the seseq above.

Now, in the general case, we consider the quotient of the integers by the subgroup $n\mathbb{Z}$. There is a seseq

$$0 \longleftarrow \mathbb{Z}/n \xleftarrow{q} \mathbb{Z} \xleftarrow{n \cdot \_} \mathbb{Z} \longleftarrow 0,$$

and we follow the process *mutatis mutandis*. A set-theoretic section is chosen so that $s([0]) = 0$ and $s([x]) = x$. In general, except for $[0]$, the section could be any element that is equivalent to $x \pmod{n}$. For convenience, we chose the standard representative between 0 and $n-1$. Then a 2-cocycle ψ is defined as

$$\psi_n(x,y) = \frac{s(x) + s(y) - s(x+y)}{n}.$$

The numerator of the expression on the right is divisible by n by the definition of a set-theoretic section ($q(s([x])) = [x]$), the exactness property ($\mathrm{Im}(n \cdot \_) = \ker(q)$), and because the quotient map q is a homomorphism.

Now, we are ready to state the condition that the function ψ is a 2-cocycle.

A function $K \xleftarrow{\xi} G \times G$ from the product of a group G with itself to an abelian group K is said to be a *2-cocycle* if it satisfies the condition

$$\xi(y, z) - \xi(xy, z) + \xi(x, yz) - \xi(x, y) = 0$$

for all $x, y, z \in G$. Some 2-cocycles are more useful than others. A 2-cocycle ξ is said to be a *coboundary* if there is a function $K \xleftarrow{\gamma} G$ such that

$$\xi(x, y) = \gamma(y) - \gamma(xy) + \gamma(x).$$

Furthermore, two 2-cocycles, ξ_1 and ξ_2, are said to be *cohomologous* if they differ by a coboundary. That is to say,

$$\xi_2(x, y) - \xi_1(x, y) = \gamma(y) - \gamma(xy) + \gamma(x)$$

for some $K \xleftarrow{\gamma} G$. We'll also say that $\xi_1(x, y)$ and $\xi_2(x, y)$ are *equivalent cocycles*. This relationship between cocycles is, indeed, an equivalence relation. It is reflexive, symmetric, and transitive. We hope you will take some time to verify these properties. Our doing so within the text will only take us on a further tangent. Cocycles that are not coboundaries are the more important ones.

These definitions are unmotivated at the moment, but soon (Section 8.1.2), we give them some geometric content. Before that, let us at least show that ψ_n is a 2-cocycle that is not a coboundary.

Proposition 2. *The function*

$$\psi_n(x, y) = \frac{(s(x) + s(y) - s(x + y))}{n}$$

is a 2-cocycle. So,

$$\psi_n(y, z) - \psi_n(x + y, z) + \psi_n(x, y + z) - \psi_n(x, y) = 0.$$

Furthermore, ψ_n is not a coboundary. There is no function $\mathbb{Z} \xleftarrow{\gamma} \mathbb{Z}/n$ such that

$$\psi_n(x, y) = \gamma(y) - \gamma(x + y) + \gamma(x).$$

Proof. The proof that ψ_n is a 2-cocycle is straightforward:

$$\psi_n(y, z) = \frac{s(y) + s(z) - s(y + z)}{n},$$

$$-\psi_n(x + y, z) = \frac{s((x + y) + z) - s(z) - s(x + y)}{n},$$

$$\psi_n(x, y + z) = \frac{s(x) + s(y + z) - s(x + (y + z))}{n},$$

$$-\psi_n(x, y) = \frac{s(x + y) - s(y) - s(x)}{n}.$$

Add the quantities on the right and cancel like terms. Since the quotient group $\mathbb{Z}/n$ is a group, it is associative and $s((x + y) + z) - s(x + (y + z)) = 0$. Note also that the proof did not depend upon the nature of the set-theoretic section s. Indeed, rather than dividing by n, one could instead compute the inverse image under the inclusion map of the kernel of the quotient map. A function of the form $\psi(x, y) = i^{-1}(s(x) + s(y) - s(x + y))$ is still a 2-cocycle.

Suppose that there is a function γ such that

$$\psi_n(x, y) = \gamma(y) - \gamma(x + y) + \gamma(x)$$

for all $x, y \in \mathbb{Z}/n$. The dependence of γ upon the modulus n is suppressed in the notation. The method by which a contradiction arises in that assumption falls into two cases that depend upon the parity of n. We'll write down the argument as a condensation of how we found it. But please be aware, working this out was not an easy task. In particular, we first worked this through on a case-by-case basis for small values of n (i.e. $n = 2, 3, 4, 5, 6, 7$) and often had to redo the calculations before we developed a systematic method.

First Table 8.1 evaluates the cocycles ψ_n, for $n = 2, 3, 4$.

Table 8.1. A collection of 2-cocycle values for similarly defined cocycles.

ψ_2	0	1
0	0	0
1	0	1

ψ_3	0	1	2
0	0	0	0
1	0	0	1
2	0	1	1

ψ_4	0	1	2	3
0	0	0	0	0
1	0	0	0	1
2	0	0	1	1
3	0	1	1	1

In general, the table of cocycle values for ψ_n consists of an upper triangular block of $\frac{n(n+1)}{2}$ zeroes and a lower triangular block of $\frac{(n-1)n}{2}$ ones since

$$\psi_n(x,y) = \begin{cases} 0 & \text{if } x+y < n, \\ 1 & \text{if } n \le x+y. \end{cases}$$

It is always good to start from the most elementary example. Here, that is when $n = 2$. In general, we'll write $\gamma_\ell = \gamma(\ell)$, but also switch between the notations as context dictates. The condition

$$\gamma(y) - \gamma(x+y) + \gamma(x) = \psi_2(x,y)$$

translates to four linear equations:

$$\gamma_0 - \gamma_0 + \gamma_0 = 0,$$
$$\gamma_1 - \gamma_1 + \gamma_0 = 0,$$
$$\gamma_0 - \gamma_1 + \gamma_1 = 0,$$
$$\gamma_1 - \gamma_0 + \gamma_1 = 1.$$

The first three of these give that $\gamma_0 = \gamma(0) = 0$. The last one can be read as $2\gamma(1) = 1$. But the function γ takes values in $\mathbb{Z}$. So, no solution exists.

Case $n = 2m$: Then $\psi(m,m) = 1$. If ψ is a coboundary,

$$\gamma_m - \gamma_0 + \gamma_m = \psi(m,m) = 1.$$

So, $2\gamma(m) = 1$, but γ only takes integral values. So, ψ_{2m} is not a coboundary.

Case $n = 2m+1$: In general, a system of n^2 equations in n unknowns results. But the equations are symmetric in x, y. Furthermore, along the first row (or column) every equation results in the conclusion that $\gamma(0) = 0$. So, if the cocycle is a coboundary, then the function γ of which ψ is a coboundary always satisfies $\gamma(0) = 0$. We can restrict our considerations to the equations that correspond to the lower-right triangular block of the table of cocycle values.

It turns out that a contradiction occurs directly when only the equations that correspond to the second row of the cocycle table are considered.

Let us begin at the most elementary level: $n = 3$.

The equations that must be satisfied include $\gamma_0 = 0$ and the following:

$$\gamma_1 - \gamma_2 + \gamma_1 = 0,$$
$$\gamma_2 - \gamma_0 + \gamma_1 = 1,$$
$$\gamma_2 - \gamma_1 + \gamma_2 = 1.$$

These will be written as the (3×3)-matrix

$$\begin{bmatrix} 2 & -1 & | & 0 \\ 1 & 1 & | & 1 \\ -1 & 2 & | & 1 \end{bmatrix}$$

that is to be row reduced. Since $\gamma_0 = 0$, only the possible values of γ_1 and γ_2 require consideration. The process will be generalized, and so, we indicate the row operations.

$$\begin{bmatrix} 2 & -1 & | & 0 \\ 1 & 1 & | & 1 \\ -1 & 2 & | & 1 \end{bmatrix} \xrightarrow{R_1 \leftrightarrow R_2} \begin{bmatrix} 1 & 1 & | & 1 \\ 2 & -1 & | & 0 \\ -1 & 2 & | & 1 \end{bmatrix} \xrightarrow{2R_1 - R_2 \mapsto R_2} \begin{bmatrix} 1 & 1 & | & 1 \\ 0 & 3 & | & 2 \\ -1 & 2 & | & 1 \end{bmatrix}.$$

In the middle equation, we obtain $3\gamma(2) = 2$ which is a contradiction because $\gamma(2)$ should be an integer.

Note that the equations involved in the row reduction were only those read from the second row of the cocycle table.

In the general case, the first operation is a cyclic permutation. Rather than writing $R_1 \mapsto R_2 \mapsto \cdots R_{n-1} \mapsto R_1$, we'll write it in permutation cycle notation $(12\ldots(n-1))$.

While the case $n = 5$ is also instructive, the inductive case is more easily seen by way of the case $n = 7$. First, observe that $\psi_7(1, \ell) = 0$ when $\ell = 0, 1, \ldots, 5$ and $\psi_7(1, 6) = 1$ since the section values $s(1) + s(\ell) = s(1 + \ell)$ when $\ell < 6$.

The linear system of equations that will be considered is

$$\gamma_1 - \gamma_2 + \gamma_1 = 0,$$
$$\gamma_2 - \gamma_3 + \gamma_1 = 0,$$
$$\gamma_3 - \gamma_4 + \gamma_1 = 0,$$

$$\gamma_4 - \gamma_5 + \gamma_1 = 0,$$
$$\gamma_5 - \gamma_6 + \gamma_1 = 0,$$
$$\gamma_6 - \gamma_0 + \gamma_1 = 1$$

and as before $\gamma_0 = 0$. So, six linear equations in six unknowns will be examined for a solution in the integers $\mathbb{Z}$. In matrix form, we row-reduce

$$S = \begin{bmatrix} 2 & -1 & 0 & 0 & 0 & 0 & | & 0 \\ 1 & 1 & -1 & 0 & 0 & 0 & | & 0 \\ 1 & 0 & 1 & -1 & 0 & 0 & | & 0 \\ 1 & 0 & 0 & 1 & -1 & 0 & | & 0 \\ 1 & 0 & 0 & 0 & 1 & -1 & | & 0 \\ 1 & 0 & 0 & 0 & 0 & 1 & | & 1 \end{bmatrix}.$$

We are fond of the last row. So, let's put that one on top, by reordering the equations using the cyclic permutation (123456), and then perform row reductions to the result.

$$S' = \begin{bmatrix} 1 & 0 & 0 & 0 & 0 & 1 & | & 1 \\ 2 & -1 & 0 & 0 & 0 & 0 & | & 0 \\ 1 & 1 & -1 & 0 & 0 & 0 & | & 0 \\ 1 & 0 & 1 & -1 & 0 & 0 & | & 0 \\ 1 & 0 & 0 & 1 & -1 & 0 & | & 0 \\ 1 & 0 & 0 & 0 & 1 & -1 & | & 0 \end{bmatrix}$$

$$\xrightarrow{2R_1 - R_2 \mapsto R_2} \begin{bmatrix} 1 & 0 & 0 & 0 & 0 & 1 & | & 1 \\ 0 & 1 & 0 & 0 & 0 & 2 & | & 2 \\ 1 & 1 & -1 & 0 & 0 & 0 & | & 0 \\ 1 & 0 & 1 & -1 & 0 & 0 & | & 0 \\ 1 & 0 & 0 & 1 & -1 & 0 & | & 0 \\ 1 & 0 & 0 & 0 & 1 & -1 & | & 0 \end{bmatrix}$$

$$\xrightarrow{R} \begin{bmatrix} 1 & 0 & 0 & 0 & 0 & 1 & | & 1 \\ 0 & 1 & 0 & 0 & 0 & 2 & | & 2 \\ 0 & 1 & -1 & 0 & 0 & -1 & | & -1 \\ 0 & 0 & 1 & -1 & 0 & -1 & | & -1 \\ 0 & 0 & 0 & 1 & -1 & -1 & | & -1 \\ 0 & 0 & 0 & 0 & 1 & -2 & | & -1 \end{bmatrix} = U,$$

where the row operation R is the composition

$$(R_3 - R_1 \mapsto R_3) \circ (R_4 - R_1 \mapsto R_4)$$
$$\circ (R_5 - R_1 \mapsto R_5) \circ (R_6 - R_1 \mapsto R_6).$$

Use the bottom row to clear the $(5,5)$ entry and continue upward:

$$\begin{bmatrix} 1 & 0 & 0 & 0 & 0 & 1 & 1 \\ 0 & 1 & 0 & 0 & 0 & 2 & 2 \\ 0 & 1 & -1 & 0 & 0 & -1 & -1 \\ 0 & 0 & 1 & -1 & 0 & -1 & -1 \\ 0 & 0 & 0 & 1 & -1 & -1 & -1 \\ 0 & 0 & 0 & 0 & 1 & -2 & -1 \end{bmatrix}$$

$$\xrightarrow{R_5+R_6 \mapsto R_5} \begin{bmatrix} 1 & 0 & 0 & 0 & 0 & 1 & 1 \\ 0 & 1 & 0 & 0 & 0 & 2 & 2 \\ 0 & 1 & -1 & 0 & 0 & -1 & -1 \\ 0 & 0 & 1 & -1 & 0 & -1 & -1 \\ 0 & 0 & 0 & 1 & 0 & -3 & -2 \\ 0 & 0 & 0 & 0 & 1 & -2 & -1 \end{bmatrix}$$

$$\xrightarrow{R_4+R_5 \mapsto R_4} \begin{bmatrix} 1 & 0 & 0 & 0 & 0 & 1 & 1 \\ 0 & 1 & 0 & 0 & 0 & 2 & 2 \\ 0 & 1 & -1 & 0 & 0 & -1 & -1 \\ 0 & 0 & 1 & 0 & 0 & -4 & -3 \\ 0 & 0 & 0 & 1 & 0 & -3 & -2 \\ 0 & 0 & 0 & 0 & 1 & -2 & -1 \end{bmatrix}$$

$$\xrightarrow{R_3+R_4 \mapsto R_3} \begin{bmatrix} 1 & 0 & 0 & 0 & 0 & 1 & 1 \\ 0 & 1 & 0 & 0 & 0 & 2 & 2 \\ 0 & 1 & 0 & 0 & 0 & -5 & -4 \\ 0 & 0 & 1 & 0 & 0 & -4 & -3 \\ 0 & 0 & 0 & 1 & 0 & -3 & -2 \\ 0 & 0 & 0 & 0 & 1 & -2 & -1 \end{bmatrix}$$

$$\xrightarrow{R_2-R_3 \mapsto R_2} \begin{bmatrix} 1 & 0 & 0 & 0 & 0 & 1 & 1 \\ 0 & 0 & 0 & 0 & 0 & 7 & 6 \\ 0 & 1 & 0 & 0 & 0 & -5 & -4 \\ 0 & 0 & 1 & 0 & 0 & -4 & -3 \\ 0 & 0 & 0 & 1 & 0 & -3 & -2 \\ 0 & 0 & 0 & 0 & 1 & -2 & -1 \end{bmatrix}.$$

The second row corresponds to the equation $7\gamma_6 = 6$ which does not have a solution in the integers.

In the general case when n is an odd integer, 2-cocycle values are

$$\psi_n(1, \ell) = \begin{cases} 0 & \text{if } \ell < n-1, \\ 1 & \text{if } \ell = n-1. \end{cases}$$

and the equation $\gamma_{n-1} + \gamma_1 = 1$ can be used to write the values of the remaining γs in terms of γ_{n-1}. In particular,

$$\gamma_{n-2} + 2\gamma_{n-1} = -1.$$

All of γ_3 through γ_{n-2} are written in terms of γ_{n-1}, and the originally first equation (that corresponded to the value $\psi_n(1,1)$) becomes $n\gamma_{n-1} = n - 1$. Therein lies the contradiction. This completes the proof. □

At this stage, you should be wondering, "Why do we care?" To be fair, the proposition and the conceptual context of 2-cocycles is still hidden. We also promised, and as of yet haven't delivered, that there is a geometric context in which 2-cocycles can be conceptualized. Before we address your concerns, we'd like to suggest an exercise. Please be patient.

Exercise 18. Careless section. In the case $n = 3$, play around with the value of the set-theoretic section. For example, let $s_b([0]) = 0$, as always, but let $s_b([1]) = 7$ and $s_b([2]) = 20$ or something else that is correct but a careless choice.[3] Show that the resulting cocycle $\psi_b(x, y) = \frac{1}{3}[s_b(x) + s_b(y) - s_b(x+y)]$ differs from ψ_3 by a coboundary.

8.1.1. Classifying seseqs

The principal idea about 2-cocycles is that they classify seseqs up to an equivalence. We recall that equivalent cocycles differ by a coboundary. Specifically, 2-cocycles $K \xleftarrow{\psi_1} G \times G$ and $K \xleftarrow{\psi_2} G \times G$ that take values in a commutative group K are said to be *equivalent*

---
[3]The subscript b indicates something "bad."

if there is a function $K \xleftarrow{\gamma} G$ such that

$$\psi_2(x,y) - \psi_1(x,y) = \gamma(y) - \gamma(xy) + \gamma(x)$$

for all group elements $x, y \in G$.

A 2-cocycle is said to be trivial if it is equivalent to the constant function $\psi_0(x,y) = 0$ for all $x, y \in G$.

In this section of the chapter, we'll examine the differences between the seseqs

$$0 \longleftarrow \mathbb{Z}/3 \xleftarrow{q} \mathbb{Z}/9 \xleftarrow{3\cdot} \mathbb{Z}/3 \longleftarrow 0$$

and

$$0 \longleftarrow \mathbb{Z}/3 \xleftarrow{p_2} (\mathbb{Z}/3 \times \mathbb{Z}/3) \xleftarrow{x} \mathbb{Z}/3 \longleftarrow 0$$

by means of constructing and comparing 2-cocycles. In the notation above, $q(0) = q(3) = q(6) = 0$, $q(1) = q(4) = q(7) = 1$, $q(2) = q(5) = q(8) = 2$, $p_2(t,u) = u$, and $x(t) = (t,0)$.

In general, if there is a seseq

$$1 \longleftarrow H = G/K \xleftarrow{q} G \xleftarrow{i} K \longleftarrow 0$$

in which the kernel K is an abelian group (written additively), then the underlying set of the group G can be identified with the cartesian product of $(H \times K)$. In fact, we have been using these identifications when we represented group elements as strings-with-quipu.

Let

$$K \xleftarrow{\xi} H \times H$$

denote the 2-cocycle that is defined by the equation

$$\xi(x,y) = i^{-1}[s(x) + s(y) - s(xy)],$$

where $G/K \xrightarrow{s} G$ is a set-theoretic section (so, $s(1) = 1$ and $q(s(z)) = z$ for all $z \in G/K$). The "ambiguous name" section can define a section $s([x]) = x$, where $x \in G$ is a coset representative, but some care will be needed, namely, we will want that

$s(x^{-1}) = -s(x)$ — the value of the section at the inverse of an element should be the negative of the value at the given element. So, suppose in addition that the cocycle ξ satisfies the condition

$$\xi(x, x^{-1}) = \xi(x^{-1}, x) = 0.$$

In the case of

$$0 \longleftarrow \mathbb{Z}/n \xleftarrow{q} \mathbb{Z} \xleftarrow{n \cdot} \mathbb{Z} \longleftarrow 0,$$

we chose coset representatives $0, 1, \ldots, n-1$ and defined the section as $s([\ell]) = \ell$. It can be shown[4] in a manner similar to the proof of Proposition 2 that the cocycle $\psi_n = (s(x)+s(y)-s(x+y))/n$ is equivalent to a cocycle ξ_n defined by the same formula but for which the section takes integer values in the interval $\left[-n + \lfloor \frac{n}{2} \rfloor + 1, n - \lceil \frac{n}{2} \rceil\right]$. This interval is symmetric around 0 when n is odd, and it leans toward the positive side when n is even. Then ξ_n satisfies the second condition.

Suppose that a function $K \xleftarrow{\xi} H \times H$ satisfies the conditions

$$\xi(y, z) - \xi(xy, z) + \xi(x, yz) - \xi(x, y) = 0,$$
$$\xi(x, x^{-1}) = \xi(x^{-1}, x) = 0,$$

and

$$\xi(1, h) = \xi(h, 1) = 0 \quad \text{for all } h \in H.$$

So, that ξ is a 2-cocycle defined on H with values in the abelian group K. Use the cocycle ξ to define a product structure on $(K \times H)$ by the equation

$$(k_1, h_1) \times_\xi (k_2, h_2) = (k_1 + k_2 + \xi(h_1, h_2), h_1 h_2).$$

Then $(K \times_\xi H)$ is a group that happens to be isomorphic to G.

---

[4]The desired proof also involves row reduction of some augmented matrices in which the right-hand side of the corresponding system of equations reflects the differences between the cocycle values. Giving it here would be even more of a distraction than this long footnote is.

First, we demonstrate that $(0,1)$ is an identity for the operation $\times_\xi$:

$$(k_1, h_1) \times_\xi (0,1) = (k_1 + 0 + \xi(h_1, 1), h_1 \cdot 1) = (k_1, h_1).$$

$$(0,1) \times_\xi (k_2, h_2) = (0 + k_2 + \xi(1, h_2), 1 \cdot h_2) = (k_2, h_2).$$

Next, we check that inverses behave as expected.

$$(k_1, h_1) \times_\xi (-k_1, h_1^{-1}) = (k_1 - k_1 + \xi(h_1, h_1^{-1}), h_1 \cdot h_1^{-1}) = (0,1),$$

$$(-k_2, h_2^{-1}) \times_\xi (k_2, h_2) = (-k_2 + k_2 + \xi(h_2^{-1}, h_2), h_2^{-1} \cdot h_2) = (0,1).$$

Finally, we check that the product is associative.

$$[(k_1, h_1) \times_\xi (k_2.h_2)] \times_\xi (k_3, h_3)$$
$$= ((k_1 + k_2 + \xi(h_1, h_2)) + k_3 + \xi(h_1 h_2, h_3), (h_1 h_2) h_3).$$

On the other hand,

$$(k_1, h_1) \times_\xi [(k_2.h_2) \times_\xi (k_3, h_3)]$$
$$= (k_1 + \xi(h_1, h_2 h_3) + (k_2 + k_3 + \xi(h_2, h_3)), h_1(h_2 h_3)).$$

These two expressions are equal by means of the 2-cocycle condition. In particular, by letting

$$\xi(x, y) = i^{-1}[s(x) + s(y) - s(xy)],$$

where s is a set-theoretic section with $s(x^{-1}) = -s(x)$ for all $x \in H$, we obtain a cocycle that defines the product structure on the group G in the seseq

$$1 \longleftarrow H = G/K \xleftarrow{q} G \xleftarrow{i} K \longleftarrow 0.$$

The proof that $(K \times_\xi H)$ is isomorphic to G will be given in more generality in Theorem 5. Also, we are going to postpone the definition of equivalent seseqs until the end of the section.

First, we want to use quipu representations for the elements in $\mathbb{Z}/9$ and the elements in $(\mathbb{Z}/3 \times \mathbb{Z}/3)$ to compute the cocycles that take values in $\mathbb{Z}/3$ and that are associated to the seseqs that contain these groups flanked by $\mathbb{Z}/3$ — those for which the kernel and the image are $\mathbb{Z}/3$.

We invite you to examine Figs. 2.15 and 2.16 for visual context. Let us start from the seseq

$$0 \longleftarrow \mathbb{Z}/3 \xleftarrow{q} \mathbb{Z}/9 \xleftarrow{3 \cdot \_} \mathbb{Z}/3 \longleftarrow 0.$$

Define a (normalized[5]) set-theoretic section $\mathbb{Z}/3 \xrightarrow{s} \mathbb{Z}/9$ by

$$s(0) = 0, \quad s(1) = 1, \quad s(-1) = -1.$$

The elements of $\mathbb{Z}/3$ are represented as $-1, 0, 1$ and those of $\mathbb{Z}/9$ are represented as $-4, -3, -2, -1, 0, 1, 2, 3, 4$. In particular, $-1 \equiv 8$ mod 9 — a fact that helps when considering the 3-strings-with-$(\mathbb{Z}/3)$-quipu representation given in Fig. 2.15. A 2-cocycle $\mathbb{Z}/3 \xleftarrow{\xi} \mathbb{Z}/3 \times \mathbb{Z}/3$ is defined by

$$\xi(x, y) = \frac{1}{3}[s(x) + s(y) - s(x + y)].$$

Clearly, $\xi(x, y) = \xi(y, x)$. It's also not too difficult to see that $\xi(0, y) = \xi(x, 0) = 0$ for all $x, y \in \mathbb{Z}/3$. We are left to compute three values of ξ:

$$\xi(1, 1) = \frac{1}{3}[s(1) + s(1) - s(-1)] = \frac{1}{3}[3] = 1,$$

$$\xi(1, -1) = \frac{1}{3}[s(1) + s(-1) - s(0)] = 0,$$

and

$$\xi(-1, -1) = \frac{1}{3}[s(-1) + s(-1) - s(1)] = -1.$$

The values of ξ are given in Table 8.2, where the two notations $\{-1, 0, 1\}$ and $\{0, 1, 2\}$ are intentionally mixed.

---

[5]The set-theoretic section is *normalized* when $s(0) = 0$ and $s(-x) = -s(x)$.

Table 8.2. The values of the cocycle $\mathbb{Z}/3 \xleftarrow{\xi} \mathbb{Z}/3 \times \mathbb{Z}.3$.

ξ	0	1	2
0	0	0	0
1	0	1	0
2	0	0	−1

Let us explain our intentional mixing of the notations.

Exercise 19. Interpret the cocycle values of ξ as being integers. Then demonstrate that ξ and the integral cocycle ψ_3 given in Table 8.1 are equivalent. That is, find a function $\mathbb{Z} \xleftarrow{\gamma} \mathbb{Z}/3$ so that

$$\psi_3(x,y) - \xi(x,y) = \gamma(y) - \gamma(x+y) + \gamma(x).$$

The context of this exercise is to address a gap in our presentation, namely, the 2-cocycles ψ_n had not been normalized. That is, $\psi_n(x, -x) \neq 0$. Yet, it is cohomologous to a normalized cocycle. This exercise addresses the case $n = 3$ with the general case following from arguments similar to Proposition 2.

Next, we follow the same process and determine a cocycle that defines the seseq

$$0 \longleftarrow \mathbb{Z}/3 \xleftarrow{p_2} (\mathbb{Z}/3 \times \mathbb{Z}/3) \xleftarrow{x} \mathbb{Z}/3 \longleftarrow 0.$$

The section that we choose for the projection $\mathbb{Z}/3 \xleftarrow{p_2} (\mathbb{Z}/3 \times \mathbb{Z}/3)$ appears in Fig. 2.16 as reading up the left most column. If $u \in \mathbb{Z}/3$, then let $s(u) = (0, u)$. In general,

$$s(u) + s(v) - s(u+v) = (0, u) + (0, v) - (0, u+v) = 0.$$

So, a 2-cocycle $\mathbb{Z}/3 \xleftarrow{\eta} \mathbb{Z}/3 \times \mathbb{Z}/3$ defined by

$$\eta(x, y) = \frac{1}{3}[s(u) + s(v) - s(u+v)]$$

will constantly be 0. That is, $\eta(x, y) = 0$ for all $x, y \in \mathbb{Z}/3$.

The groups $\mathbb{Z}/3 \times \mathbb{Z}/3$ and $\mathbb{Z}/9$ are not isomorphic. An *equivalence* between seseqs

$$1 \longleftarrow H_\ell = G_\ell/K_\ell \xleftarrow{q_\ell} G_\ell \xleftarrow{i_\ell} K_\ell \longleftarrow 0$$

for $\ell = 1, 2$ includes isomorphisms $K_1 \approx K_2$, $G_1 \approx G_2$, and $H_1 \approx H_2$ that commute with the inclusions and projections. We demonstrated that ξ is cohomologous to ψ_3 and the latter is not a coboundary. So, the cocycles η and ξ are not equivalent since the former is trivial.

8.1.2. A geometric interpretation

This section is a digression on the geometric nature of the formulas for cocycles and coboundaries. It is part of a much longer discussion about group cohomology that will only be mentioned in passing here. Some of the more general aspects are given in [2]. The material of this section begins to address the situation in which the groups G and $H = G/K$ in the seseq

$$1 \longleftarrow H = G/K \xleftarrow{q} G \xleftarrow{i} K \longleftarrow 0 \quad [\star]$$

are not commutative. As we continue, we will also address the possible non-commutativity of K. Particularly, we are interested in $K = Q_8$, but it is better to move to that case gradually.

For a moment, let us reconsider the cases

$$0 \longleftarrow \mathbb{Z}/3 \xleftarrow{p_2} (\mathbb{Z}/3 \times \mathbb{Z}/3) \xleftarrow{x} \mathbb{Z}/3 \longleftarrow 0$$

and

$$0 \longleftarrow \mathbb{Z}/3 \xleftarrow{q} \mathbb{Z}/9 \xleftarrow{3\cdot} \mathbb{Z}/3 \longleftarrow 0.$$

In the case that $G = \mathbb{Z}/3 \times \mathbb{Z}/3$, the set-theoretic section satisfies the equation

$$s(u) + s(v) - s(u + v) = 0.$$

So, in fact, the function s is a group homomorphism. Meanwhile, in the case that $G = \mathbb{Z}/9$, the section $s(0) = 0$, $s(1) = 1$, and $s(-1) = -1$ is a choice of coset representatives of the subgroup

$\mathbb{Z}/3 = \{0, 3, 6\}$. The 2-cocycle measures the failure of that set-theoretic section to be a group homomorphism in the sense that

$$\xi(x, y) = s(x) + s(y) - s(x + y) \neq 0.$$

That is, there are some values of x and y for which $\xi(x, y) \neq 0$, so that $s(x + y) \neq s(y) + s(y)$ for such values.

In the seseq [⋆], the idea when a normalized set-theoretic section is a group homomorphism will be examined further. Then some geometric context will be added.

In general, there are set-theoretic sections $H \xrightarrow{s} G$ that are obtained by picking coset representatives of G/K. We choose such representatives so that

- $s(1) = 1$;
- $[s(h)]^{-1} = s(h^{-1})$;
  since these are coset representatives,
- $q(s(h)) = h$ for all $h \in H$.

We recall that such a function is a normalized set-theoretic section.

Since the kernel K is a normal subgroup of G, for any $g \in G$, conjugation of K by elements of the group G yield an equality of sets: $gKg^{-1} = K$. In particular, the quotient group H acts upon K by means of the section. We state this as a proposition (see also [1, p. 87]).

Proposition 3. *Let a normalized set-theoretic section $H \xrightarrow{s} G$ for the seseq [⋆] be given. For every $h \in H$ and for every $k \in K$, there is a $k' \in K$ such that $s(h)i(k)s(h)^{-1} = i(k')$.*

Write $k' = k^h$ to indicate that the element is the result of a conjugation action of H upon K. Then the above relation can be rewritten as

$$i(k^h)s(h) = s(h)i(k).$$

Since the group G is not abelian, the action of H upon K may not be trivial. We will use this action to refine the idea of a 2-cocycle.

Computing Group 2-cocycles 371

The geometric context of the section is encapsulated in how the boundaries of simplices were computed in Section 4.1.2 of Chapter 4. Only the cases of a a triangle and a tetrahedron are important to the text, but be aware that there are higher-dimensional versions.

The binary operation that defines a group structure interests us. It is schematized via labelings of a Y shape as on the left of Fig. 8.1. On the right of Fig. 8.1, the Y is surrounded by a triangle Δ whose edges are oriented. The edge of positive slope is oriented upward; the edge that has negative slope is oriented downward, and the horizontal edge is oriented from left to right. Let us write oriented boundary of the triangle Δ as a formal product

$$(x)(y)(xy)^{-1},$$

where the group elements are labels, but they are merely juxtaposed and not to be thought of as being multiplied. The orientations upon the edges of the triangles agree with the directions of the short red arrows that decorate the Y.

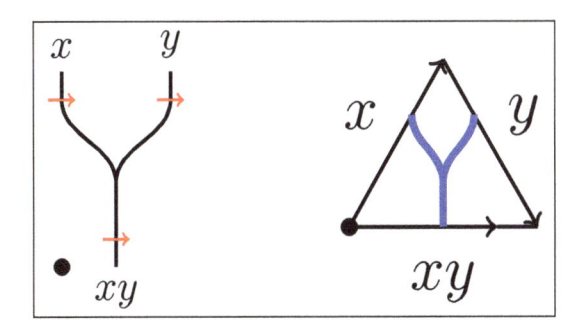

Figure 8.1. Multiplication and the boundary of a triangle.

The triangle on the right of Fig. 8.1 is oriented in the opposite direction from that illustrated in Fig. 4.3, but observe that it has two contiguous edges oriented coherently with the remaining edge running oppositely just as that in Fig. 4.3. When we read that the boundary is the formal product $(x)(y)(xy)^{-1}$, we are reading around the boundary of the triangle in a clockwise fashion starting at the lower left vertex that is indicated by •.

Now, associate a normalized set-theoretic section $H \xleftarrow{s} K$ to the segments of the boundary of the triangle, and define a function $K \xrightarrow{\xi} H \times H$ by the equation

$$\xi(x,y) = s(x)s(y)[s(xy)]^{-1}.$$

We think of $\xi(x,y)$ as having been assigned to the triangle on the right of Fig. 8.1.

These ideas are extended to a labeling of the edges of a tetrahedron Δ_3, where the labels are group elements. Three contiguous edges that do not bound a triangle on a tetrahedron are labeled by means of group elements $x, y, z \in H$, as illustrated in Fig. 8.2.

The boundary of the tetrahedron is envisioned as a front and back. Each of these is a pair of triangles glued upon an edge. The back is illustrated on the left of Fig. 8.2. The edge labeled x is imagined to be oriented upward. The edges labeled y and xyz are oriented left to right. The edge labeled z is oriented downward. The upper left triangle on the left of the figure is the triangle that may be denoted as $[x, y]$. The other triangle on the left is denoted $[xy, z]$ since the orientation on the edge labeled xy is upward and left to right. These two triangles appear on the left of the illustration since they are on the back of the tetrahedron as it has been depicted.

On the right of the figure, the diagonal that is labeled by the product yz is downward pointing. That orientation is induced from the directions upon y and z. The two triangles on the right are denoted $[x, yz]$ and $[y, z]$. Compare these orientations with those in Fig. 4.4.

Each triangle will be associated to a value of the function ξ. The group operation in the kernel K is written multiplicatively. The geometric relation that will be imposed upon ξ is that

$$\xi(x,y)\xi(xy,z) = [\xi(y,z)]^x \xi(x,yz).$$

Here, we recall that ξ is defined upon $H \times H$, it takes values in K, and H is acting upon K via conjugation that is induced by the set-theoretic section s.

Computing Group 2-cocycles 373

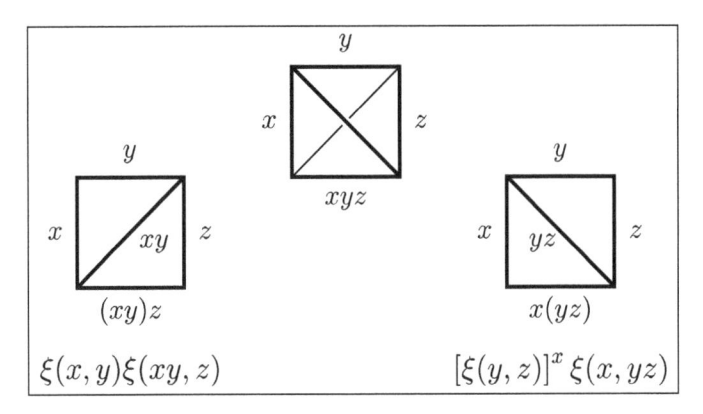

Figure 8.2. The edges of a tetrahedron labeled by group elements.

In general, a *2-cocycle that takes values in a group K upon which a group H acts* is a function ξ that satisfies

$$\xi(x,y)\xi(xy,z) = [\xi(y,z)]^x \xi(x,yz),$$

where $[\xi(y,z)]^x$ is indicative of the group action. The reason that this is a "cocycle" is that it is evaluated upon the boundary of a tetrahedron which is a "cycle." We recall from Section 4.1.2 that a cycle is an object whose boundary, ∂, is 0, and moreover, $\partial^2 = 0$. So, the boundary of the tetrahedron is a 2-cycle.

It's time now to combine the ideas that (a) ξ measures the failure of a set-theoretic section to be a homomorphism, (b) the 2-cocycle condition is associated to the boundary of a tetrahedron, and (c) the quotient group, H, acts upon the kernel, K, via conjugation and by means of the set-theoretic section s.

Theorem 5 is an expansion of the material in [1], at roughly pp. 90–92. There are some notational complexities in its statement and in its proof that are, perhaps, unavoidable. Some of these are more pronounced with the proof of Proposition 4.

Theorem 5. *Consider a seseq*

$$1 \longleftarrow H = G/K \xleftarrow{q} G \xleftarrow{i} K \longleftarrow 1,$$

in which the kernel K is a group that is written multiplicatively. Let a function $H \xrightarrow{s} G$ that satisfies

- $s(1) = 1$;
- $[s(h)]^{-1} = s(h^{-1})$;
- $q(s(h)) = h$ for all $h \in H$

be given. Define a function $K \xleftarrow{\xi} H \times H$ by the equation

$$\xi(x, y) = s(x)s(y)[s(xy)]^{-1}.$$

Then ξ satisfies the 2-cocycle condition

$$\xi(x, y)\xi(xy, z) = [\xi(y, z)]^x \xi(x, yz).$$

Moreover, as a set, the group G can be identified with $K \times H$, and under this operation, its group operation can be given by

$$(\alpha, x) \times_\xi (\beta, y) = (\alpha \cdot \beta^x \cdot \xi(x, y), xy).$$

Proof. The idea of the proof is to evaluate the set-theoretic section s of the H-labeled edges of the tetrahedron and associate to the triangular faces of the tetrahedron the values of ξ. So, (L)

$$\xi(x, y)\xi(xy, z) = s(x)s(y)[s(xy)]^{-1}s(xy)s(z)[s(xyz)]^{-1}$$

and (R)

$$[\xi(y, z)]^x \xi(x, yz)$$
$$= s(x)s(y)s(z)[s(yz)]^{-1}[s(x)]^{-1}s(x)s(yz)[s(xyz)]^{-1}.$$

These two expressions are equal.

Read around the boundary of the square on the left-hand side of Fig. 8.2 from the lower left corner, but evaluate s at the successive edges. When you get to the upper right corner, go back to the corner along the diagonal. In this way, you will have circumnavigated the $[x, y]$ triangle. Then circumnavigate the $[xy, z]$ triangle. These circumnavigations are successively evaluating $\xi(x, y)$ and $\xi(xy, z)$. By reading s at the successive labels on this tour, you will have voiced the right-hand side of the equation (L).

On the square on the right, perform a similar reading, but think about the upper right triangle as having a "tail" that is labeled by x. To circumnavigate the $[y, z]$ triangle, you have to follow the x edge out, and return along x before circumnavigating the $[x, yz]$ triangle. In this way, you will have voiced the right side of equation (R).

Next, let us demonstrate that
$$(\alpha, x) \cdot (\beta, y) = (\alpha \cdot \beta^x \cdot \xi(x, y), xy)$$
defines a group structure. First, consider the element $(1, 1)$:
$$(1, 1) \cdot (\beta, y) = (1 \cdot \beta \cdot \xi(1, y), 1y) = (\beta, y)$$
and
$$(\alpha, x) \cdot (1, 1) = (\alpha \cdot 1^x \cdot \xi(x, 1), x) = (\alpha, x).$$
As far as inverses are concerned,
$$(\alpha, x) \cdot \left([\alpha^{-1}]^{x^{-1}}, x^{-1} \right) = \left(\alpha \cdot \left[[\alpha^{-1}]^{x^{-1}} \right]^x \cdot \xi(x, x^{-1}), xx^{-1} \right)$$
$$= (\alpha \cdot (\alpha^{-1}) \cdot 1, 1) = (1, 1).$$
This follows since
$$\left[[\alpha^{-1}]^{x^{-1}} \right]^x = [s(x^{-1})\alpha^{-1} s(x)]^x$$
$$= s(x) s(x^{-1}) \alpha^{-1} s(x) s(x)^{-1} = \alpha^{-1}.$$
Furthermore,
$$\left([\alpha^{-1}]^{x^{-1}}, x^{-1} \right) \cdot (\alpha, x)$$
$$= \left([\alpha^{-1}]^{x^{-1}} \alpha^{x^{-1}} \xi(x^{-1}, x), x^{-1} x \right)$$
$$= \left(s(x^{-1}) \alpha^{-1} s\left(x^{-1}\right)^{-1} s(x^{-1}) \alpha s \left(x^{-1}\right)^{-1}, 1 \right) = (1, 1)$$
since all the rigamarole in the center on the left cancels.

To show that the product is associative takes a bit more work:
$$[(\alpha, x)(\beta, y)](\gamma, z) = (\alpha \beta^x \xi(x, y), xy) \, (\gamma, z)$$
$$(\alpha \beta^x \xi(x, y) \gamma^{xy} \xi(xy, z), (xy)z) \, .$$
Meanwhile,
$$(\alpha, x)[(\beta, y)(\gamma, z)] = (\alpha, x) \, (\beta \gamma^y \xi(y, z), yz)$$
$$= (\alpha \, (\beta \gamma^y \xi(y, z))^x \, \xi(x, yz), x(yz)) \, .$$
To match these expressions, we consider
$$(\beta \gamma^y \xi(y, z))^x = s(x) \beta s(y) \gamma s(y)^{-1} \xi(y, z) s(x)^{-1}.$$

There is a technique that is commonly used when faced with a conjugation, say by $s(x)$, on the outside of a multiplicative expression, namely, insert expressions $[s(x)]^{-1}s(x)$ between the factors. Then we rewrite things using the exponential notation for conjugation:

$$(\beta\gamma^y\xi(y,z))^x = (\beta^x\,(\gamma^y)^x\,\xi(y,z)^x).$$

That conjugation distributes over group multiplication is one reason that the exponential notation is used. In ordinary arithmetic, exponents distribute over multiplication.

We must show that

$$L = \alpha\beta^x\xi(x,y)\gamma^{xy}\xi(xy,z)$$

and

$$R = \alpha\beta^x\,(\gamma^y)^x\,[\xi(y,z)]^x\xi(x,yz)$$

are the same. First, cancel the common factors of $\alpha\beta^x$, and then rewrite $[\xi(y,z)]^x\xi(x,yz) = \xi(x,y)\xi(xy,z)$ using the cocycle condition. Then we ask

$$¿\ L' = \xi(x,y)\gamma^{xy} \stackrel{?}{=} (\gamma^y)^x\,\xi(x,y) = R'\ ?$$

Here, we will be careful about the conjugation actions and rewrite ξ in terms of sections. Then

$$R' = (\gamma^y)^x\,\xi(x,y)$$
$$= s(x)s(y)\gamma[s(y)]^{-1}[s(x)]^{-1}s(x)s(y)[s(xy)]^{-1}$$
$$= s(x)s(y)\gamma[s(xy)]^{-1}.$$

Meanwhile,

$$L' = \xi(x,y)\gamma^{xy}$$
$$= s(x)s(y)[s(xy)]^{-1}s(xy)\gamma[s(xy)]^{-1}$$
$$= s(x)s(y)\gamma[s(xy)]^{-1}.$$

These expressions are equal, and so, the multiplication is associative.

As a final step, we want to show that the multiplication on $H \times K$ that is twisted by the cocycle ξ agrees with the multiplication in G. First, define a bijection between $K \times H$ and G by means of

$$(\alpha, x) \leftrightarrow i(\alpha)s(x).$$

Then consider the product in G

$$i(\alpha)s(x) \cdot i(\beta)s(y) = i(\alpha)[s(x)i(\beta)]s(y) = i(\alpha)[i(\beta^x)s(x)]s(y)$$

because of the relation given immediately below Proposition 3. We continue as follows:

$$i(\alpha)s(x) \cdot i(\beta)s(y) = i(\alpha)[i(\beta^x)s(x)]s(y)$$
$$= i(\alpha\beta^x)i(\xi(x,y))s(xy) = i(\alpha\beta^x \xi(x,y))s(xy)$$
$$\leftrightarrow (\alpha\beta^x \xi(x,y), xy).$$

Therefore, the correspondence is a bijective homomorphism or an isomorphism. This completes the proof. □

Remark. Please note that in the proof the order of multiplication was maintained. In this way, the kernel K did not need to be a commutative group.

In the above proof, the 2-cocycle condition insured that the product is **associative**. In general, the associativity condition is encoded within the tetrahedron, as indicated in Fig. 8.2. There are higher-dimensional cocycle conditions, and these can be read from the boundaries of higher-dimensional simplices. The cocycle conditions, then, are expressions about the so-called higher associativities. The volume [2] contains some of the details of those theories.

We now turn to computing some 2-cocycles explicitly. In the process, we'll demonstrate how to extend the definition of the cocycle condition so that the group K in which the cocycle takes values is the quaternions Q_8.

8.2. Example 1. The symmetric group Σ_4

We recall from Section 5.1 that there is a seseq

$$1 \longleftarrow \Sigma_3 \xleftarrow{p} \Sigma_4 \xleftarrow{i} K_4 \longleftarrow 0$$

that is the prototype for similar sequences in which the group in the middle is an extension of the symmetric group Σ_4.

Our goal in this section is to define a 2-cocycle $K_4 \xleftarrow{\xi} \Sigma_3 \times \Sigma_3$ that classifies the seseq above. Since K_4 is given to be a subgroup of Σ_4, the value of ξ will be determined by means of a normalized set-theoretic section $s : \Sigma_3 \to \Sigma_4$.

$$\xi(g,h) = s(g)s(h)[s(gh)]^{-1}.$$

Such a function has the following properties:

- $p(s(g)) = g$ for all $g \in \Sigma_3$;
- $s(1) = 1$;
- $s(g^{-1}) = [s(g)]^{-1}$.

The illustrations in Figs. 5.1–5.3 will be used to choose such a section s. For each of the transpositions (12), (23), and (13), there are two choices for the value of s because among the four preimages, two are 4-cycles as permutations. Since $s(g^{-1}) = [s(g)]^{-1}$, the transpositions must lift to transpositions. For all three, we will chose the element that has only one (mod 2)-quipu upon its 3-strings picture. Thus, $s(12) = (24)$ (Fig. 5.2), $s(23) = (34)$ (Fig. 5.1), and $s(13) = (14)$ (Fig. 5.3). For the permutation 3-cycles, we choose $s(123) = (123)$ (top of Fig. 5.2) and $s(132) = (132)$ (top of Fig. 5.3). since neither of these elements has quipu.

For more visual clarity, the elements in Σ_3 will often be written with their parentheses removed, especially when they are arguments of functions. That is, the elements will be written as indicated: $1 := (1)$; $12 := (12)$; $23 := (23)$; $123 := (123)$; and $132 := (132)$.

We will use the 3-strings-with-(mod 2)-quipu representations of the elements in Σ_4 to compute $\xi(g,h) = s(g)s(h)[s(gh)]^{-1}$. Unlike the cases above, the group Σ_4 is not abelian, so the table of cocycle values will not be symmetric. However,

$$\xi(1,g) = s(1)s(g)[s(g)]^{-1} = s(1) = 1 = s(g)s(1)[s(g)]^{-1} = \xi(g,1)$$

and

$$\xi(g,g) = s(g)s(g)[s(g^2)]^{-1} = 1$$

when g is a transposition. Meanwhile,

$$\xi(123,123) = s(123)s(123)[s(132)]^{-1} = (123)^3 = 1$$

and

$$\xi(132,132) = s(132)s(132)[s(123)]^{-1} = (132)^3 = 1.$$

So, there are only 20 values of ξ that remain to be computed. These computations are indicated in Figs. 8.3–8.7

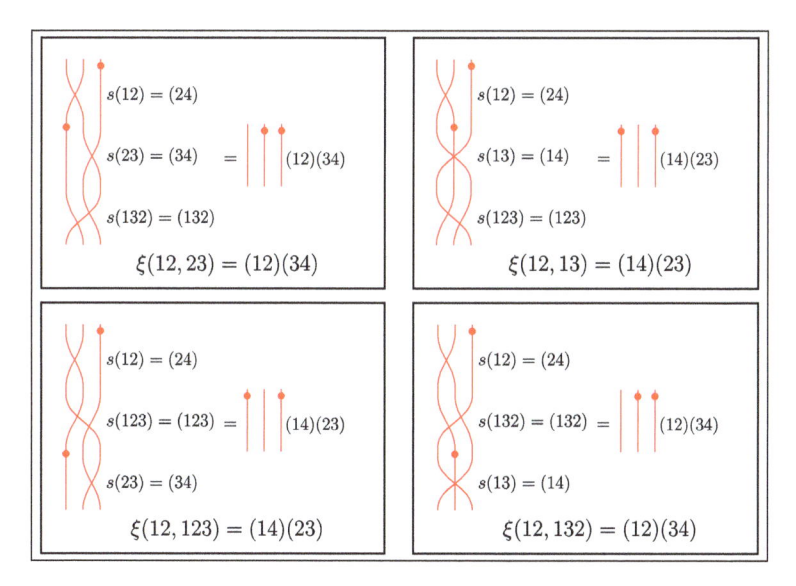

Figure 8.3. The cocycle values $\xi(12, \_)$.

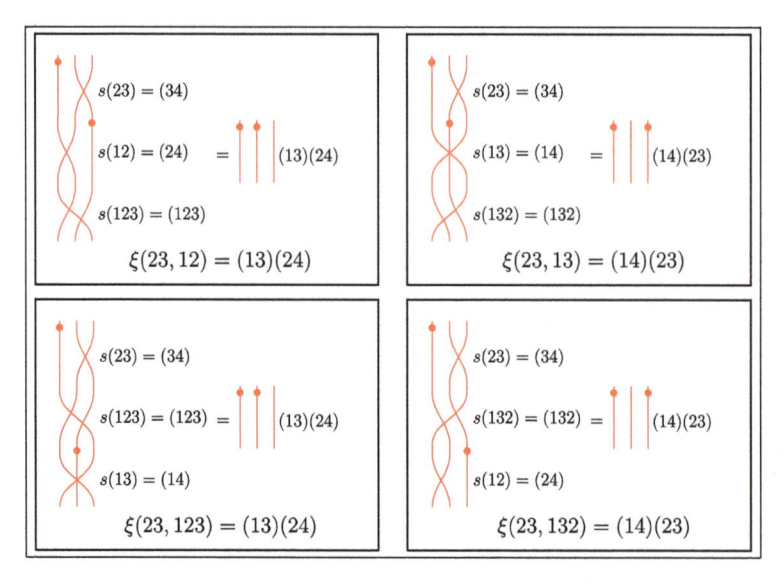

Figure 8.4. The cocycle values $\xi(23, \_)$.

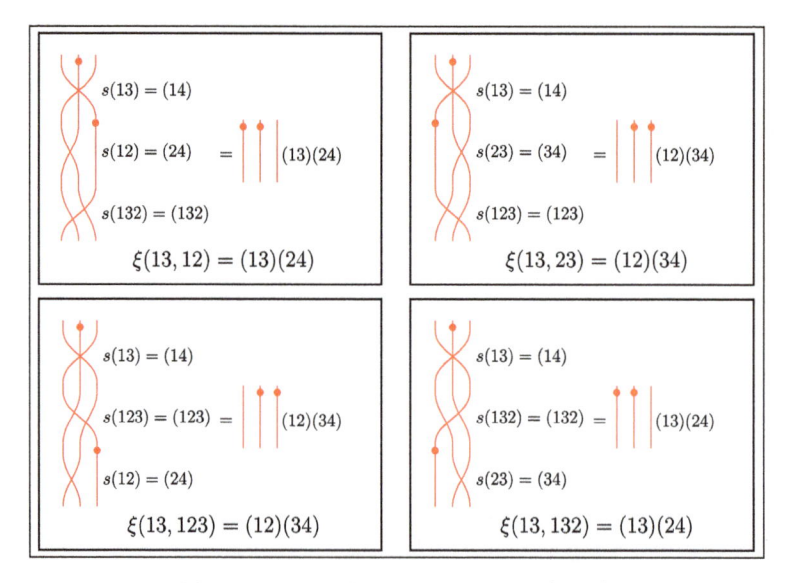

Figure 8.5. The cocycle values $\xi(13, \_)$.

Computing Group 2-cocycles

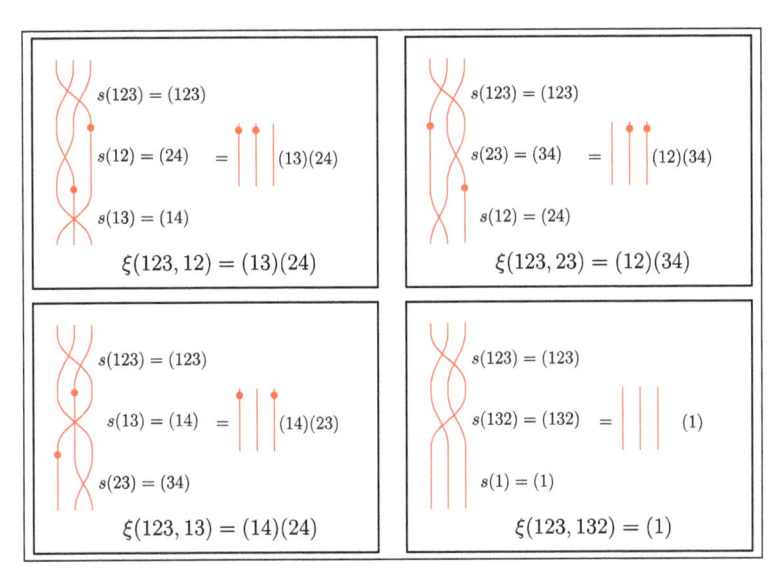

Figure 8.6. The cocycle values $\xi(123, \_)$.

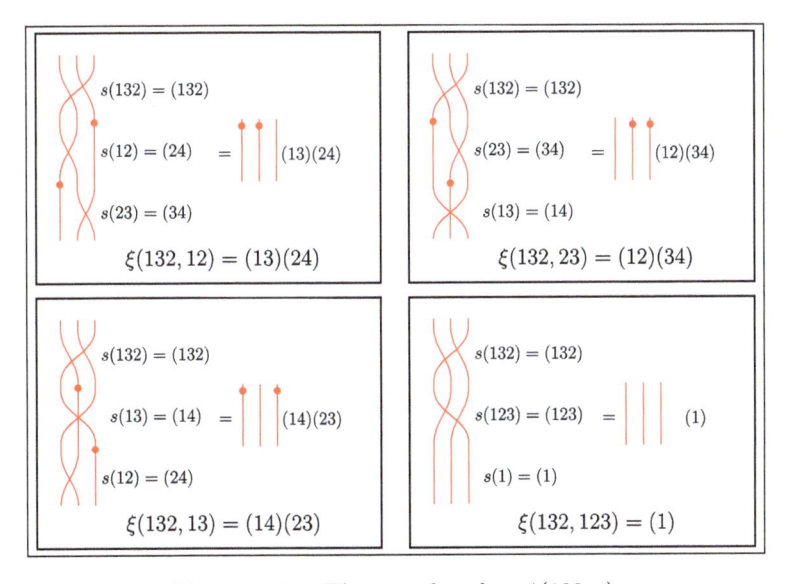

$s(123) = (123)$
$s(12) = (24)$
$s(13) = (14)$
$\xi(123, 12) = (13)(24)$

$s(123) = (123)$
$s(23) = (34)$
$s(12) = (24)$
$\xi(123, 23) = (12)(34)$

$s(123) = (123)$
$s(13) = (14)$
$s(23) = (34)$
$\xi(123, 13) = (14)(24)$

$s(123) = (123)$
$s(132) = (132)$
$s(1) = (1)$
$\xi(123, 132) = (1)$

Figure 8.7. The cocycle values $\xi(132, \_)$.

$s(132) = (132)$
$s(12) = (24)$
$s(23) = (34)$
$\xi(132, 12) = (13)(24)$

$s(132) = (132)$
$s(23) = (34)$
$s(13) = (14)$
$\xi(132, 23) = (12)(34)$

$s(132) = (132)$
$s(13) = (14)$
$s(12) = (24)$
$\xi(132, 13) = (14)(23)$

$s(132) = (132)$
$s(123) = (123)$
$s(1) = (1)$
$\xi(132, 123) = (1)$

All of the values of the function $K_4 \xleftarrow{\xi} \Sigma_3 \times \Sigma_3$ are presented in Table 8.3.

Table 8.3. A table of 2-cocycle values $K_4 \xleftarrow{\xi} \Sigma_3 \times \Sigma_3$.

$\xi(r,c)$	(1)	(123)	(132)	(12)	(23)	(13)
(1)	(1)	(1)	(1)	(1)	(1)	(1)
(123)	(1)	(1)	(1)	(13)(24)	(12)(34)	(14)(23)
(132)	(1)	(1)	(1)	(13)(24)	(12)(34)	(14)(23)
(12)	(1)	(14)(23)	(12)(34)	(1)	(12)(34)	(14)(23)
(23)	(1)	(13)(24)	(14)(23)	(13)(24)	(1)	(14)(23)
(13)	(1)	(12)(34)	(13)(24)	(13)(24)	(12)(34)	(1)

By Theorem 5, the function $K_4 \xleftarrow{\xi} \Sigma_3 \times \Sigma_3$ that is presented in Table 8.3 satisfies

$$\xi(x,y)\xi(xy,z) = [\xi(y,z)]^x \xi(x,yz).$$

The function ξ is a 2-cocycle.

Remark. The action of Σ_3 upon K_4 can be given in a clever way. Let us rename the elements of K_4 as

$$\mathbf{0} \leftrightarrow (1), \; \mathbf{1} \leftrightarrow (12)(34), \; \mathbf{2} \leftrightarrow (14)(23), \mathbf{3} \leftrightarrow (13)(24).$$

The conjugation action of Σ_3 upon K_4 is via the set-theoretic section s. The images of the elements in Σ_3 are strings-with-(mod 2)-quipu with the fixed string in the transpositions being decorated with a quipu, and the section values of the 3-cycles have no quipu. The non-trivial elements in K_4 are straight strings with two quipu among the three strings.

In the relabeling, we observe that the conjugation action of any element of Σ_3 fixes $\mathbf{0}$, (12) interchanges $\mathbf{1}$ and $\mathbf{2}$ while it fixes $\mathbf{3}$, (23) interchanges $\mathbf{2}$ and $\mathbf{3}$ while it fixes $\mathbf{1}$, and this tautological phenomenon holds for each element of Σ_3. If you wish to check directly that ξ satisfies the cocycle condition that appears immediately above this remark, then you will find this labeling and this tautological action helpful.

For example, let's check that

$$¿\xi(12,23)\xi((12)(23),13) \stackrel{?}{=} [\xi(23,13)]^{12}\xi(12,(23)(13))?$$

We compute $(12)(23) = (123) = (23)(13)$ and rewrite the cocycle values:

$$\begin{aligned} \text{LHS} &= \xi(12,23)\xi((12)(23),13) \\ &= \xi(12,23)\xi(123,13) \\ &= (12)(34) \cdot (14)(23) = (13)(24) \end{aligned}$$

and

$$\begin{aligned} \text{RHS} &= [\xi(23,13)]^{12}\xi(12,(23)(13)) \\ &= \mathbf{2}^{12}\xi(12,123) = \mathbf{1}(14)(23) \\ &= (12)(34)(13)(23) = (13)(24). \end{aligned}$$

Perhaps, it is impractical to check all 216 possible equations in this way, but you might find some satisfaction in checking a few values of x, y, and z directly.

As a final item of this section, the cocycle ξ whose values are presented in Table 8.3 is not a coboundary. To demonstrate this, suppose that

$$\xi(x,y) = \delta g(x,y) = g(\partial(x,y)) = g(y)[g(x,y)]^{-1}g(x)$$

for all $x, y \in \Sigma_3$ where g is a function $K_4 \xleftarrow{g} \Sigma_3$ in order to find a contradiction. The Klein 4-group K_4 is abelian and each of its elements is of order 2. Let $x = y = (123)$. We have that $\xi((123),(123)) = 1$, and therefore,

$$g(123)g(132)g(123) = g(132) = 1.$$

Similarly, if $x = y = (132)$,

$$g(132)g(123)g(132) = g(123) = \xi(132,132) = 1.$$

Next, compare some values of $g(y)g((12)y)g(12)$ with the values $\xi(x, 12)$.

$$g(23)g((12)(23))g(12) = g(23)g(12),$$
$$g(123)g((12)(123))g(12) = g(23)g(12).$$

Meanwhile,

$$\xi(12, 23) = (12)(34),$$
$$\xi(12, 123) = (14)(23).$$

According to the assumption, $(12)(34) = (14)(23)$. So, the assumption that ξ is a coboundary is incorrect.

A 2-cocycle that defines a seseq

$$1 \leftarrow G/K = H \xleftarrow{q} G \xleftarrow{i} K \leftarrow 1$$

is a coboundary if and only if G is a semi-direct product $G \approx K \rtimes H$. Therefore, the symmetric group Σ_4 is not a semi-direct product of K_4 and Σ_3.

8.3. Example 2. The group $SL_2(\mathbb{Z}/4)$

From Section 5.3 and in particular Sections 5.3.1 and 5.3.2, there is a seseq

$$1 \longleftarrow \Sigma_3 \xleftarrow{q} SL_2(\mathbb{Z}/4) \xleftarrow{i} (\mathbb{Z}/2)^3 \longleftarrow 0 \quad [\text{Ses}(1)].$$

The illustrations Figs. 5.54–5.59 and 5.62–5.67 indicate the elements of the matrix group as 3-strings-with-dihedral-quipu. By ignoring the (mod 2)-quipu upon the teal-colored strings in Figs. 5.68–5.73, it may be easy to recognize the projection q. But please recall that the labels on those illustrations are not elements in Σ_3 but rather in Σ_4 that are induced from the blue 4-strings-with-dicyclic-quipu representations. Moreover, this projection from Σ_4 to Σ_3 disagrees with the homomorphism that was used in Section 8.2.

The goal of this section is to use these illustrations to define a 2-cocycle $(\mathbb{Z}/2)^3 \xleftarrow{\eta} \Sigma_3 \times \Sigma_3$ that classifies Ses(1). To that end, a set-theoretic section $\Sigma_3 \xrightarrow{s} SL_2(\mathbb{Z}/4)$ will be defined by means of examining Figs. 5.68–5.73.

To construct the 2-cocycle η_4, a *not entirely normal* set-theoretic section $\Sigma_3 \xrightarrow{s} SL_2(\mathbb{Z}/4)$ will be defined. It is not entirely normal since it is impossible that $s(x^{-1}) = [s(x)]^{-1}$ when x is a transposition ($x = (12), (23)$, or (13)). Consider, for example, Fig. 5.69, the possible values for $s(12)$ are $\pm j$, $\pm Aj$, $\pm Bj$, or $\pm Cj$. Yet all of these elements have order 4 in $SL_2(\mathbb{Z}/4)$.

The choices for the values of the (not entirely normal) set-theoretic section s at the transpositions will be made for convenience. We choose those with homogenous quipu:

$$s(1) = I = \begin{bmatrix} 1 & 0 \\ 0 & 1 \end{bmatrix}; \quad s(12) = -Cj = \begin{bmatrix} 2 & 1 \\ -1 & 2 \end{bmatrix},$$

$$s(23) = j\rho = \begin{bmatrix} -1 & 1 \\ 2 & 1 \end{bmatrix}; \quad s(13) = Bj\rho^2 = \begin{bmatrix} -1 & 2 \\ -1 & 1 \end{bmatrix},$$

$$s(123) = -C\rho = \begin{bmatrix} 0 & 1 \\ -1 & -1 \end{bmatrix}; \quad s(132) = -B\rho^2 = \begin{bmatrix} -1 & -1 \\ 1 & 0 \end{bmatrix}.$$

After finding the cocycle η and proving that it is not a coboundary, the issue of inverses will be discussed.

Several values of η are computed to be trivial. We write $I = \begin{bmatrix} 1 & 0 \\ 0 & 1 \end{bmatrix}$ and $-I = \begin{bmatrix} -1 & 0 \\ 0 & -1 \end{bmatrix}$. Across the first row of the cocycle table, we have

$$\eta(1,1) = s(1)s(1)[s(1)]^{-1} = I,$$
$$\eta(1,12) = s(1)s(12)[s(12)]^{-1} = I,$$
$$\eta(1,23) = s(1)s(23)[s(23)]^{-1} = I,$$
$$\eta(1,13) = s(1)s(13)[s(13)]^{-1} = I,$$
$$\eta(1,123) = s(1)s(123)[s(123)]^{-1} = I,$$
$$\eta(1,132) = s(1)s(132)[s(132)]^{-1} = I.$$

Down the first column, we have

$$\eta(12,1) = s(12)s(1)[s(12)]^{-1} = I,$$
$$\eta(23,1) = s(23)s(1)[s(23)]^{-1} = I,$$
$$\eta(13,1) = s(13)s(1)[s(13)]^{-1} = I,$$
$$\eta(123,1) = s(123)s(1)[s(123)]^{-1} = I,$$
$$\eta(132,1) = s(132)s(1)[s(132)]^{-1} = I.$$

Along the diagonals, we have

$$\eta(123,123) = s(123)s(123)[s(132)]^{-1} = I,$$
$$\eta(132,132) = s(132)s(132)[s(123)]^{-1} = I,$$
$$\eta(12,12) = s(12)s(12)[s(1)]^{-1} = -I,$$
$$\eta(23,23) = s(23)s(23)[s(1)]^{-1} = -I,$$
$$\eta(13,13) = s(13)s(13)[s(1)]^{-1} = -I.$$

Two other values can be computed without using matrix multiplication or the graphical calculus.

$$\eta(123,132) = s(123)s(132)[s(1)]^{-1} = I,$$
$$\eta(132,123) = s(132)s(123)[s(1)]^{-1} = I.$$

The rest of the values of the cocycle η are computed in Figs. 8.8–8.16.

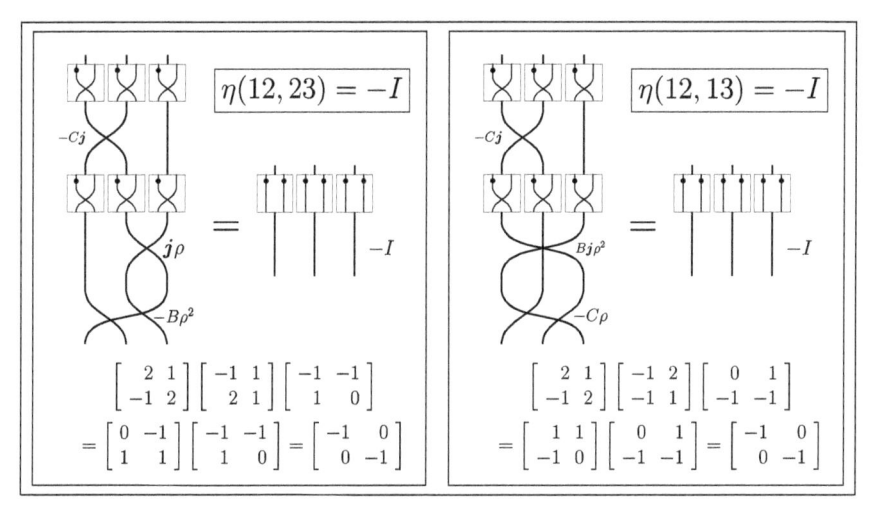

Figure 8.8. Computing $\eta(12, 23)$ and $\eta(12, 13)$.

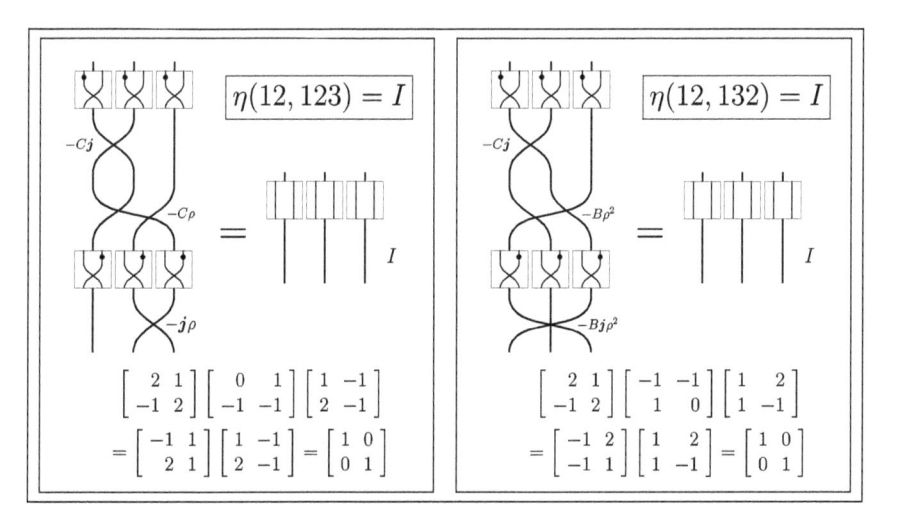

Figure 8.9. Computing $\eta(12, 123)$ and $\eta(12, 132)$.

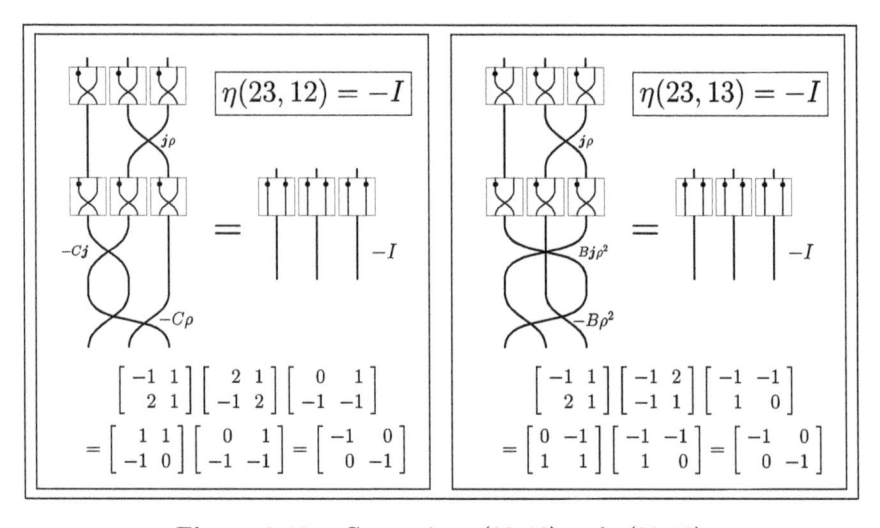

Figure 8.10. Computing $\eta(23, 12)$ and $\eta(23, 13)$.

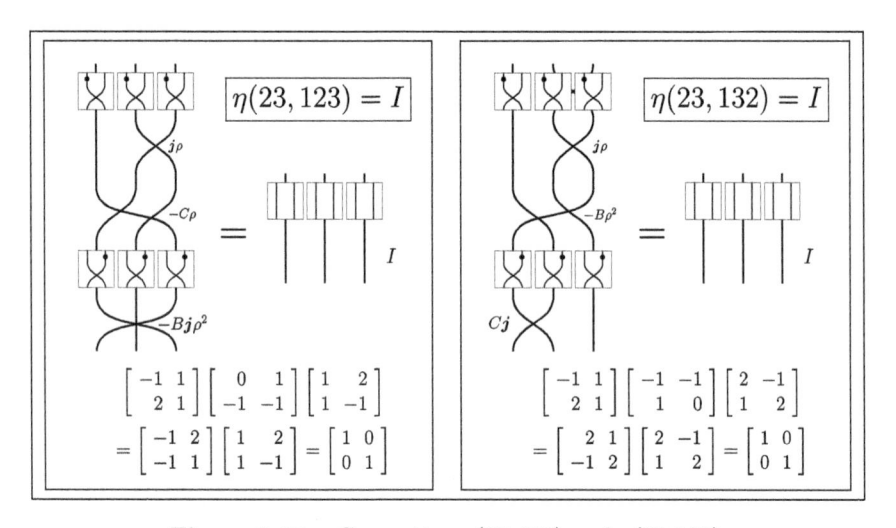

Figure 8.11. Computing $\eta(23, 123)$ and $\eta(23, 132)$.

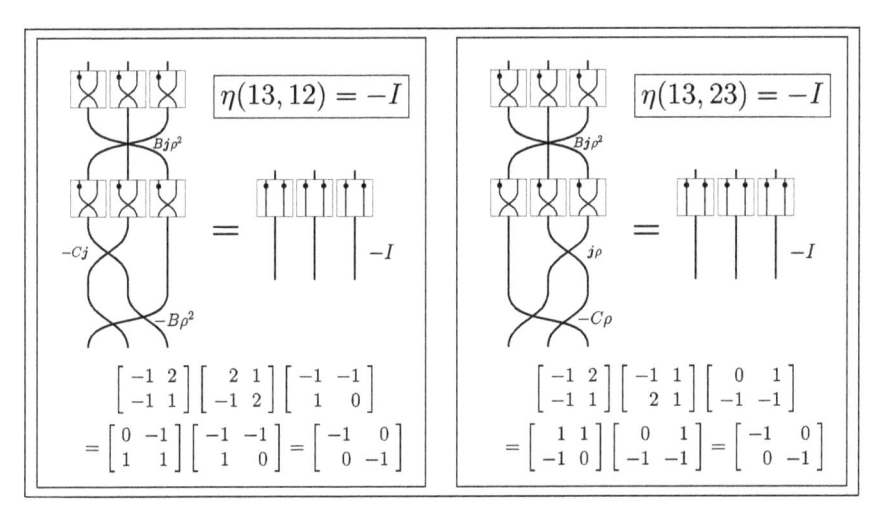

Figure 8.12. Computing $\eta(13,12)$ and $\eta(13,23)$.

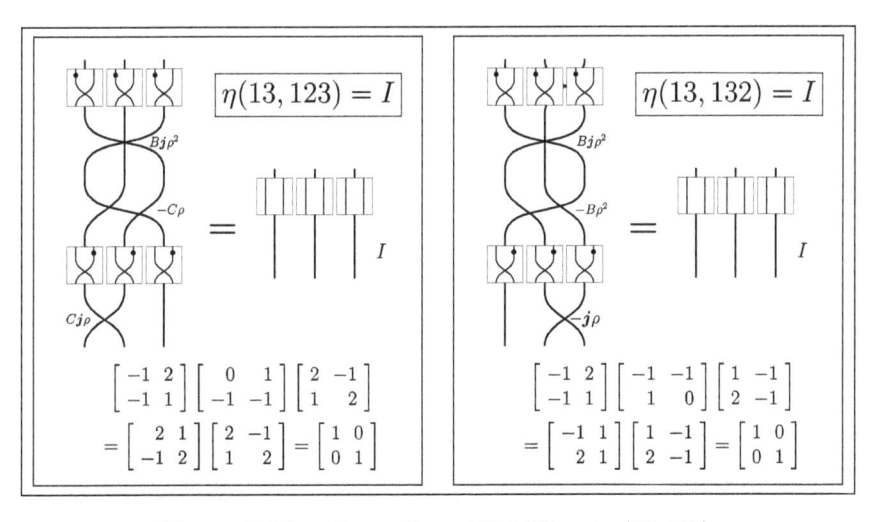

Figure 8.13. Computing $\eta(13,123)$ and $\eta(13,132)$.

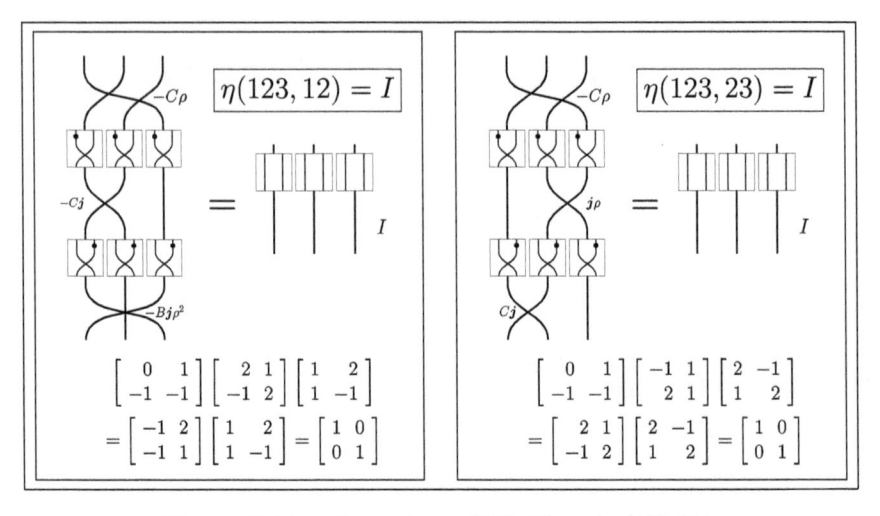

Figure 8.14. Computing $\eta(123,12)$ and $\eta(123,23)$.

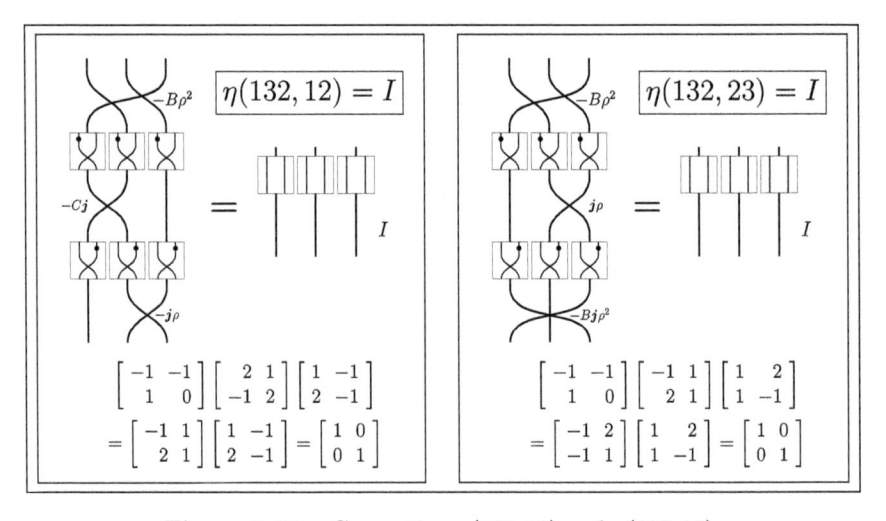

Figure 8.15. Computing $\eta(132,12)$ and $\eta(132,23)$.

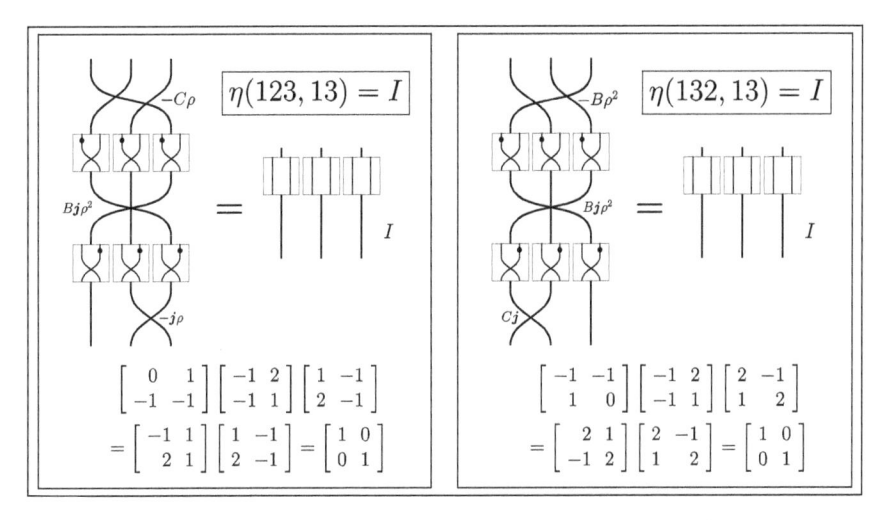

Figure 8.16. Computing $\eta(123, 13)$ and $\eta(132, 13)$.

Table 8.4 compiles the cocycle values for η.

Table 8.4. The collection of 2-cocycle values for η.

$\eta(r,c)$	(1)	(123)	(132)	(12)	(23)	(13)
(1)	I	I	I	I	I	I
(123)	I	I	I	I	I	I
(132)	I	I	I	I	I	I
(12)	I	I	I	$-I$	$-I$	$-I$
(23)	I	I	I	$-I$	$-I$	$-I$
(13)	I	I	I	$-I$	$-I$	$-I$

Please observe that $\eta(x, y) = -I$ when both x and y are transpositions, and $\eta(z, w) = I$ when either z or w is a 3-cycle.

For your convenience, we summarize the action of Σ_3 upon the subgroup $K = \{\pm I, \pm A, \pm B, \pm C\}$. Recall from Section 5.3.1 that the quipu descriptions of the elements of K are as indicated in Fig. 8.17. In this illustration, these have been reordered and partitioned into four subsets.

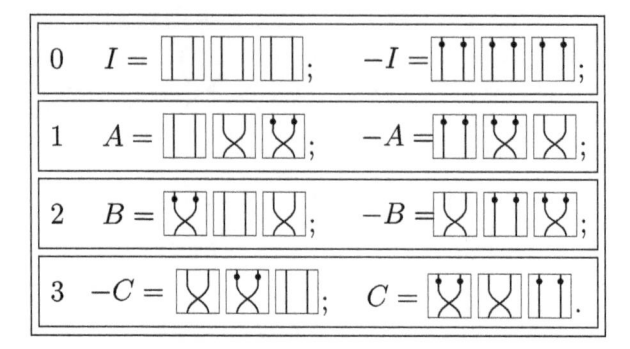

Figure 8.17. Partitioning $(\mathbb{Z}/3)^3$ into four subsets upon which Σ_3 acts.

Recall that the action of an element $x \in \Sigma_3$ upon K is given via the set-theoretic section: $k^x = s(x)k[s(x)]^{-1}$. The section is not entirely normal, and $[s(x)]^{-1} = -s(x^{-1})$ when x is a transpositions. So, it is helpful to compute the conjugation actions that appear in Fig. 8.18. We hope that it is straightforward to see that $\pm I$ are fixed by any of the elements of Σ_3. Of course, this can be computed algebraically as well. Let $x \in \Sigma_3$, then $(\pm I)^x = s(x)(\pm I)[s(x)]^{-1} = \pm I$.

These two values are sufficient to compute the action of Σ_3 upon $\eta(x, y)$ since the values of the 2-cocycle, η are all $\pm I$. Nevertheless, it is amusing to compute the actions upon the remaining elements in K.

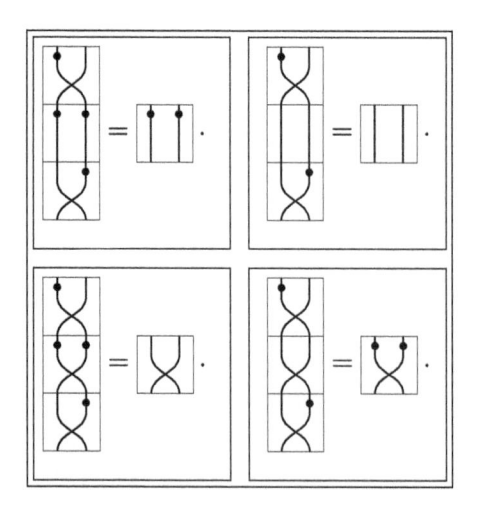

Figure 8.18. Conjugating the elements of K_4.

The three K_4 quipu that define each of $\pm A, \pm B,$ and $\pm C$ will slide up the permutations. For example, the action of (12) upon A switches the 1 and the X quipu before the quipu of $s(12)$ and $[s(12)]^{-1}$ pre and post multiplying them. This gives that $A^{(12)} = \pm B$. But from Fig. 8.18, we can see that $A^{(12)} = B$. Similar computations work for the remaining elements.

One can compute that each element of Σ_3 permutes the three non-trivial cosets of $(I, -I)$, as the notation suggests. For example, $(1, 2, 3)^{(12)} = (2, 1, 3)$.

It is worth taking a moment to consider some alternate, and perhaps more natural, notation for the elements of $K = (\mathbb{Z}/2)^3$. The elements are $(\pm 1, \pm 1, \pm 1)$. Then $I = (1, 1, 1)$ and $-I = (-1, -1, -1)$. We make the correspondences:

$$A \leftrightarrow (-1, 1, 1), \quad -A \leftrightarrow (1, -1, -1),$$
$$B \leftrightarrow (1, -1, 1), \quad -B \leftrightarrow (-1, 1, -1),$$
$$-C \leftrightarrow (1, 1, -1), \quad C \leftrightarrow (-1, -1, 1).$$

The action of Σ_3 is to permute the coordinates.

Next, we show that η is not a coboundary. To that end, suppose that there is a function $(\mathbb{Z}/2)^3 \xleftarrow{\gamma} \Sigma_3$ such that

$$\eta(x, y) = \gamma(y)[\gamma(xy)]^{-1}\gamma(x) \quad \text{for all } x, y \in \Sigma_3.$$

We note, under the assumption that we are trying to disprove, that $I = \eta(1, 1) = \gamma(1)[\gamma(1)]^{-1}\gamma(1)$ so that $\gamma(1) = I$. Furthermore, if x is a transposition $(x = (12), (23),$ or $(13))$,

$$-I = \eta(x, x) = \gamma(x)[\gamma(1)]^{-1}\gamma(x).$$

So,

$$[\gamma(x)]^2 = -I,$$

But γ takes values in $\{\pm I, \pm A, \pm B, \pm C\}$ and each of these eight elements squares to I. So, there is the contradiction.

The conclusion of this section is that the seseq (Ses(1)) does not split. That is,
$$\mathrm{SL}_2(\mathbb{Z}/4) \neq (\mathbb{Z}/2)^3 \rtimes \Sigma_3.$$
The matrix group is a non-trivial extension of Σ_3.

8.3.1. Computing inverses

Since the set-theoretic section that was used to define η was not entirely normal, the formula for inverses will need to be modified by appropriately adding factors similar to one of the following: $\eta(x, x^{-1})^{\pm 1}$ or $\eta(x^{-1}, x)^{\pm 1}$. Which factor and when is the subject of Proposition 4.

The problem for inverses exists in general, and we choose to examine the issue even in case K is not abelian. Consequently, the notation in Proposition 4 and in its proof is more intense than we would like.

Proposition 4. *Suppose*

$$1 \xleftarrow{} H \xleftarrow{q} G \xleftarrow{i} K \xleftarrow{} 1 \quad (\mathrm{Ses}(*))$$

is a seseq. Suppose that $H \xrightarrow{s} G$ is a not entirely normal cocycle. So, it is not necessarily true that $[s(x)]^{-1} = s(x^{-1})$. However, $s(1) = 1$. Let

$$\eta(x, y) = i^{-1}\left(s(x)s(y)\,[s(xy)]^{-1}\right).$$

The multiplication in $K \times H$ that is defined via η, given by

$$(\alpha, x) \cdot_\eta (\beta, y) = (\alpha \beta^x \eta(x, y), xy),$$

yields an isomorphism between $(K \times H, \cdot_\eta)$ and G.
The formula for the left inverse in $(\cdot_\eta)$ is given by

$$L_\eta(\beta, y) = (\beta, y)^{-1} = \left([\eta(y^{-1}, y)]^{-1}\,[\beta^{-1}]^{y^{-1}}, y^{-1}\right)$$

so that
$$L_\eta(\beta, y) \cdot_\eta (\beta, y) = (1, 1).$$

In addition, the left inverse serves as a right inverse:
$$R_\eta(\alpha, x) = L_\eta(\alpha, x).$$

Proof. The isomorphism $(K \times H, \cdot_\eta) \approx G$ was presented in Theorem 5. The context of the proposition is merely to address the issues of inverses.

The first step of the proof of the proposition is to simplify the product

$$L_\eta(\beta, y^{-1}) \cdot_\eta (\beta, y) = \left([\eta(y^{-1}, y)]^{-1} [\beta^{-1}]^{y^{-1}}, y^{-1} \right) \cdot_\eta (\beta, y)$$
$$= \left([\eta(y^{-1}, y)]^{-1} [\beta^{-1}]^{y^{-1}} \beta^{y^{-1}} \eta(y^{-1}, y), y^{-1} y \right)$$
$$= \left([\eta(y^{-1}, y)]^{-1} [\beta^{-1}]^{y^{-1}} \beta^{y^{-1}} \eta(y^{-1}, y), 1 \right).$$

The product
$$[\beta^{-1}]^{y^{-1}} \beta^{y^{-1}} = s(y^{-1})\beta^{-1} [s(y^{-1})]^{-1} s(y^{-1})\beta [s(y^{-1})]^{-1}$$
$$= s(y^{-1})\beta^{-1}\beta [s(y^{-1})]^{-1} = s(y^{-1}) [s(y^{-1})]^{-1} = 1$$

simplifies as indicated, and so

$$[\eta(y^{-1}, y)]^{-1} [\beta^{-1}]^{y^{-1}} \beta^{y^{-1}} \eta(y^{-1}, y)$$
$$= [\eta(y^{-1}, y)]^{-1} \eta(y^{-1}, y) = 1.$$

Therefore,
$$L_\eta(\beta, y) \cdot_\eta (\beta, y) = (1, 1),$$

as desired.

We wish to show that

$$(\alpha, x) \cdot_\eta \left([\eta(x^{-1}, x)]^{-1} [\alpha^{-1}]^{x^{-1}}, x^{-1} \right) = (1, 1).$$

To that end, consider

$$\alpha \left([\eta(x^{-1}, x)]^{-1} [\alpha^{-1}]^{x^{-1}} \right)^x \eta(x, x^{-1}).$$

Since $\eta(x^{-1}, x) = s(x^{-1})s(x)[s(1)]^{-1} = s(x^{-1})s(x)$, the first factor in parentheses can be rewritten as

$$[\eta(x^{-1}, x)]^{-1} = [s(x^{-1})s(x)]^{-1} = [s(x)]^{-1} [s(x^{-1})]^{-1}.$$

The next factor is expanded:

$$[\alpha^{-1}]^{x^{-1}} = s(x^{-1})\alpha^{-1} [s(x^{-1})]^{-1}.$$

Then

$$[\eta(x^{-1}, x)]^{-1} [\alpha^{-1}]^{x^{-1}} = [s(x)]^{-1} [s(x^{-1})]^{-1} s(x^{-1}) \alpha^{-1} [s(x^{-1})]^{-1}$$
$$= [s(x)]^{-1} \alpha^{-1} [s(x^{-1})]^{-1}.$$

Conjugate by $s(x)$ as follows:

$$\left([\eta(x^{-1}, x)]^{-1} [\alpha^{-1}]^{x^{-1}} \right)^x$$
$$= s(x) [s(x)]^{-1} \alpha^{-1} [s(x^{-1})]^{-1} [s(x)]^{-1}$$
$$= \alpha^{-1} [\eta(x, x^{-1})]^{-1}.$$

Therefore,

$$\alpha \left([\eta(x^{-1}, x)]^{-1} [\alpha^{-1}]^{x^{-1}} \right)^x \eta(x, x^{-1})$$
$$= \alpha \alpha^{-1} [\eta(x, x^{-1})]^{-1} \eta(x, x^{-1}) = 1.$$

We obtain,

$$(\alpha, x) \cdot_\eta \left([\eta(x^{-1}, x)]^{-1} [\alpha^{-1}]^{x^{-1}}, x^{-1} \right) = (1, 1),$$

as desired. This completes the proof. □

8.4. Example 3. The binary tetrahedral group

The binary tetrahedral group $\widetilde{A_4}$ is a subgroup of both $GL_2(\mathbb{Z}/3)$ and the binary octahedral group $\widetilde{\Sigma_4}$. In this section, we prove the following results.

Theorem 6.

(1) *The binary tetrahedral group $\widetilde{A_4}$ can be written as a semi-direct product*:

$$\widetilde{A_4} \approx Q_8 \rtimes \mathbb{Z}/3$$

of the cyclic group of order 3, $\mathbb{Z}/3$ and the unit quaternions Q_8.

(2) *The alternating group A_4 can be written as a semi-direct product*:

$$A_4 \approx K_4 \rtimes \mathbb{Z}/3$$

of $\mathbb{Z}/3$ and the Klein 4-group K_4.

The result is known. The first statement can be found by a web search of the term "binary tetrahedral group." The second statement can be found by looking up the classification of groups of order 12. Both Wikipedia sources are devoid of details. So, they will be provided here by means of demonstrating that cocycles $Q_8 \xleftarrow{\eta_1} \mathbb{Z}/3\times\mathbb{Z}/3$ and $K_4 \xleftarrow{\eta_3} \mathbb{Z}/3 \times \mathbb{Z}/3$ that are derived from set-theoretic sections for the seseqs

$$1 \longleftarrow \mathbb{Z}/3 \xleftarrow{q_1} \widetilde{A_4} \xleftarrow{i_1} Q_8 \longleftarrow 1 \quad (\text{Ses}(1))$$

and

$$1 \longleftarrow \mathbb{Z}/3 \xleftarrow{q_3} A_4 \xleftarrow{i_3} K_4 \longleftarrow 1 \quad (\text{Ses}(3))$$

are equivalent to trivial cocycles. One is trivial. The other is a coboundary. In the process, the sections will be used to define an action of $\mathbb{Z}/3$ upon the kernels.

These seseqs fit into a large diagram (Fig. 8.19) that also involves the seseq

$$1 \longleftarrow K_4 \xleftarrow{q_2} Q_8 \xleftarrow{i_2} \mathbb{Z}/2 \longleftarrow 1 \quad (\text{Ses}(2)),$$

which is classified by means of a 2-cocycle $\mathbb{Z}/2 \xleftarrow{\eta_4} K_4 \times K_4$, and the seseq,

$$1 \longleftarrow A_4 \xleftarrow{q_4} \widetilde{A_4} \xleftarrow{i_3} \mathbb{Z}/2 \longleftarrow 1 \quad (\text{Ses}(4)),$$

which is classified by means of a 2-cocycle $\mathbb{Z}/2 \xleftarrow{\eta_4} A_4 \times A_4$.

These seseqs are combined in a diagram in Fig. 8.19.

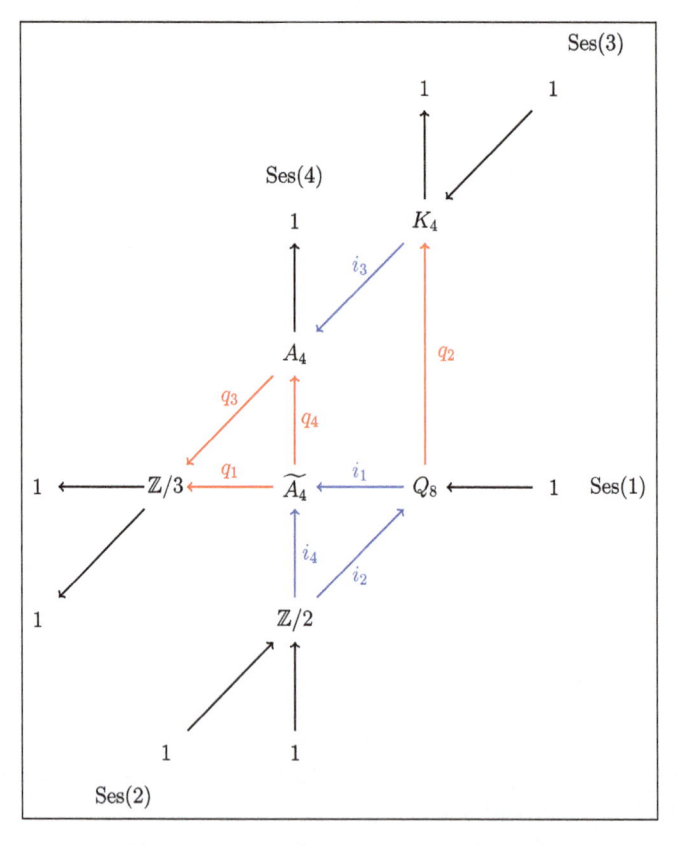

Figure 8.19. Combining several seseqs.

Computations of all of these cocycles will be given via various set-theoretic sections and within the various subsections of the text. The non-sequential numbering of the seseqs is chosen to aid the authors in their writing of the proof of Proposition 5.

Proposition 5. *The diagram that is depicted in Fig. 8.19 is commutative. Specifically, equalities hold among the group homomorphisms that are depicted:*

$$i_3 \circ q_2 = q_4 \circ i_1; \quad q_3 \circ q_4 = q_1; \quad i_1 \circ i_2 = i_4.$$

Proof. Each of the seseqs in the figure is informed by quipu representations of the groups involved. The sequences Ses(2) and Ses(4) are constructed by means of 4-strings-with-(mod 2)-quipu. The sequence Ses(1) is constructed using 3-strings-with-Q_8-quipu, and the sequence Ses(3) is constructed by means of 3-strings-with-(mod 2)-quipu.

Figures that help visualize the result are scattered throughout the text, and consequently, you may need to flip among several sections to fully understand the proof. Let us start from Fig. 2.3. Therein the elements (1), $(12)(34)$, $(13)(24)$, and $(14)(23)$ in the Klein 4 group, K_4, are found among the elements of Σ_4 in their 4-strings representations.

To understand Ses(4), and in particular, the quotient map $K_4 \xleftarrow{q_2} Q_8$, consider Fig. 4.5. The quotient map q_2 ignores the (mod 2)-quipu upon the strings. In this way, $\pm 1 \mapsto 1$, $\pm i \mapsto (12)(34)$, $\pm j \mapsto (13)(24)$, and $\pm k \mapsto (14)(23)$.

Ses(1) can be imagined via the illustrations on the right of Figs. 6.3–6.5. Therein 3-strings-with-Q_8-quipu represent elements in the binary tetrahedral group $\widetilde{A_4}$. The projection q_1 to $\mathbb{Z}/3$ removes the quipu, and we observe that the elements project to either the identity or to 3-cycles in Σ_3. In particular, the elements in Q_8 project to the identity element in $\mathbb{Z}/3$. The integers modulo 3 is now considered to be the subgroup of Σ_3 that is generated by the 3-cycle (123).

Among these illustrations, Fig. 6.5 is most relevant to the comparison between $q_4 \circ i_1$ and $i_3 \circ q_2$. The red-stringed illustrations of -1, i, j, and k that appear in the center-left column correspond to the 3-strings pictures that are depicted on the right. The quotient map $K_4 \xleftarrow{q_2} Q_8$ strips the (mod 2)-quipu from the 4-strings pictures.

The injection $\widetilde{A_4} \xleftarrow{i_1} Q_8$ is the inclusion of the elements of Q_8 as the depictions that have three straight strings. The representations of the Q_8-quipu within Fig. 6.5 differ from those presented in Figs. 5.35 and 5.36 only because the the quaternions are also elements in the semi-dihedral group and so the cyclic quipu in that subgroup of $GL_2(\mathbb{Z}/3)$ are (mod 8)-quipu.

Yet, Figs. 5.35 and 5.36 better explain the composition $i_3 \circ q_2$. In particular, an alternative description of the quotient map q_2 is to map the quipu that correspond to $\pm \boldsymbol{j}$ and $\pm \boldsymbol{k}$ to non-trivial (mod 2)-quipu within the 3-strings-with-(mod 2)-quipu descriptions of K_4. Thus, the elements of K_4 are found on the right-hand side of Figs. 5.35 and 5.36 as every other entry. Moreover, $K_4 \subset A_4 \subset \Sigma_4$ is a sequence of subgroup inclusions, and the elements of Σ_4 are represented in these figures as 3-strings-with-(mod 2)-quipu. More specifically, Σ_4 is the group of symmetries of the 3-cube, and these representations are given by means of (3×3)-signed permutation matrices that have determinant 1.

In a similar manner, the quotient map $A_4 \xleftarrow{q_2} \widetilde{A_4}$ from the binary tetrahedral group to the alternating group is also described by means of mapping the elements in the coset $\{\pm \boldsymbol{j}, \pm \boldsymbol{k}\}$ to -1 so that those quipu which are represented with crossings are mapped to dots in their 3-strings representatives.

Both of the quotient homomorphisms q_1 and q_2 are described by means of mapping the quipu that represent elements in the coset $\{\pm \boldsymbol{j}, \pm \boldsymbol{k}\}$ to -1 and mapping the elements in the subgroup $\{\pm 1, \pm \boldsymbol{i}\}$ to the identity element. In this way, $i_3 \circ q_2 = q_4 \circ i_1$.

Note further that $q_1 = q_3 \circ q_4$ since both quotient homomorphisms q_3 and q_1 are obtained by stripping the quipu from the 3-strings-with-quipu representatives. The quotient q_4 converts quaternions to elements in $\mathbb{Z}/2$: specifically, $\pm 1, \pm \boldsymbol{i} \mapsto 1$ and $\pm \boldsymbol{j}, \pm \boldsymbol{k} \mapsto -1$.

Both of the inclusions i_2 and i_4 map ± 1 into same named elements in either the quaternions, Q_8 or the binary tetrahedral group, $\widetilde{A_4}$. As such, $i_1 \circ i_2 = i_4$.

This completes the proof. □

8.4.1. The cocycle η_1

In this section, we prove the first statement of Theorem 6, namely, the binary tetrahedral group is a semi-direct product

$$\widetilde{A_4} \approx Q_8 \rtimes \mathbb{Z}/3.$$

Starting from the seseq

$$1 \longleftarrow \mathbb{Z}/3 \xleftarrow{q_1} \widetilde{A_4} \xleftarrow{i_1} Q_8 \longleftarrow 1 \quad (\text{Ses}(1)),$$

we construct a cocycle

$$Q_8 \xleftarrow{\eta_1} \mathbb{Z}/3 \times \mathbb{Z}/3.$$

Let $\mathbb{Z}/3 \xrightarrow{s} \widetilde{A_4}$ denote the normalized section

$$s(1) = 1, \quad s(123) = a, \quad s(132) = -a^2.$$

To be sure, the cyclic group $\mathbb{Z}/3$ is being considered as a subgroup of Σ_3 and its elements are written as cycles. The element $a = (1 + i + j + k)/2$ as usual, and its inverse is given as $-a^2 = (1 - i - j - k)/2$. Graphically, the section s is represented as in Fig. 8.20.

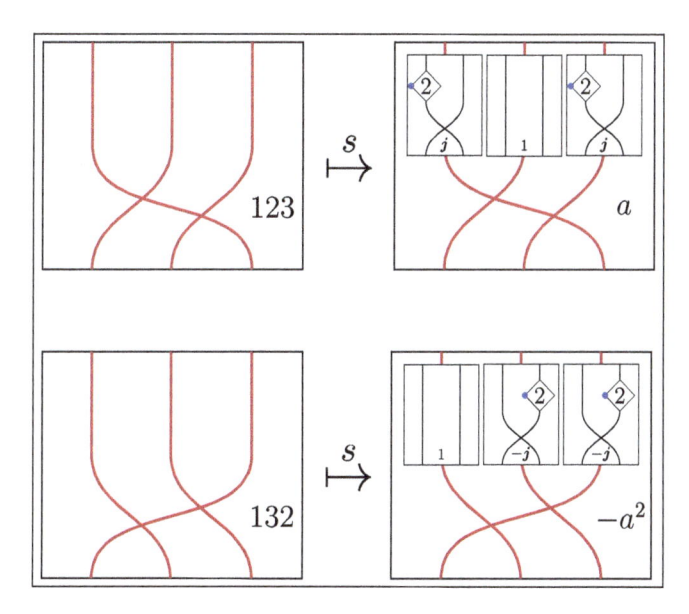

Figure 8.20. The section values of the units of $\mathbb{Z}/3$.

The values of the cocycle $\eta_1(x,y) = s(x)s(y)[s(xy)]^{-1}$ are very easy to compute since the section values are powers of a. Along the top row of the table of cocycle values, we have

$$\eta(1,1) = 1, \quad \eta(1,123) = 1 \cdot a(-a^2) = 1, \quad \eta(1,132) = 1 \cdot (-a^2)a = 1.$$

Similarly, down the first column,

$$\eta(123,1) = a \cdot 1(-a^2) = 1, \quad \eta(132,1) = (-a^2) \cdot 1 \cdot a = 1.$$

Along the diagonals, we compute

$$\eta(123,123) = a \cdot a(s(132))^{-1} = a^3 = -1,$$
$$\eta(132,132) = (-a^2) \cdot (-a^2) \cdot (-a^2) = -1.$$

The remaining entries

$$\eta(123,132) = a \cdot (-a^2) \cdot 1 = 1,$$
$$\eta(132,123) = (-a^2) \cdot a \cdot 1 = 1.$$

So, the table of cocycle values is as follows:

$\eta_1(r,c)$	1	123	132
1	1	1	1
123	1	-1	1
132	1	1	-1

One might think that η_1 defines a non-trivial extension. However, let $Q_8 \xleftarrow{\gamma} \mathbb{Z}/3$ be defined by

$$\gamma(1) = 1, \quad \gamma(123) = \gamma(132) = -1.$$

Tabulate the values

$$\gamma(y)[\gamma(xy)]^{-1}\gamma(x)$$

for all $x, y \in \mathbb{Z}/3$. We have

$$\gamma(y)(\gamma(1y))^{-1}\gamma(1) = \gamma(1)(\gamma(x1))^{-1}\gamma(x) = 1,$$

and if $x \neq y$ and neither is 1,

$$\gamma(y)[\gamma(1)]^{-1}\gamma(x) = 1.$$

Meanwhile, if $x \neq 1$,
$$\gamma(x)[\gamma(x^2)]^{-1}\gamma(x) = -1.$$
So,
$$\eta_1(x,y) = \gamma(y)[\gamma(xy)]^{-1}\gamma(x),$$
and the cocycle η_1 is a coboundary.

If η' is a 2-cocycle that is defined by $s'(123) = -a = a^4$ and $s'(132) = a^2$, so that $s'(x) = \gamma(x)s(x)$, then one can show (and you should show) that $\eta'(x,y) = 1$ for all $x, y \in \mathbb{Z}/3$.

Perhaps, the most expeditious manner to describe the action of $\mathbb{Z}/3$ upon Q_8 is by means of the matrix representations

$$a \leftrightarrow \begin{bmatrix} 0 & 0 & j \\ 1 & 0 & 0 \\ 0 & j & 0 \end{bmatrix}, \quad -a^2 \leftrightarrow \begin{bmatrix} 0 & 1 & 0 \\ 0 & 0 & -j \\ -j & 0 & 0 \end{bmatrix},$$

$$i \leftrightarrow \begin{bmatrix} i & 0 & 0 \\ 0 & k & 0 \\ 0 & 0 & j \end{bmatrix}, \quad j \leftrightarrow \begin{bmatrix} j & 0 & 0 \\ 0 & i & 0 \\ 0 & 0 & -k \end{bmatrix}, \quad k \leftrightarrow \begin{bmatrix} k & 0 & 0 \\ 0 & j & 0 \\ 0 & 0 & -i \end{bmatrix}.$$

We compute

$$i^{(123)} = \begin{bmatrix} 0 & 0 & j \\ 1 & 0 & 0 \\ 0 & j & 0 \end{bmatrix} \begin{bmatrix} i & 0 & 0 \\ 0 & k & 0 \\ 0 & 0 & j \end{bmatrix} \begin{bmatrix} 0 & 1 & 0 \\ 0 & 0 & -j \\ -j & 0 & 0 \end{bmatrix}$$

$$= \begin{bmatrix} 0 & 0 & -1 \\ i & 0 & 0 \\ 0 & i & 0 \end{bmatrix} \begin{bmatrix} 0 & 1 & 0 \\ 0 & 0 & -j \\ -j & 0 & 0 \end{bmatrix} = \begin{bmatrix} j & 0 & 0 \\ 0 & i & 0 \\ 0 & 0 & -k \end{bmatrix}.$$

So, $i^{(123)} = j$. Similarly,

$$j^{(123)} = \begin{bmatrix} 0 & 0 & j \\ 1 & 0 & 0 \\ 0 & j & 0 \end{bmatrix} \begin{bmatrix} j & 0 & 0 \\ 0 & i & 0 \\ 0 & 0 & -k \end{bmatrix} \begin{bmatrix} 0 & 1 & 0 \\ 0 & 0 & -j \\ -j & 0 & 0 \end{bmatrix}$$

$$= \begin{bmatrix} 0 & 0 & -i \\ j & 0 & 0 \\ 0 & -k & 0 \end{bmatrix} \begin{bmatrix} 0 & 1 & 0 \\ 0 & 0 & -j \\ -j & 0 & 0 \end{bmatrix} = \begin{bmatrix} k & 0 & 0 \\ 0 & j & 0 \\ 0 & 0 & -i \end{bmatrix},$$

and

$$k^{(123)} = \begin{bmatrix} 0 & 0 & j \\ 1 & 0 & 0 \\ 0 & j & 0 \end{bmatrix} \begin{bmatrix} k & 0 & 0 \\ 0 & j & 0 \\ 0 & 0 & -i \end{bmatrix} \begin{bmatrix} 0 & 1 & 0 \\ 0 & 0 & -j \\ -j & 0 & 0 \end{bmatrix}$$

$$= \begin{bmatrix} 0 & 0 & k \\ k & 0 & 0 \\ 0 & -1 & 0 \end{bmatrix} \begin{bmatrix} 0 & 1 & 0 \\ 0 & 0 & -j \\ -j & 0 & 0 \end{bmatrix} = \begin{bmatrix} i & 0 & 0 \\ 0 & k & 0 \\ 0 & 0 & j \end{bmatrix}.$$

So, $j^{(123)} = k$, and $k^{(123)} = i$. Negatives commute with all the elements in $\widetilde{A_4}$. We leave it as an exercise for you to show that

$$i \mapsto k \mapsto j \mapsto i$$

under the conjugation action of (132). The cocycle η_1 is equivalent to the trivial cocycle η' that was defined above. The cyclic group $\mathbb{Z}/3$ acts upon Q_8 via the section $s' = \gamma s$ in the obvious fashion as a cyclic permutation of i, j, and k. The added factors of -1 will not affect the conjugation. Since η' is trivial, the section s' is a homomorphism. Thus, the seseq Ses(1) splits, and the binary tetrahedral group is a semi-direct product:

$$\widetilde{A_4} \approx Q_8 \rtimes \mathbb{Z}/3,$$

as claimed in part 1 of Theorem 6.

8.4.2. The cocycle η_2

Consider the seseq

$$1 \longleftarrow K_4 \xleftarrow{q_2} Q_8 \xleftarrow{i_2} \mathbb{Z}/2 \longleftarrow 1 \quad (\text{Ses}(2)).$$

A not entirely normal set-theoretic section $K_4 \xrightarrow{s} Q_8$ is defined by

$$s(1) = 1, \quad s(12.34) = i, \quad s(13.24) = j, \quad s(14.23) = k.$$

It should be apparent from Fig. 8.21 that s is indeed a section. That it is not entirely normal follows because every non-trivial element in K_4 has order 2, but the images have order 4.

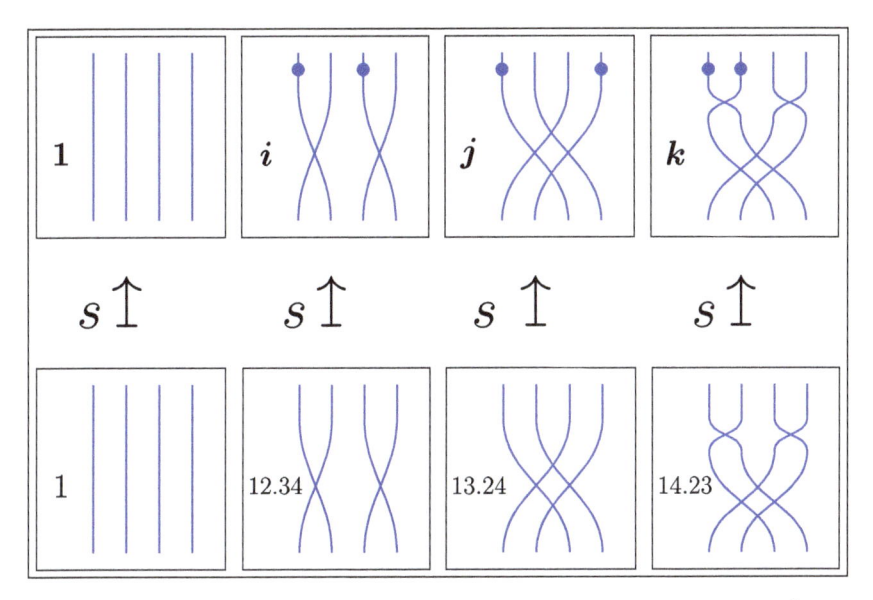

Figure 8.21. Defining a not entirely normal set-theoretic section $K_4 \xrightarrow{s} Q_8$.

As usual, define a 2-cocycle $\mathbb{Z}/2 \xleftarrow{\eta_2} K_4 \times K_4$ by the equation

$$\eta(x, y) = s(x)s(y)[s(xy)]^{-1}.$$

It won't be too difficult to compute the values of η_2 once we make the following observations:

$$\boldsymbol{ijk} = \boldsymbol{jki} = \boldsymbol{kij} = -1, \quad \boldsymbol{ikj} = \boldsymbol{jik} = \boldsymbol{kji} = 1.$$

The cocycle values will be computed row-by-row as follows:

$$\eta(1, 1) = s(1)s(1)[s(1 \cdot 1)]^{-1} = 1,$$

$$\eta(1, 12.34) = s(1)s(12.34)[s(1 \cdot 12.34)]^{-1} = 1 \cdot \boldsymbol{i} \cdot (-\boldsymbol{i}) = 1,$$

$$\eta(1, 13.24) = s(1)s(13.24)[s(1 \cdot 13.24)]^{-1} = 1 \cdot \boldsymbol{j} \cdot (-\boldsymbol{j}) = 1,$$

$$\eta(1, 14.23) = s(1)s(14.23)[s(1 \cdot 14.23)]^{-1} = 1 \cdot \boldsymbol{k} \cdot (-\boldsymbol{k}) = 1.$$

$$\eta(12.34, 1) = s(12.34)s(1)[s(12.34 \cdot 1)]^{-1} = 1,$$

$$\eta(12.34, 12.34) = s(12.34)s(12.34)[s(1)]^{-1} = \boldsymbol{i} \cdot \boldsymbol{i} = -1,$$

$\eta(12.34, 13.24) = s(12.34)s(13.24)[s(14, 23)]^{-1} = \boldsymbol{i} \cdot \boldsymbol{j} \cdot (-\boldsymbol{k}) = 1,$

$\eta(12.34, 14.23) = s(12.34)s(14.23)[s(13.24)]^{-1} = \boldsymbol{i} \cdot \boldsymbol{k} \cdot (-\boldsymbol{j}) = -1.$

$\eta(13.24, 1) = s(13.24)s(1)[s(13.24 \cdot 1)]^{-1} = 1,$

$\eta(13.24, 12.34) = s(13.24)s(12.34)[s(14.23)]^{-1} = \boldsymbol{j} \cdot \boldsymbol{i} \cdot (-\boldsymbol{k}) = -1,$

$\eta(13.24, 13.24) = s(13.24)s(13.24)[s(1)]^{-1} = \boldsymbol{j} \cdot \boldsymbol{j} = -1,$

$\eta(13.24, 14.23) = s(13.24)s(14.23)[s(12.34)]^{-1} = \boldsymbol{j} \cdot \boldsymbol{k} \cdot (-\boldsymbol{i}) = 1.$

$\eta(14.23, 1) = s(14.23)s(1)[s(14.23 \cdot 1)]^{-1} = 1,$

$\eta(14.23, 12.34) = s(14.23)s(12.34)[s(13.24)]^{-1} = \boldsymbol{k} \cdot \boldsymbol{i} \cdot (-\boldsymbol{j}) = 1,$

$\eta(14.23, 13.24) = s(14.23)s(13.24)[s(12.34)]^{-1} = \boldsymbol{k} \cdot \boldsymbol{j} \cdot (-\boldsymbol{i}) = -1,$

$\eta(14.23, 14.23) = s(14.23)s(14.23)[s(1)]^{-1} = \boldsymbol{k} \cdot \boldsymbol{k} = -1.$

The cocycle values are provided in Table 8.5. Before that, it is worthwhile to note a couple of things about these values. Each of the six permutations of $\{\boldsymbol{i}, \boldsymbol{j}, \boldsymbol{k}\}$ appears in the off-diagonal and non-identity entries among these values. Again, forget about the identity entries, then the nine remaining values in the rows are $(-1, 1, -1)$, $(-1, -1, 1)$, and $(1, -1, -1)$. These coincide with the 3-strings-with-(mod 2)-quipu representations of the non-trivial elements in K_4. This is probably a coincidence or might be something more profound.

Table 8.5. The cocycle values for $\mathbb{Z}/2 \xleftarrow{\eta_2} K_4 \times K_4$.

$\eta_2(r, c)$	1	(12)(34)	(13)(24)	(14)(23)
1	1	1	1	1
(12)(34)	1	-1	1	-1
(13)(24)	1	-1	-1	1
(14)(23)	1	1	-1	-1

If there is a function $\mathbb{Z}/2 \xleftarrow{\gamma} K_4$ such that $\eta_1(x,y) = \gamma(y)[\gamma(xy)]^{-1}\gamma(x)$, then $-1 = \eta_1(12.34, 12.34) = [\gamma(12.34)]^2$, but the square of any element in $\mathbb{Z}/2$ is 1. Therefore, the cocycle η_1 is not a coboundary.

8.4.3. The cocycle η_3

In this section, we prove the second statement of Theorem 6, namely, the alternating group A_4 can be written as a semi-direct product:

$$A_4 \approx K_4 \rtimes \mathbb{Z}/3.$$

The current study begins with the seseq

$$1 \longleftarrow \mathbb{Z}/3 \xleftarrow{q_3} A_4 \xleftarrow{i_3} K_4 \longleftarrow 1 \quad (\text{Ses}(3)).$$

The elements in $\mathbb{Z}/3$ will be expressed multiplicatively, and it is convenient to imagine the non-trivial elements as 3-cycles in Σ_3. Meanwhile, we consider the 3-strings-with-(mod 2)-quipu representation of A_4. The elements have either two (mod 2)-quipu or no (mod 2)-quipu and are either straight strings or 3-cycles.

For your convenience, we recall from Section 5.2.4 that the representative of K_4 in the current context are as illustrated in Fig. 8.22.

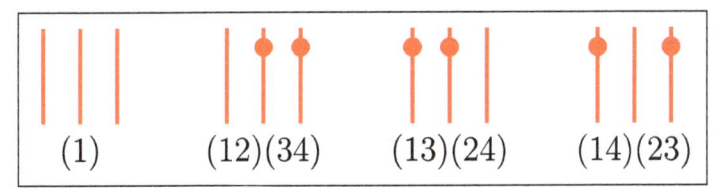

Figure 8.22. The elements in K_4 as 3-strings-with-(mod 2)-quipu from the current perspective.

As in Section 8.2, a section $\mathbb{Z}/3 \xrightarrow{s} A_4$ is chosen so that

$$s(1) = 1; \quad s(123) = (123); \quad s(132) = s(132).$$

Be aware that the values of s are permutations on four letters. So, it would be more accurate to write $s(132) = (132)(4)$.

Exercise 20. See Section 8.2 for comparison. The 2-cocycle $K_4 \xleftarrow{\eta_3} \mathbb{Z}/3 \times \mathbb{Z}/3$ that is defined by

$$\eta_3(x,y) = s(x)s(y)[s(xy)]^{-1},$$

— where $\mathbb{Z}/3 \xrightarrow{s} A_4$ is the normalized section

$$s(1) = (1)(2)(3)(4), \quad s(123) = (123)(4), \quad s(132) = s(132)(4)$$

— takes constant values:

$$\eta_3(x,y) = 1 \quad \text{for all } x,y \in Z.3.$$

The conjugation action of $\mathbb{Z}/3$ upon K_4 is given by

$$[(12)(34)]^{(123)} = (13)(24), \quad [(13)(24)]^{(123)} = (14)(23),$$
$$[(14)(23)]^{(123)} = (12)(34), \quad [(12)(34)]^{(132)} = (14)(23),$$
$$[(13)(24)]^{(132)} = (12)(34), \quad [(14)(23)]^{(123)} = (13)(24).$$

This completes the proof of part 2 of Theorem 6.

8.4.4. The cocycle η_4

In this section, we tabulate the values of a 2-cocycle

$$\mathbb{Z}/2 \xleftarrow{\eta_4} A_4 \times A_4$$

that is associated to the seseq

$$1 \longleftarrow A_4 \xleftarrow{q_4} \widetilde{A_4} \xleftarrow{i_4} \mathbb{Z}/2 \longleftarrow 1 \quad (\text{Ses}(4))$$

and which is defined by a not entirely normal set-theoretic section $A_4 \xrightarrow{s} \widetilde{A_4}$ via the usual formula.

$$\eta_4(x,y) = s(x)s(y)[s(xy)]^{-1}.$$

The function s will be defined in a moment.

It turns out that away from Q_8, we can choose $s(x^{-1}) = [s(x)]^{-1}$. From considerations in Section 8.4.1, we guess that it is better to choose $s(123) = -a$ and $s(132) = a^2$ than the choices in Fig. 8.20. There are more reasons that involve the 4-strings-with-(mod 2)-quipu for these two choices, namely, neither of the representatives of $-a$ nor of a^2 have any quipu. So, the diagrams for these elements are more easily multiplied.

Since there are 12 elements in A_4, the table of values will have 144 entries. The decomposition, $A_4 \approx K_4 \rtimes \mathbb{Z}/3$, of the alternating group A_4 into a semi-direct product will help us organize that table into nine blocks — each of size (4×4). We order the subgroup K_4 and its cosets $(123)K_4$ and $(132)K_4$ as three blocks:

$$K_4 = (1, 12.34, 13.24, 14.23),$$
$$(123)K_4 = (123, 134, 243, 142),$$

and

$$(132)K_4 = (132, 234, 124, 143).$$

For more visual ease, parentheses will often be dropped and in the Klein 4-group, $(12)(34) = 12.34$, for example, is a more succinct notation.

The 4-strings-with-(mod 2)-quipu representations will be modified from Figs. 6.3–6.5 or Figs. 5.17–5.22. We compile the calculations of cocycle values in Figs. 8.23–8.31.

Choosing the set-theoretic section: The values of the set-theoretic section s are given in Table 8.6.

Table 8.6. The set-theoretic section for Ses(4).

$s(1) = 1$	$s(12.34) = i$	$s(13.24) = j$	$s(14.23) = k$
$s(123) = b$	$s(134) = -c^2$	$s(243) = a^2$	$s(142) = d$
$s(132) = -b^2$	$s(234) = -a$	$s(124) = -d^2$	$s(143) = c$

410   *Quipu: Decorated Permutation Representations of Finite Groups*

These values are presented using string diagrams:

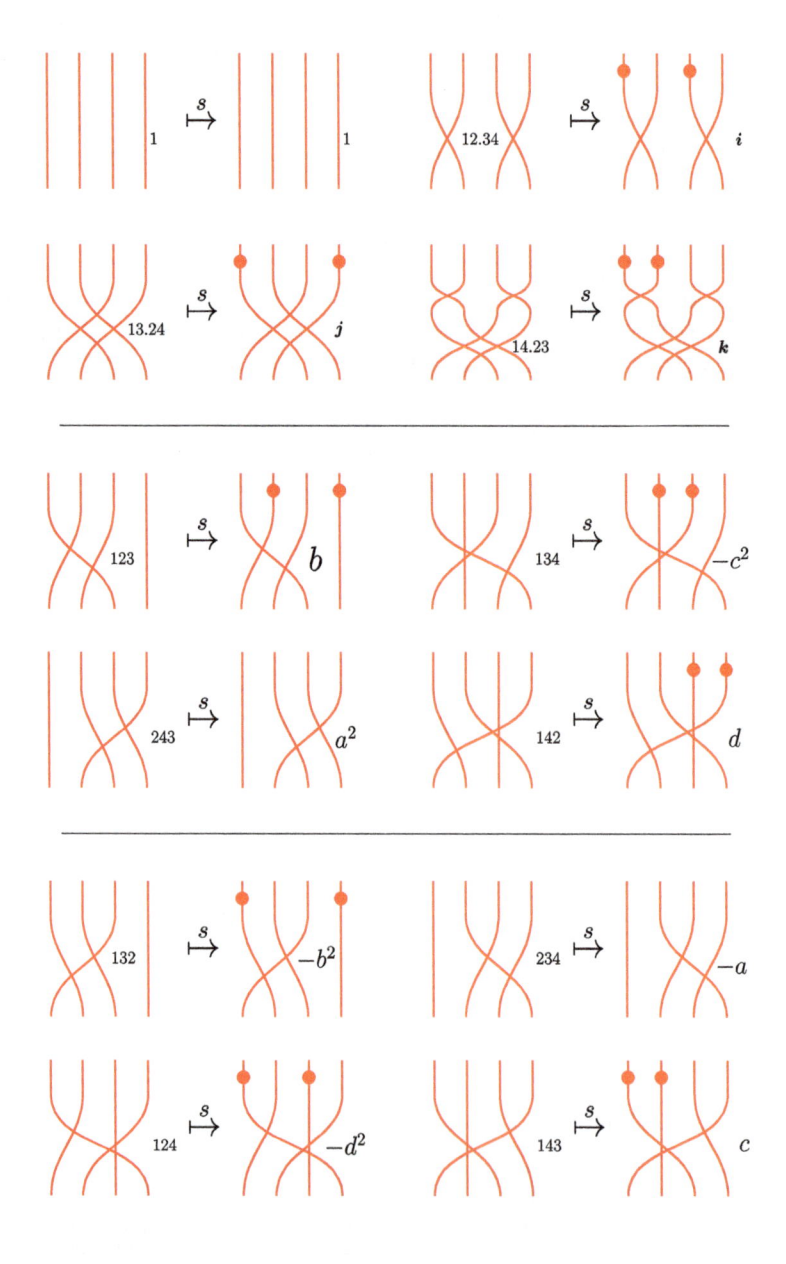

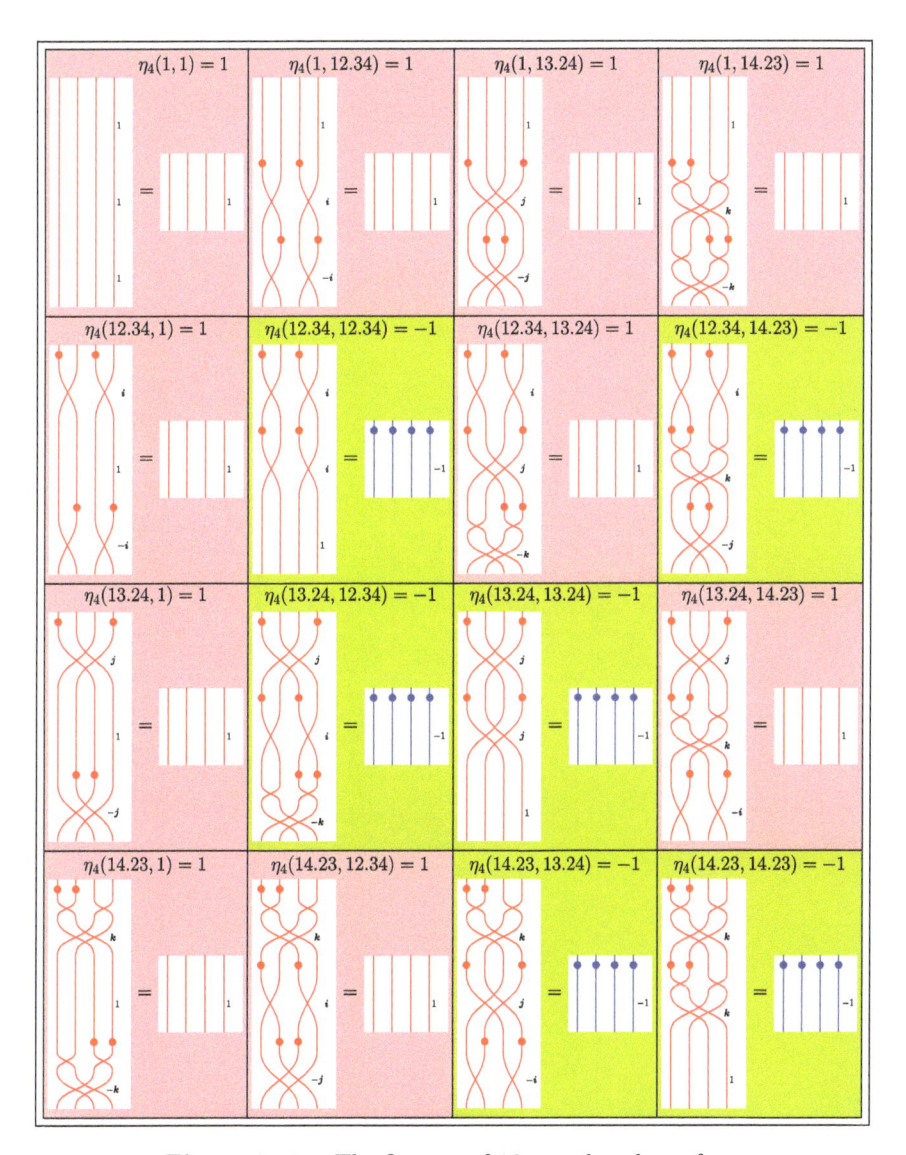

Figure 8.23. The first set of 16-cocycle values of η_4.

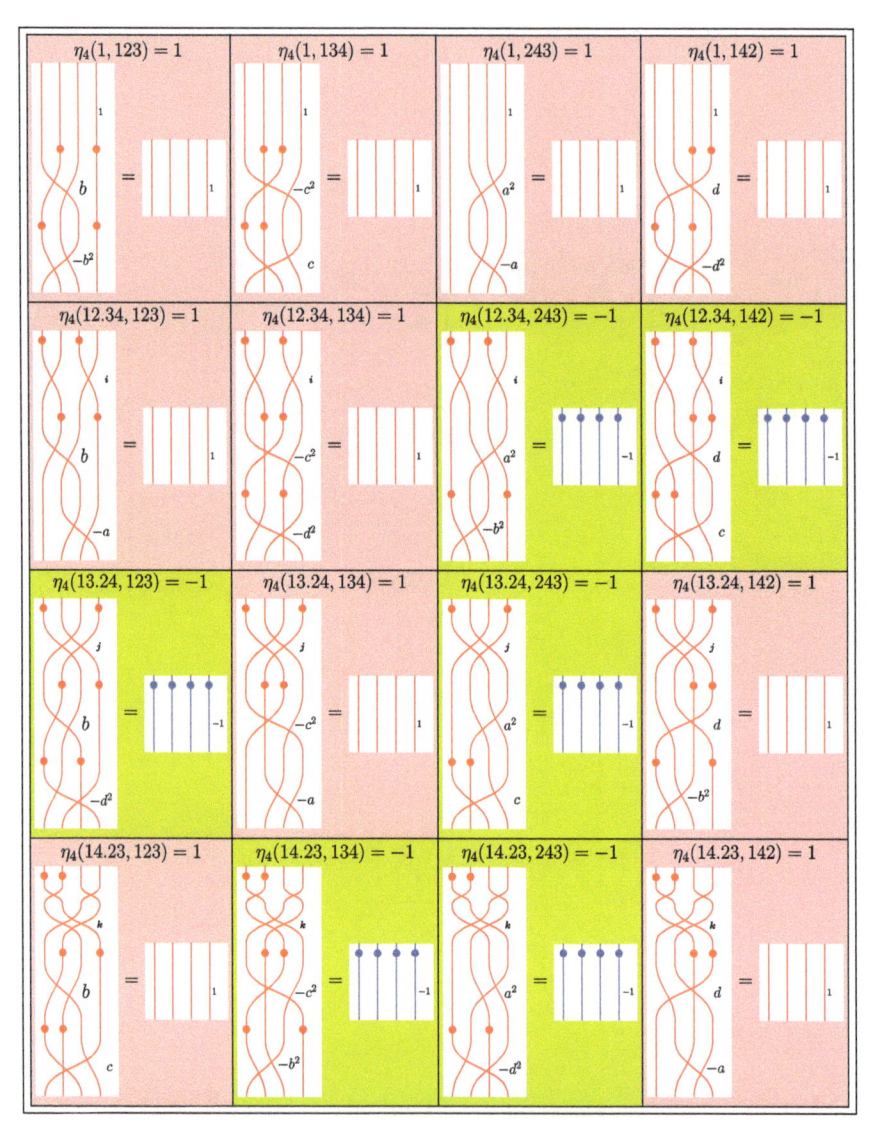

Figure 8.24. The second set of 16-cocycle values of η_4.

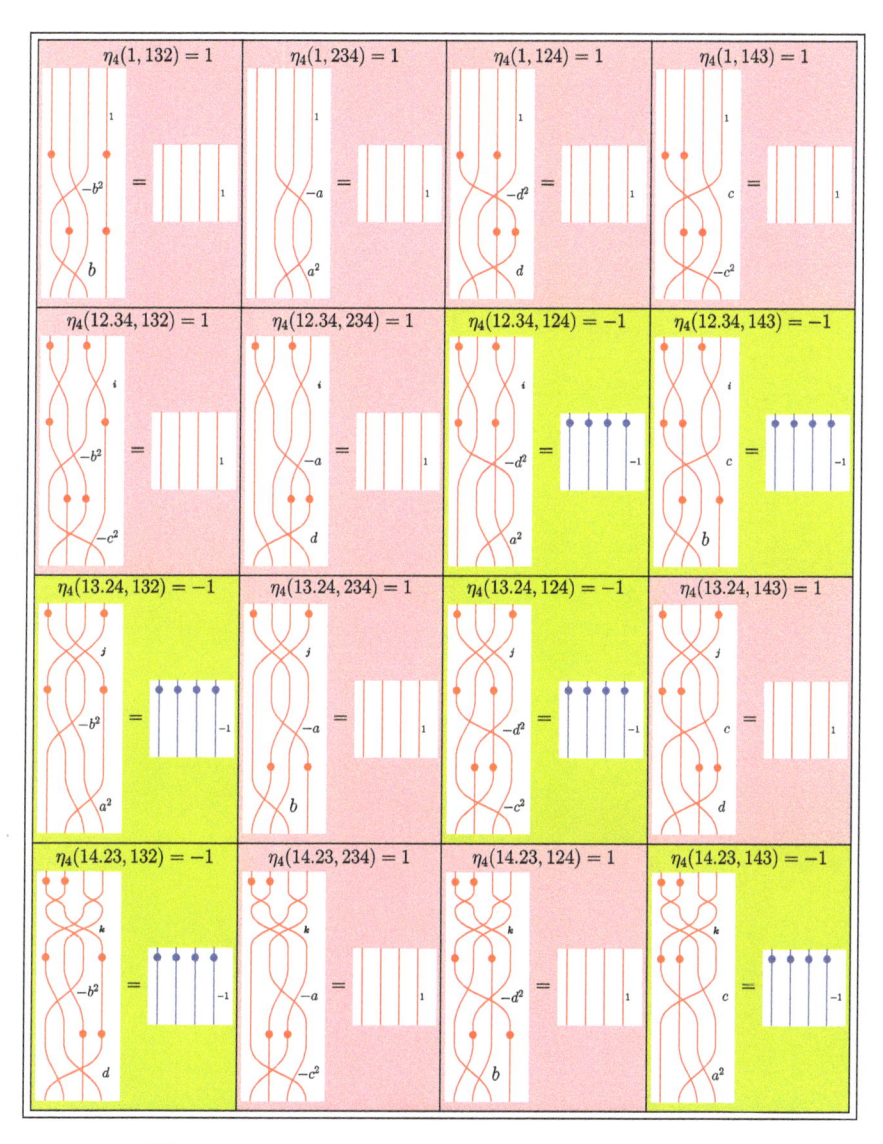

Figure 8.25. The third set of 16-cocycle values of η_4.

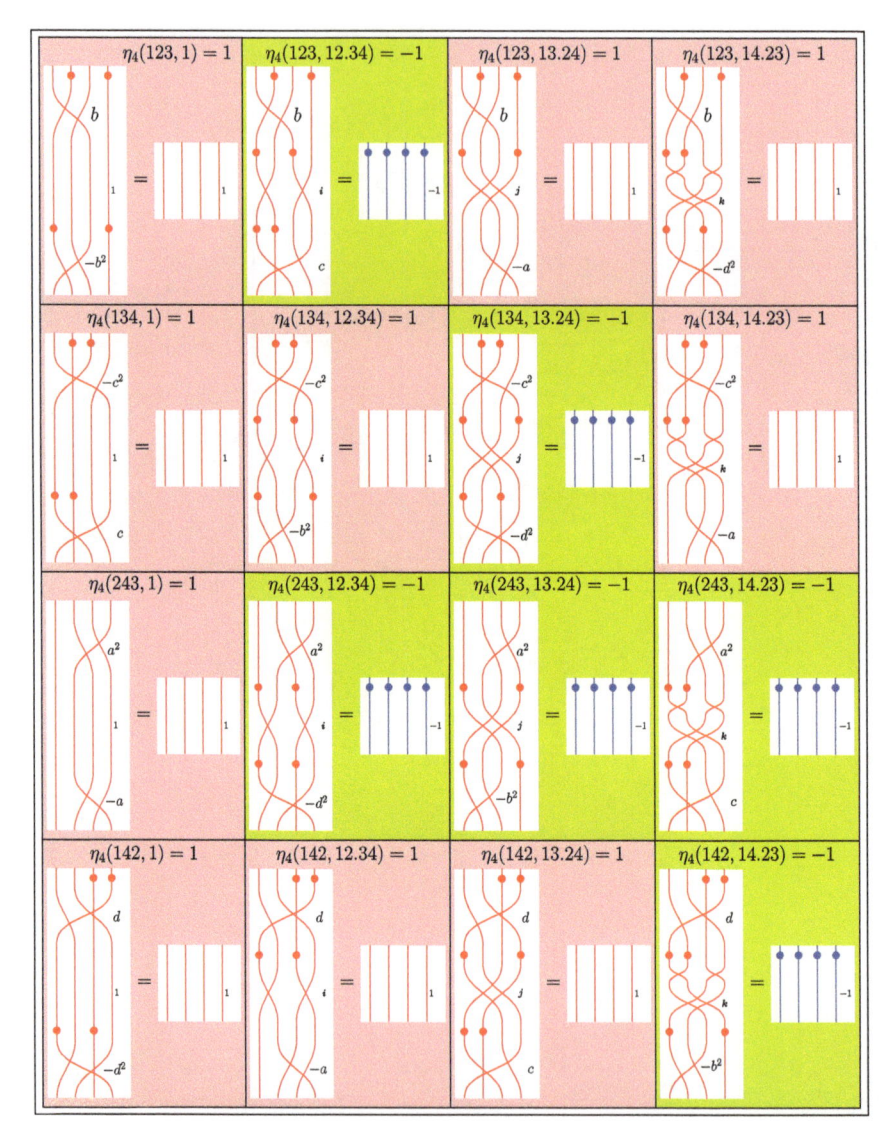

Figure 8.26. The fourth set of 16-cocycle values of η_4.

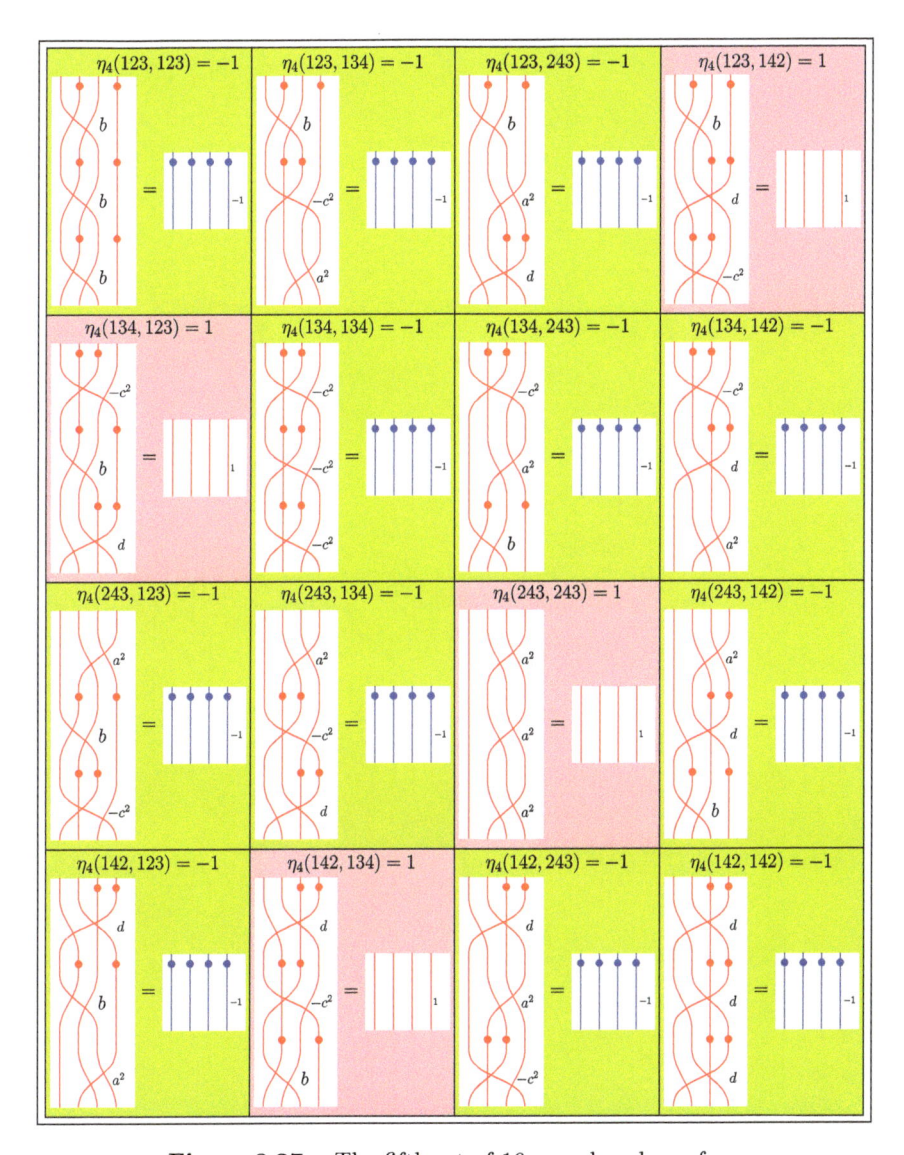

Figure 8.27. The fifth set of 16-cocycle values of η_4.

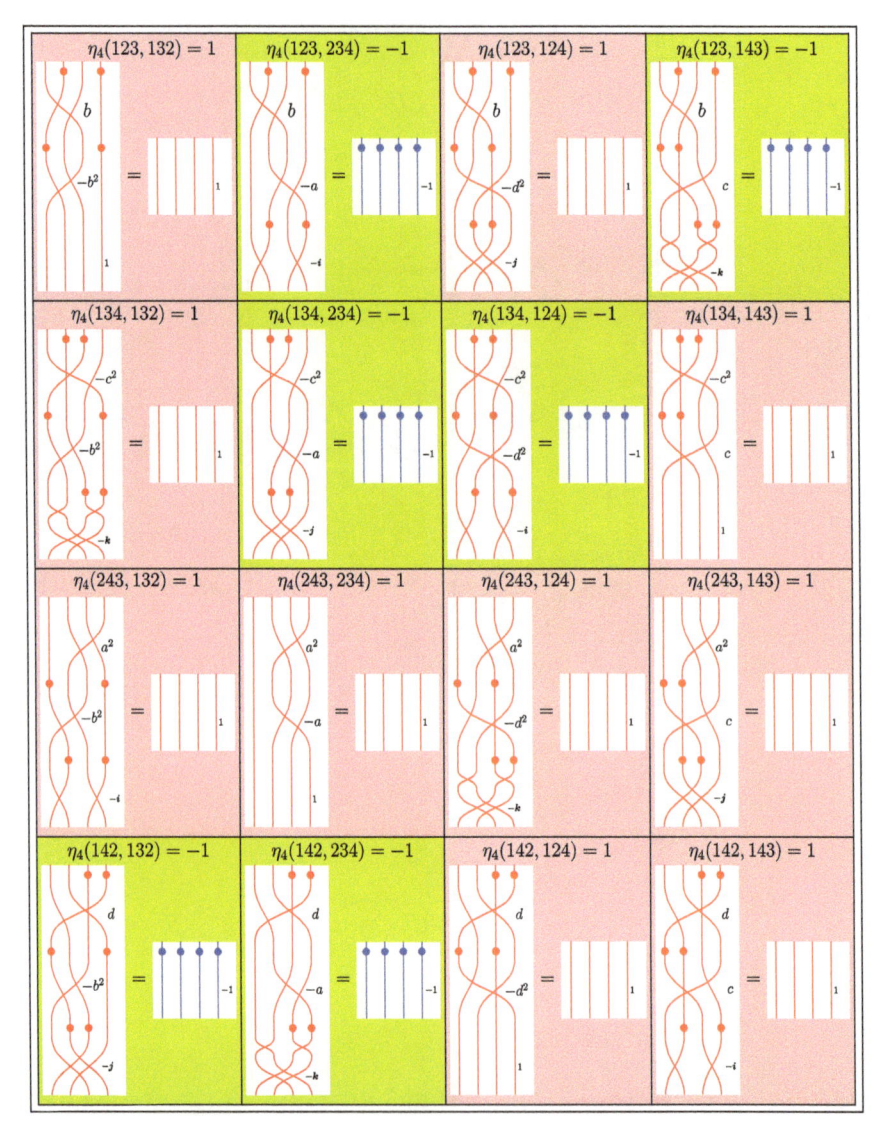

Figure 8.28. The sixth set of 16-cocycle values of η_4.

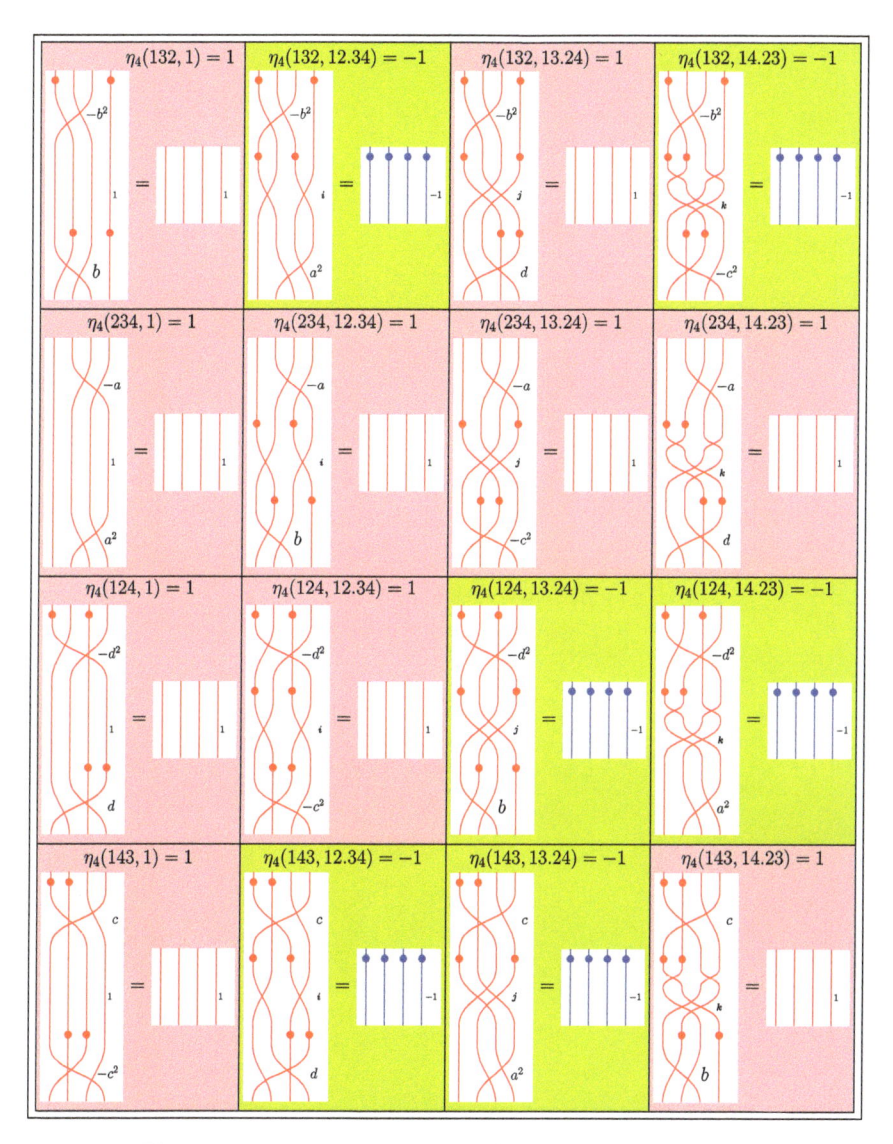

Figure 8.29. The seventh set of 16-cocycle values of η_4.

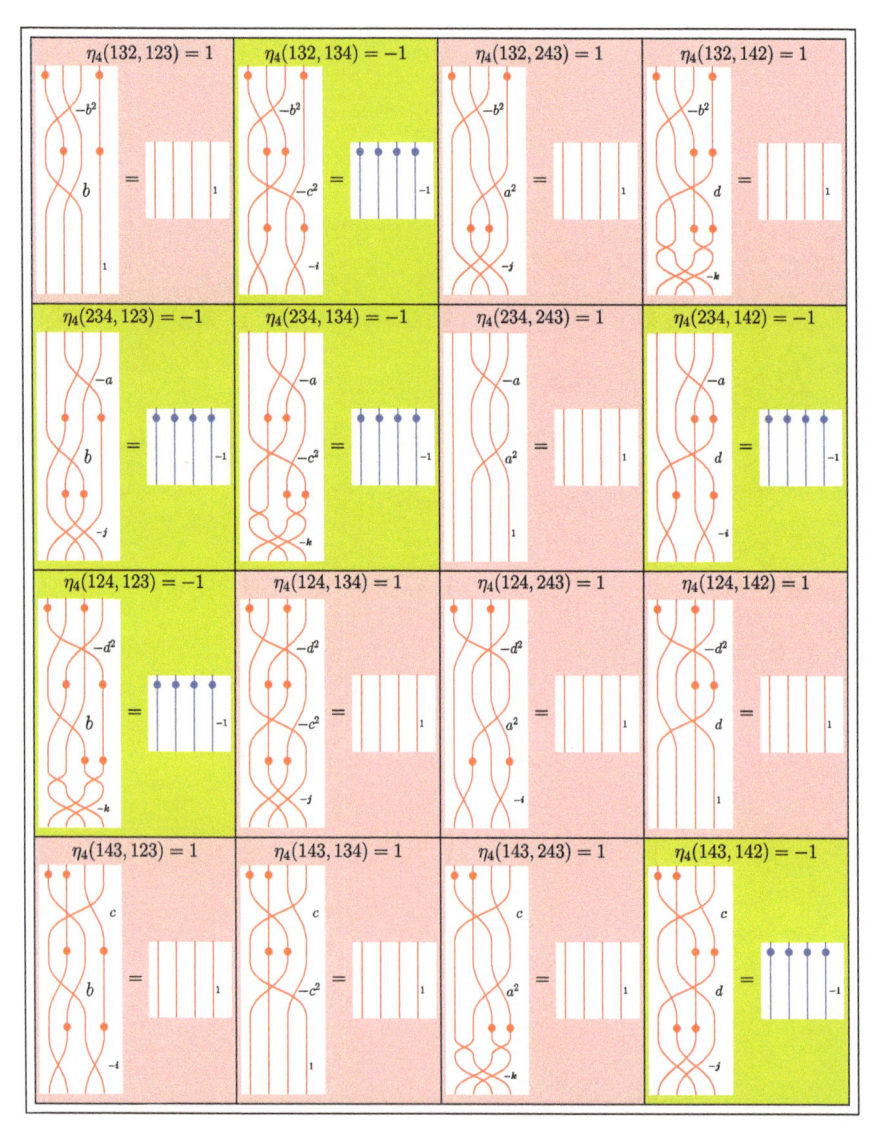

Figure 8.30. The eighth set of 16-cocycle values of η_4.

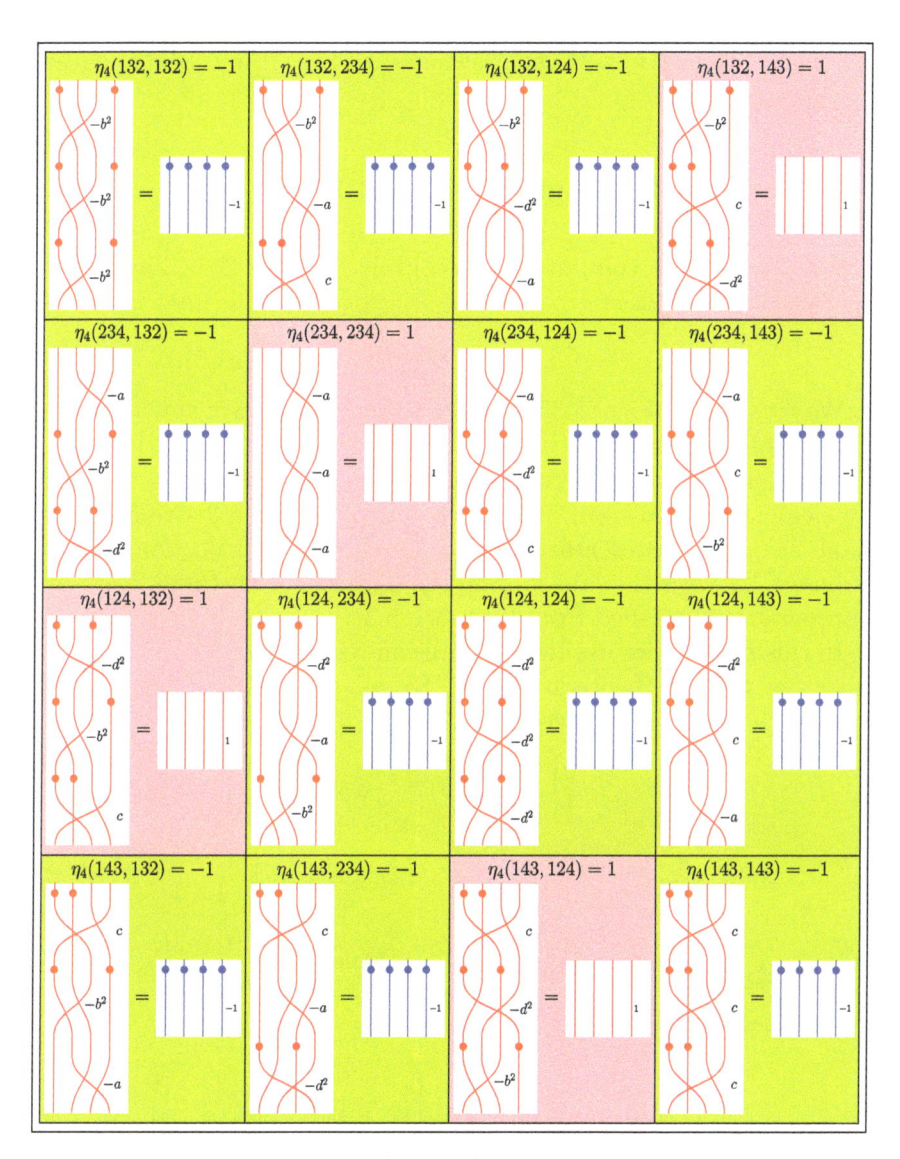

Figure 8.31. The ninth (and last) set of 16-cocycle values of η_4.

The proof that η_4 is not a coboundary is already in place because the cocycle values that appear in Fig. 8.23 coincide with the values that appear in Table 8.5. The Klein 4-group K is a subgroup of A_4, and the set-theoretic sections associated to Ses(2) and Ses(4) agree on that subgroup.

8.5. Example 4. The group $GL_2(\mathbb{Z}/3)$

In this section, we compute the 2-cocycle $Q_8 \xleftarrow{\psi} \Sigma_3 \times \Sigma_3$ that is associated to the seseq

$$1 \longleftarrow \Sigma_3 \xleftarrow{q_5} GL_2(\mathbb{Z}/3) \xleftarrow{i_5} Q_8 \longleftarrow 1 \quad \text{Ses}(5).$$

We recall from Figs. 5.35–5.40 that there are 3-strings-with-semi-dihedral-quipu representatives of the elements in $GL_2(\mathbb{Z}/3)$. By mapping the quipu in the subgroup G of the semi-dihedral group to 1 and the elements in the coset jG to -1, a projection to 3-strings-with-(mod 2)-quipu representatives is achieved. This projection has as its image Σ_4 with the labels upon the 3-strings-with-(mod 2)-quipu representatives recalled from Figs. 5.1–5.3.

In this section, we use those figures on the right of Figs. 5.35–5.40 to choose the set-theoretic section $\Sigma_3 \xrightarrow{s} GL_2(\mathbb{Z}/3)$. The section that we choose is normalized. Specifically,

$$s(1) = \begin{bmatrix} 1 & 0 \\ 0 & 1 \end{bmatrix}, \quad s(123) = -a = \begin{bmatrix} 0 & -1 \\ 1 & -1 \end{bmatrix},$$

$$s(132) = a^2 = \begin{bmatrix} -1 & 1 \\ -1 & 0 \end{bmatrix}, \quad s(12) = b^2 g = \begin{bmatrix} -1 & 0 \\ 1 & 1 \end{bmatrix},$$

$$s(23) = -kg = \begin{bmatrix} 1 & 0 \\ 1 & -1 \end{bmatrix}, \quad s(13) = c^2 g = \begin{bmatrix} 0 & 1 \\ 1 & 0 \end{bmatrix}.$$

One can check directly that

$$\begin{bmatrix} -1 & 0 \\ 1 & 1 \end{bmatrix} \cdot \begin{bmatrix} -1 & 0 \\ 1 & 1 \end{bmatrix} = \begin{bmatrix} 1 & 0 \\ 1 & -1 \end{bmatrix} \cdot \begin{bmatrix} 1 & 0 \\ 1 & -1 \end{bmatrix} = \begin{bmatrix} 0 & 1 \\ 1 & 0 \end{bmatrix} \cdot \begin{bmatrix} 0 & 1 \\ 1 & 0 \end{bmatrix}$$

$$= \begin{bmatrix} 0 & -1 \\ 1 & -1 \end{bmatrix} \cdot \begin{bmatrix} -1 & 1 \\ -1 & 0 \end{bmatrix} = \begin{bmatrix} 1 & 0 \\ 0 & 1 \end{bmatrix}.$$

Thus, the section that has been chosen is normal: $s(1) = 1$ and $s(x^{-1}) = [s(x)]^{-1}$.

Computing Group 2-cocycles 421

Among the representations that we have of the elements of $GL_2(\mathbb{Z}/3)$, it seems that products among the 4-strings-with-(mod 2)-quipu are more easily computed. In Figs. 8.32–8.37, the values of the cocycle ψ are computed both as 4-strings-with-(mod 2)-quipu and using the standard representation as (2×2)-matrices with entries in the field $\mathbb{Z}/3$.

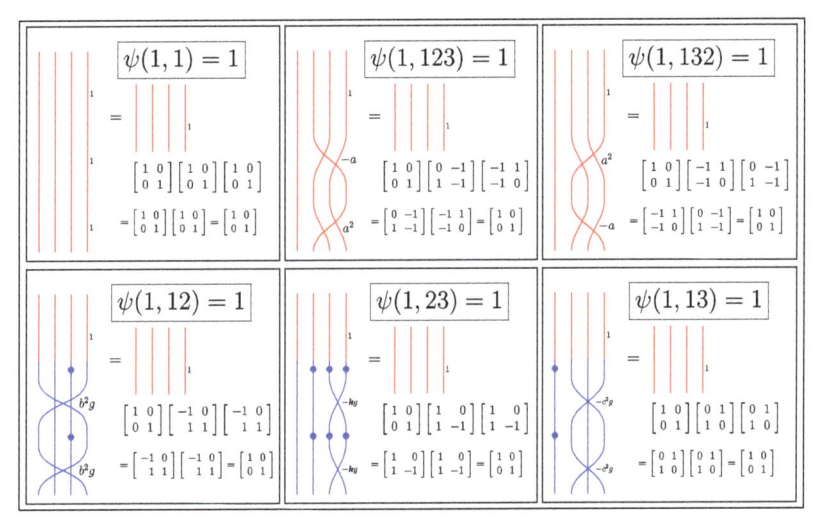

Figure 8.32. Computing cocycles for $GL_2(\mathbb{Z}/3)$ (part 1 of 6).

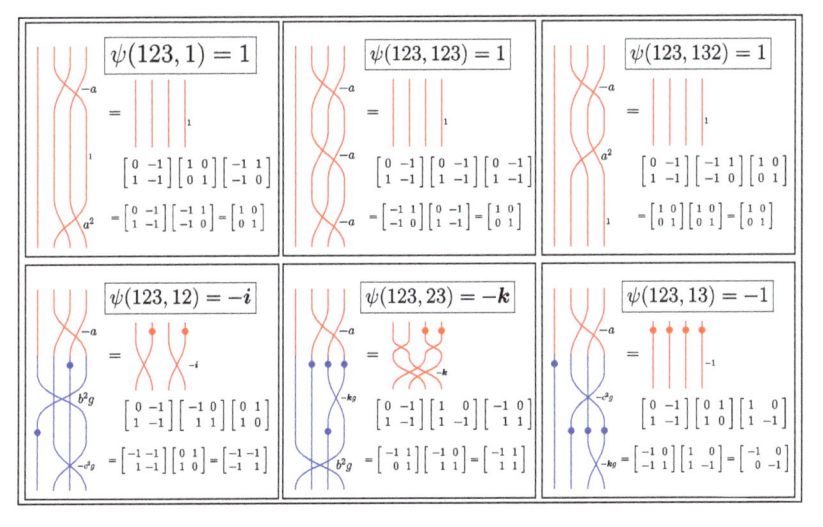

Figure 8.33. Computing cocycles for $GL_2(\mathbb{Z}/3)$ (part 2 of 6).

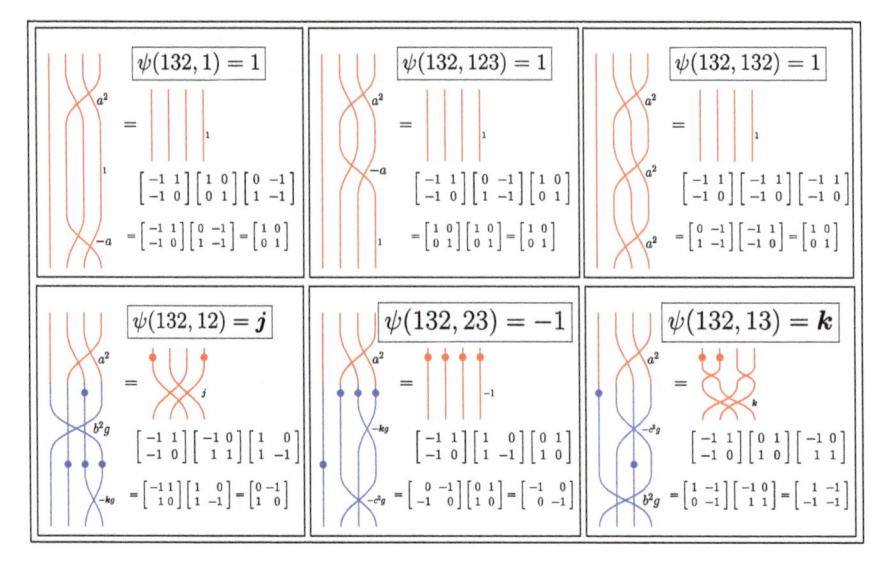

Figure 8.34. Computing cocycles for $GL_2(\mathbb{Z}/3)$ (part 3 of 6).

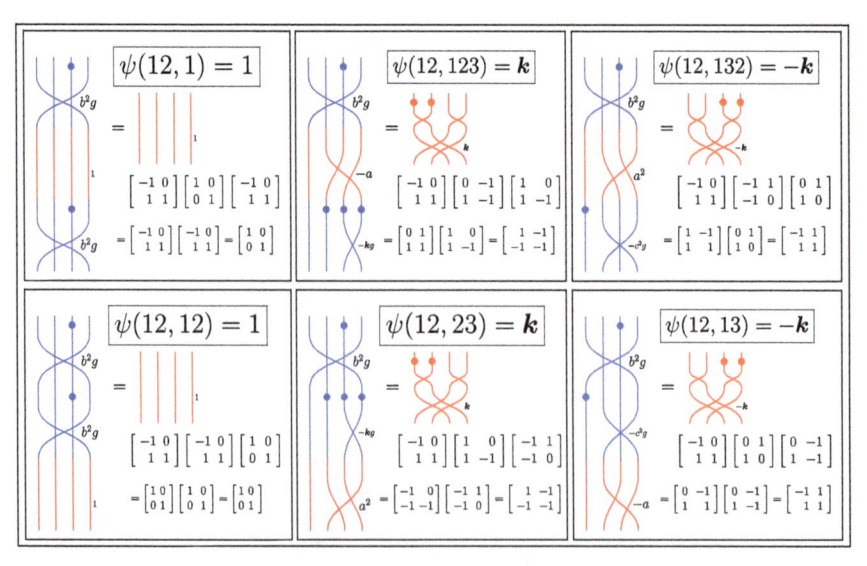

Figure 8.35. Computing cocycles for $GL_2(\mathbb{Z}/3)$ (part 4 of 6).

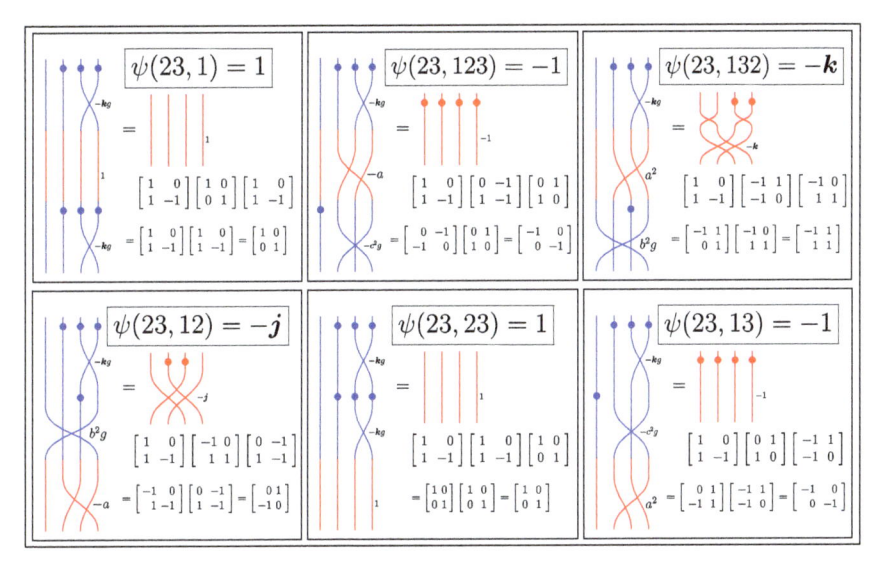

Figure 8.36. computing cocycles for $\mathrm{GL}_2(\mathbb{Z}/3)$ (part 5 of 6).

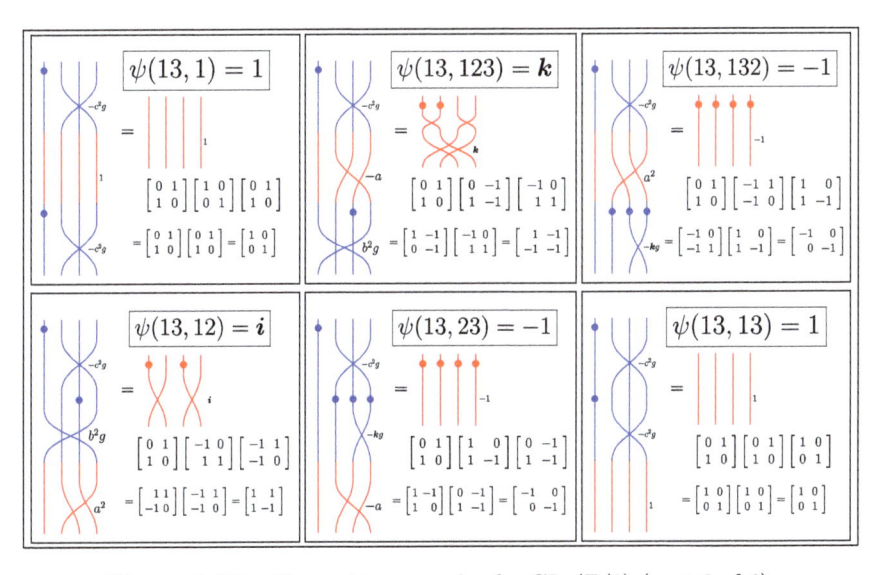

Figure 8.37. Computing cocycles for $\mathrm{GL}_2(\mathbb{Z}/3)$ (part 6 of 6).

For your convenience, these values are presented in Table 8.7.

Table 8.7. The set of cocycle values for $Q_8 \xleftarrow{\psi} \Sigma_3 \times \Sigma_3$ that is associated to Ses(5).

$\psi(r,c)$	(1)	(123)	(132)	(12)	(23)	(13)
(1)	1	1	1	1	1	1
(123)	1	1	1	$-i$	$-k$	-1
(132)	1	1	1	j	-1	k
(12)	1	k	$-k$	1	k	$-k$
(23)	1	-1	$-k$	$-j$	1	-1
(13)	1	k	-1	i	-1	1

Next, we demonstrate that the 2-cocycle ψ is not a coboundary. Suppose there were a function $Q_8 \xleftarrow{\gamma} \Sigma_3$ such that

$$\psi(x,y) = \gamma(y)[\gamma(xy)]^{-1}\gamma(x).$$

To begin, $\psi(1,1) = 1$. So,

$$\gamma(1)[\gamma(1)]^{-1}\gamma(1) = 1,$$

and

$$\gamma(1) = 1$$

Next, $\psi(123, 132) = 1$. So,

$$\gamma(132)[\gamma(1)]^{-1}\gamma(123) = 1.$$

This gives that

$$\gamma(132) = [\gamma(123)]^{-1}.$$

Since $\psi(123, 123) = 1$,

$$\gamma(123)[\gamma(132)]^{-1}\gamma(123) = [\gamma(123)]^3 = 1.$$

The only cube root of 1 in Q_8 is 1. Consequently,

$$\gamma(123) = \gamma(132) = 1.$$

Consider the transposition (12). The cocycle value $\psi(12,12) = 1$. So,

$$\gamma(12)[\gamma(1)]^{-1}\gamma(12) = [\gamma(12)]^2 = 1.$$

Similarly, for any transposition,

$$\gamma(12) = \pm 1, \quad \gamma(13) = \pm 1, \quad \gamma(13) = \pm 1.$$

We conclude this computation by recalling that $\psi(12,23) = \boldsymbol{k}$, and the function γ would satisfy the following equation:

$$\boldsymbol{k} = \gamma(23)[\gamma(123)]^{-1}\gamma(12) = (\pm 1)1(\pm 1).$$

Therein lies the contradiction.

8.6. Example 5. The binary octahedral group

In this, the last example of the chapter, a 2-cocycle

$$Q_8 \xleftarrow{\phi} \Sigma_3 \times \Sigma_3$$

that is associated to the seseq

$$1 \longleftarrow \Sigma_3 \xleftarrow{q_6} \widetilde{\Sigma_4} \xleftarrow{i_6} Q_8 \longleftarrow 1 \quad \text{Ses}(6)$$

will be computed by means of the usual method. A not entirely normal set-theoretic-section $\Sigma_3 \xrightarrow{s} \widetilde{\Sigma_4}$ will be chosen using the 3-strings-with-Dic$_8$-quipu representation of the binary octahedral group $\widetilde{\Sigma_4}$. These representatives of the elements of $\widetilde{\Sigma_4}$ are illustrated in Figs. 5.85–5.90. The section that we chose is not entirely normal since the transpositions in Σ_3, at best, lift (via the function s) to 4-cycles in the binary tetrahedral group.

On the other hand, the set-theoretic-section s can be chosen to agree with that of $\mathrm{GL}_2(\mathbb{Z}/3)$ upon the 3-cycles. Following the computation of the values $\phi(x,y)$, for $x,y \in \Sigma_3$, these will be compared to those for the cocycles η and ψ that are presented in Tables 8.4 and 8.7, respectively.

We choose

$$s(1) = 1, \quad s(123) = -a,$$

$$s(132) = a^2, \quad s(12) = af = \frac{i+j}{\sqrt{2}},$$

$$s(23) = jf = \frac{j-k}{\sqrt{2}}, \quad s(13) = bf = \frac{i-k}{\sqrt{2}}.$$

The cocycle values are determined by the usual equation

$$\phi(x,y) = s(x)s(y)[s(xy)]^{-1}.$$

We first observe that $\phi(1,y) = s(1)s(y)[s(y)]^{-1} = 1$, and $\phi(x,1) = s(x)s(1)[s(x)]^{-1} = 1$. As before, $\phi(123,123) = (-a)(-a)(a^2)^{-1} = -a^3 = 1$, and similarly, $\phi(132,132) = (a^2)(a^2)(-a)^{-1} = a^6 = 1$. Also, $\phi(123,132) = \phi(132,123) = s(123)s(132)[s(1)]^{-1} = -a \cdot a^2 = 1$. Now, let us consider the diagonal values:

$$\phi(pq,pq) = s(pq)s(pq)[s(1)]^{-1} = -1$$

for the transposition $(pq) = (12), (23)$, or (13). The remaining values $\phi(x,y)$ are computed within Figs. 8.38–8.42.

All the values are compiled in Table 8.8.

Table 8.8. The set of cocycle values for $Q_8 \xleftarrow{\phi} \Sigma_3 \times \Sigma_3$ that is associated to $\mathrm{Ses}(6)$.

$\phi(r,c)$	(1)	(123)	(132)	(12)	(23)	(13)
(1)	1	1	1	1	1	1
(123)	1	1	1	$-i$	$-k$	-1
(132)	1	1	1	j	-1	k
(12)	1	k	$-k$	-1	$-k$	k
(23)	1	-1	$-k$	j	-1	1
(13)	1	k	-1	$-i$	1	-1

Observation: Multiply the values of the cocycle η that are given in Table 8.4 with those of the cocycle ψ that are given in Table 8.7. Those products correspond, term by term, to the values of ϕ that are given in Table 8.8. We can write

$$\eta \cdot \psi = \phi.$$

Since $\mathbb{Z}/2 = \{\pm 1\}$ is a subgroup of Q_8, the values of η maybe considered to also lie within Q_8.

We have shown that both η and ψ are non-trivial. That is, neither is a coboundary. You may imitate the argument that was given in Section 8.5 to demonstrates that the 2-cocycle ϕ whose values are presented in Table 8.8 is also not a coboundary.

Suppose not. There seem to be several cases that require attention. Since $\phi(x,x) = -1$ when x is a transposition, then the function γ, for which $\phi(x,y) = \gamma(y)[\gamma(xy)]^{-1}\gamma(x)$, would have the property that $[\gamma(x)]^2 = -1$ for each of the transpositions (12), (23), and (13). Thus, the values of γ at the transpositions would be found in $\{\pm \boldsymbol{i}, \pm \boldsymbol{j}, \pm \boldsymbol{k}\}$. The lack of symmetry within Table 8.8 will induce contradictions. This last computation will be left to you, the reader, to complete at your leisure.

It is worth investigating other possible cocycles that are defined over Σ_3 with coefficients in Q_8. There are 52 groups of order 48 (groups wiki page). Among these, five are abelian. You may be able to find your own description of some of these by way of investigating the relationships among the three 2-cycles that have been described here.

It is best for us, the authors, and for you, the reader, to stop our joint investigations here since this book is already quite long. Let us conclude, then, with a summary.

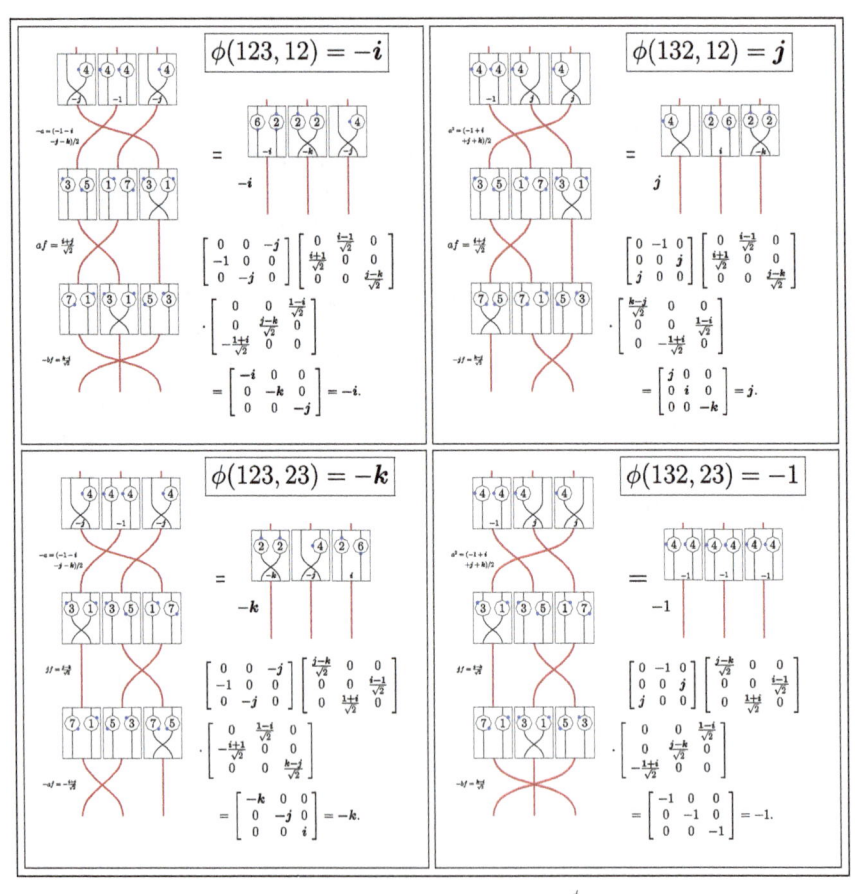

Figure 8.38. Computing some cocycle values $Q_8 \xleftarrow{\phi} \Sigma_3 \times \Sigma_3$ that are associated to Ses(6) (part 1 of 5).

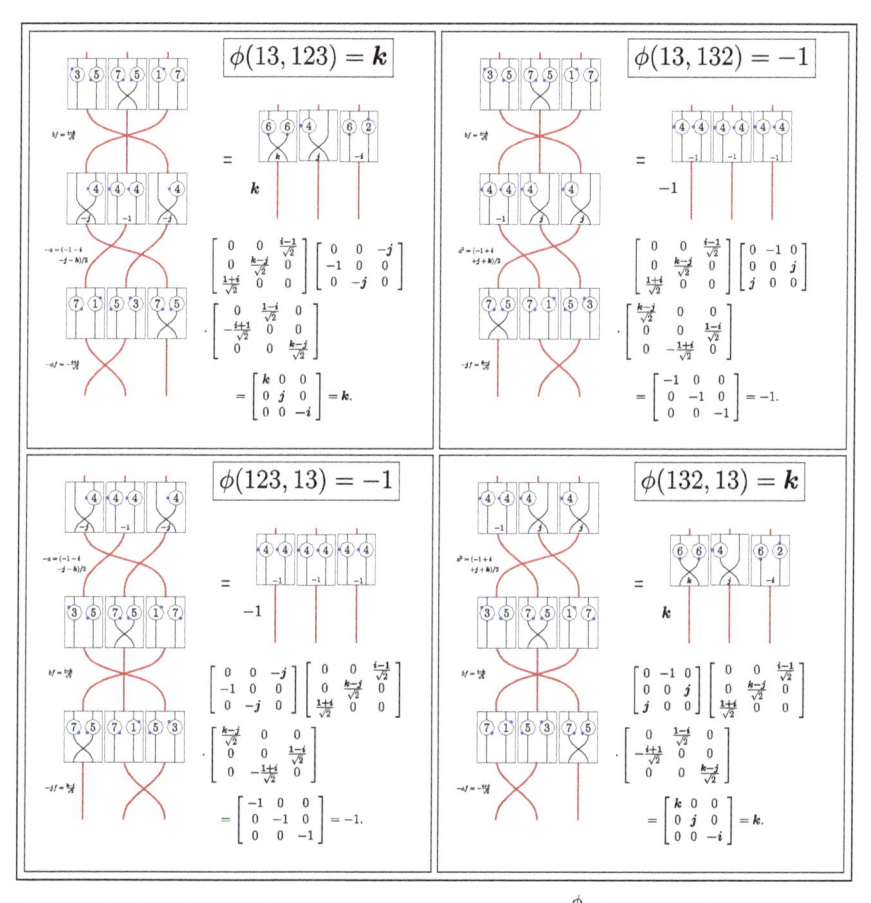

Figure 8.39. Computing some cocycle values $Q_8 \xleftarrow{\phi} \Sigma_3 \times \Sigma_3$ that are associated to Ses(6) (part 2 of 5).

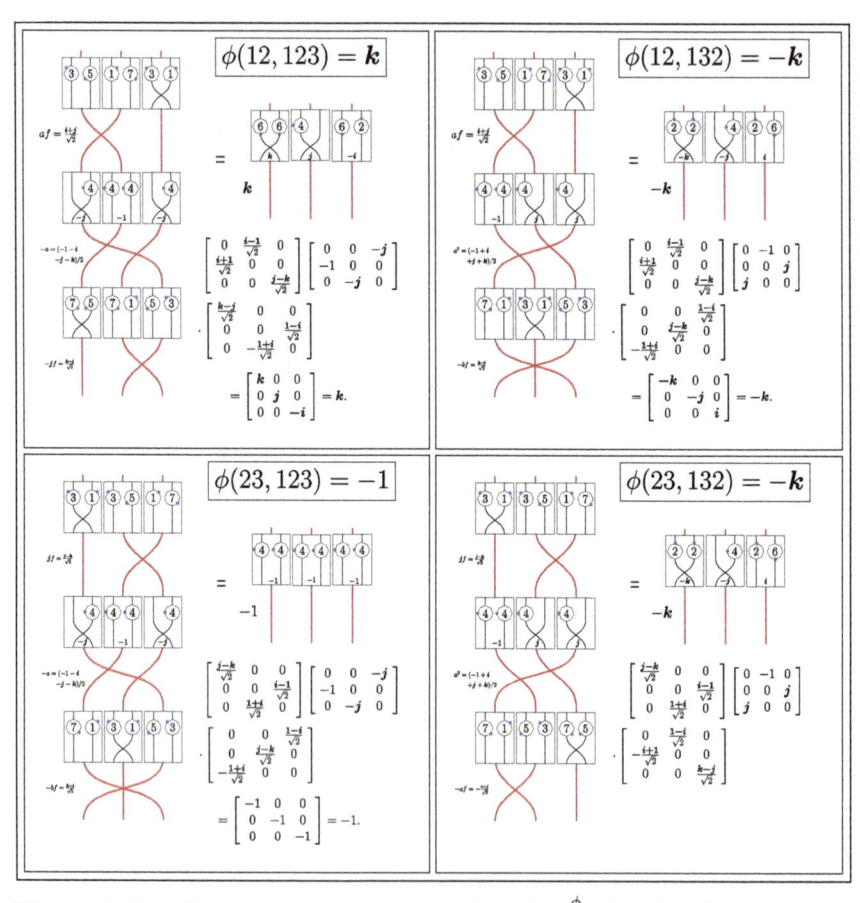

Figure 8.40. Computing some cocycle values $Q_8 \xleftarrow{\phi} \Sigma_3 \times \Sigma_3$ that are associated to Ses(6) (part 3 of 5).

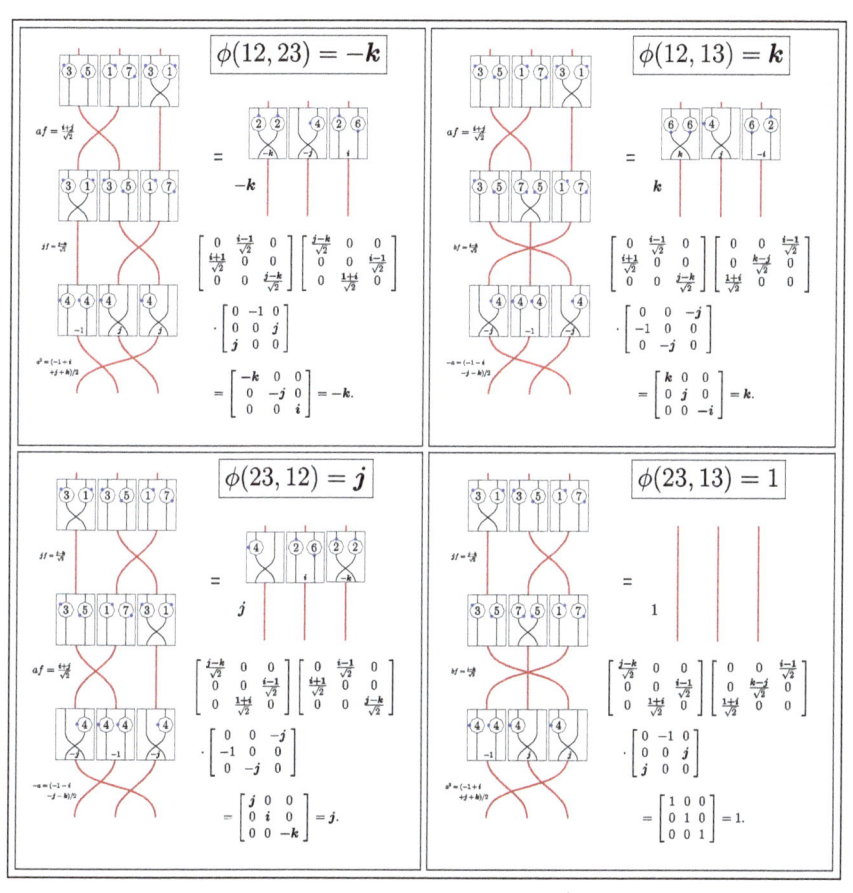

Figure 8.41. Computing some cocycle values $Q_8 \xleftarrow{\phi} \Sigma_3 \times \Sigma_3$ that are associated to Ses(6) (part 4 of 5).

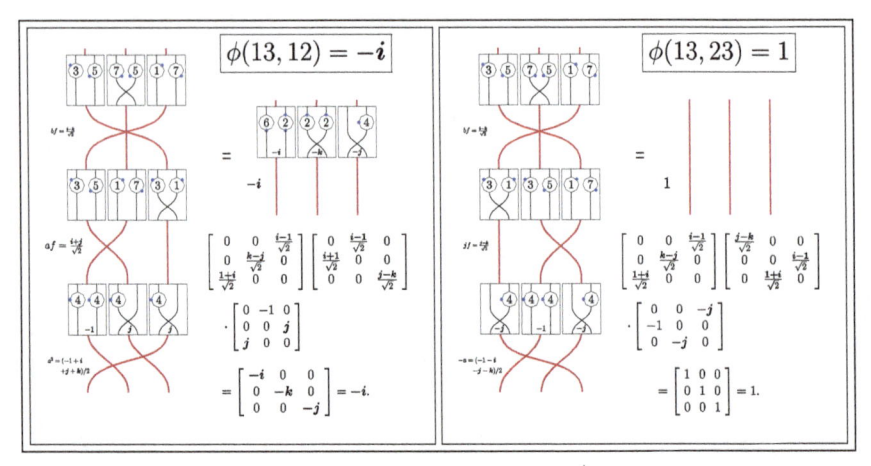

Figure 8.42. Computing some cocycle values $Q_8 \xleftarrow{\phi} \Sigma_3 \times \Sigma_3$ that are associated to Ses(6) (part 5 of 5).

8.7. Epilogue

There are several avenues down which you may explore. Certainly, the most obvious one is an understanding of a large variety of relatively small finite groups by means of the Sylow subgroups, thereby finding quipu representations of their elements. If you want to continue to study group theory from a visual perspective, we recommend [11]. Another that we only mentioned briefly was by means of some categorical aspects. Here, we give a short summary thereof.

A *category* is a mathematical structure that consists of a collection of objects for which, between any two objects, there is a set of arrows between them. There is an identity arrow from an object to itself, and the composition of arrows is associative. In our current situation, the collection of objects coincides with the set of natural numbers $\mathbb{N} = \{0, 1, 2, \ldots\}$. The identity arrow consists of straight strings that travel upward. The collection of arrows from $[n]$ to itself consists of the set of permutations in Σ_n. These are generated by the collection of adjacent transpositions of the form X.

There are mathematical identifications that can be made. A permutation corresponds to a string diagram. It can be represented as a matrix consisting of exactly one non-zero entry that is 1 in each row and column, and it is a composition of arrows that are conceived of up to an equivalence: distant crossings commute ($t_q t_r = t_r t_q$ when $1 < |r - q|$), a single crossing cancels itself ($t_q^2 = 1$), and the relation $t_q t_{q+1} t_q = t_{q+1} t_q t_{q+1}$ that factors the transposition $(q, q+1)$ in two different ways. These relationships correspond to the equivalences that were introduced for (mod 2)-quipu.

By considering (mod 2)-quipu representations, many of the groups we have studied here have representations into the collection of signed permutation matrices. We mentioned, in passing, that there are $24 \times 16 = 384$ matrices in the group of (4×4)-signed permutation matrices. Some of the subgroups therein have been found within our studies.

As cyclic subgroups are found within a given finite group, different quipu representations of the elements of the groups can be found.

One can also pursue group homology from a more abstract, and presumably more gratifying, perspective. While it is common to write, "Here be dragons," beyond the explored territory, all of the territory that we have explored herein has been explored thoroughly. The territories of group homology and homological algebra are rich with beauty and abstraction. We invite you to end your leisurely walk around the villages and begin to explore with the more advanced and abstract tools. Enjoy your studies.

References

[1] K. S. Brown. *Cohomology of Groups*, vol. 87, *Graduate Texts in Mathematics*. Springer-Verlag, New York-Berlin, 1982.

[2] J. S. Carter and S. Kamada. *Diagrammatic Algebra*, vol. 264, *Mathematical Surveys and Monographs*. American Mathematical Society, Providence, RI, [2021] ©2021.

[3] J. Dugundji. *Topology*. Allyn and Bacon, Inc., Boston, Mass., 1966.

[4] A. Hatcher. *Algebraic Topology*. Cambridge University Press, Cambridge, 2002.

[5] I. N. Herstein. *Topics in Algebra*. Xerox College Publishing, Lexington, Mass.-Toronto, Ont., second edition, 1975.

[6] C. H. Hinton. What is the fourth dimension? In *Speculations on the Fourth Dimension, Selected Writings of Charles H. Hinton*, vol. 1 (1884), pp. 1–22. Dover Publications, Inc., 1980.

[7] K. Hoffman and R. Kunze. *Linear Algebra*. Prentice-Hall, Inc., Englewood Cliffs, N.J., second edition, 1971.

[8] N. Jacobson. *Basic Algebra. I*. W. H. Freeman and Company, New York, second edition, 1985.

[9] N. Jacobson. *Basic Algebra. II*. W. H. Freeman and Company, New York, second edition, 1989.

[10] L. H. Kauffman. Non-commutative worlds and classical constraints. *Entropy*, 20(7):Paper No. 483, 25, 2018.

[11] M. Macauley. *Visual Algebra*. Forthcoming, 2025+.

[12] W. S. Massey. *A Basic Course in Algebraic Topology*, vol. 127, *Graduate Texts in Mathematics*. Springer-Verlag, New York, 1991.

[13] M. A. Peterson. The geometry of *paradise*. *Math. Intelligencer*, 30(4):14–19, 2008.

[14] Wikipedia contributors. Dicyclic group — Wikipedia, the free encyclopedia, 2018 (Accessed 23-November-2018).
[15] Wikipedia contributors. Rotation group so(3) — Wikipedia, the free encyclopedia, 2018 (Accessed 13-August-2018).
[16] Wikipedia contributors. Perelandra — Wikipedia, the free encyclopedia, 2022 (Accessed 5-April-2023).
[17] Wikipedia contributors. Jean le rond d'alembert — Wikipedia, the free encyclopedia, 2023 (Accessed 5-April-2023).

Index

$(n+1)$-dimensional space, 56
2-cocycle, 357
3-dimensional sphere, 154
3-sphere, 138
K_4-quipu, 124
n-ball, 140
n-dimensional cube, 93, 139
n-dimensional simplex, 56
n-simplex, 139
SU(2), 139
(mod 2)-quipu, 24
(mod n)-quipu, 32

A

abelian, 14
abstract presentation, 18, 60, 174, 279, 310
accidentals, 50
Alperin, Roger, 74
alternating, 83
alternating group, 97
associative, 11

B

Banchoff, Thomas, 113
Benson, David, 74
bigon, 68
bijection, 140
bijective, 13
boundary, 140

C

cartesian product, 16
category, 432
chain, 354
circle of fifths, 50
closed set, 140
closure, 140
coboundary, 357
cochain, 354
cocycle, 353–354
cohomologous, 357
commutations, 35
commutative, 14
complete graph, 61
complex numbers, 8
congruence, 12
conjugate transpose, 139
conjugation, 12
continuous function, 140
cosets, 12
covering, 177
cycle, 354

D

Daegu, 141
degree, 147
determinant, 94
dihedral group of order $2n$, 61, 175

E

ellipsis, 44
equatorial circles, 162
equivalent cocycles, 357
equivalent seseqs, 369
Eulerian, 142
exact sequence, 21

F

field, 7, 203
finite field, 203
formal difference, 143
frontier, 140
functorial pun, 354

G

Gabon, 142
generators (of a group), 27
Ghana, 142
graphical calculus, 386
group, 11
group cocycle, 353
group presentation, 18, 60, 174, 279

H

homeomorphism, 140
homomorphism, 13
Hopf fibration, 165
Hopf link, 164

I

index 2, 309
injective, 13
interior, 140
invariance of domain, 160
invertible, 11
isomorphism, 13

K

kernel, 13
key change, 51
Klein 4-group, 30, 69, 122, 158, 397, 405
Kuperberg, Greg, 74

M

metric topology, 140
modular arithmetic, 15

N

non-singular, 203
normal subgroup, 12
normalized section, 367
not entirely normal, 385

O

one-to-one, 13
onto, 13
open ball, 140
open set, 137, 140
order of a group, 11, 69
order of an element, 25, 69

P

permutations, 16
presentation, 18, 60, 174, 279

Q

quaternions, 138
quipu, 24, 32
quipu, K_4, 124
quipu, generalized, 119

R

real numbers, 7
ribbon diagrams, 29
rigid, 88
ring, 240
rook, 95

S

semi-direct product, 80, 117
seseq, 22, 27, 38, 62, 78, 117, 158, 178, 190, 231, 252, 271, 279, 309, 325, 355, 373, 378, 385, 394, 398
seseq, split, 78
set-theoretic section, 355
short exact sequence, 22
sign of a permutation, 94

signed permutations, 93
signed product rule, 147
simple group, 71
simplex, 56
slang notation, 144
space, 56
standard (coset) representatives, 15
standard unit vectors, 56
stereographic projection, 160
subgroup, 11

surjective, 13
symmetric group, 19

T

topology, 137, 140
triple point, 35
trivial group, 69

U

unitary, 11, 139

Vol. 69: *Mathematics of Harmony as a New Interdisciplinary Direction and "Golden" Paradigm of Modern Science*
Volume 3: The "Golden" Paradigm of Modern Science: Prerequisite for the "Golden" Revolution in Mathematics, Computer Science, and Theoretical Natural Sciences
by A. Stakhov

Vol. 68: *Mathematics of Harmony as a New Interdisciplinary Direction and "Golden" Paradigm of Modern Science*
Volume 2: Algorithmic Measurement Theory, Fibonacci and Golden Arithmetic's and Ternary Mirror-Symmetrical Arithmetic
by A. Stakhov

Vol. 67: On Complementarity: A Universal Organizing Principle
by J. S. Avrin

Vol. 66: *Invariants and Pictures: Low-dimensional Topology and Combinatorial Group Theory*
by V. O. Manturov, D. Fedoseev, S. Kim & I. Nikonov

Vol. 65: *Mathematics of Harmony as a New Interdisciplinary Direction and "Golden" Paradigm of Modern Science*
Volume 1: The Golden Section, Fibonacci Numbers, Pascal Triangle, and Platonic Solids
by A. Stakhov

Vol. 64: *Polynomial One-cocycles for Knots and Closed Braids*
by T. Fiedler

Vol. 63: *Board Games: Throughout the History and Multidimensional Spaces*
by J. Kyppö

Vol. 62: *The Mereon Matrix: Everything Connected through (K)nothing*
edited by J. B. McNair, L. Dennis & L. H. Kauffman

Vol. 61: *Numeral Systems with Irrational Bases for Mission-Critical Applications*
by A. Stakhov

Vol. 60: *New Horizons for Second-Order Cybernetics*
edited by A. Riegler, K. H. Müller & S. A. Umpleby

Vol. 59: *Geometry, Language and Strategy: The Dynamics of Decision Processes — Vol. 2*
by G. H. Thomas

Vol. 58: *Representing 3-Manifolds by Filling Dehn Surfaces*
by R. Vigara & A. Lozano-Rojo

Vol. 57: *ADEX Theory: How the ADE Coxeter Graphs Unify Mathematics and Physics*
by S.-P. Sirag

Vol. 56: *New Ideas in Low Dimensional Topology*
edited by L. H. Kauffman & V. O. Manturov

Vol. 55: *Knots, Braids and Möbius Strips*
by J. Avrin

Vol. 54: *Scientific Essays in Honor of H Pierre Noyes on the Occasion of His 90th Birthday*
edited by J. C. Amson & L. H. Kauffman

Vol. 53: *Knots and Physics (Fourth Edition)*
by L. H. Kauffman

Vol. 52: *Algebraic Invariants of Links (Second Edition)*
by J. Hillman

Vol. 51: *Virtual Knots: The State of the Art*
by V. O. Manturov & D. P. Ilyutko

Vol. 50: *Topological Library*
Part 3: Spectral Sequences in Topology
edited by S. P. Novikov & I. A. Taimanov

Vol. 49: *Hopf Algebras*
by D. E. Radford

Vol. 48: *An Excursion in Diagrammatic Algebra: Turning a Sphere from Red to Blue*
by J. S. Carter

Vol. 47: *Introduction to the Anisotropic Geometrodynamics*
by S. Siparov

Vol. 46: *Introductory Lectures on Knot Theory*
edited by L. H. Kauffman, S. Lambropoulou, S. Jablan & J. H. Przytycki

Vol. 45: *Orbiting the Moons of Pluto*
Complex Solutions to the Einstein, Maxwell, Schrödinger and Dirac Equations
by E. A. Rauscher & R. L. Amoroso

Vol. 44: *Topological Library*
Part 2: Characteristic Classes and Smooth Structures on Manifolds
edited by S. P. Novikov & I. A. Taimanov

Vol. 43: *The Holographic Anthropic Multiverse*
by R. L. Amoroso & E. A. Ranscher

Vol. 42: *The Origin of Discrete Particles*
by T. Bastin & C. Kilmister

Vol. 41: *Zero to Infinity: The Foundations of Physics*
by P. Rowlands

Vol. 40: *Intelligence of Low Dimensional Topology 2006*
edited by J. Scott Carter et al.

Vol. 39: *Topological Library*
Part 1: Cobordisms and Their Applications
edited by S. P. Novikov & I. A. Taimanov

Vol. 38: *Current Developments in Mathematical Biology*
edited by K. Mahdavi, R. Culshaw & J. Boucher

Vol. 37: *Geometry, Language, and Strategy*
by G. H. Thomas

Vol. 36: *Physical and Numerical Models in Knot Theory*
edited by J. A. Calvo et al.

Vol. 35: *BIOS — A Study of Creation*
by H. Sabelli

Vol. 34: *Woods Hole Mathematics — Perspectives in Mathematics and Physics*
edited by N. Tongring & R. C. Penner

Vol. 33: *Energy of Knots and Conformal Geometry*
by J. O'Hara

Vol. 32: *Algebraic Invariants of Links*
by J. A. Hillman

Vol. 31: *Mindsteps to the Cosmos*
by G. S. Hawkins

Vol. 30: *Symmetry, Ornament and Modularity*
by S. V. Jablan

Vol. 29: *Quantum Invariants — A Study of Knots, 3-Manifolds, and Their Sets*
by T. Ohtsuki

Vol. 28: *Beyond Measure: A Guided Tour Through Nature, Myth, and Number*
by J. Kappraff

Vol. 27: *Bit-String Physics: A Finite and Discrete Approach to Natural Philosophy*
by H. Pierre Noyes; edited by J. C. van den Berg

Vol. 26: *Functorial Knot Theory — Categories of Tangles, Coherence, Categorical Deformations, and Topological Invariants*
by David N. Yetter

Vol. 25: *Connections — The Geometric Bridge between Art and Science (Second Edition)*
by J. Kappraff

Vol. 24: *Knots in HELLAS '98 — Proceedings of the International Conference on Knot Theory and Its Ramifications*
edited by C. McA Gordon, V. F. R. Jones, L. Kauffman, S. Lambropoulou & J. H. Przytycki

Vol. 23: *Diamond: A Paradox Logic (Second Edition)*
by N. S. Hellerstein

Vol. 22: *The Mathematics of Harmony — From Euclid to Contemporary Mathematics and Computer Science*
by A. Stakhov (assisted by S. Olsen)

Vol. 21: *LINKNOT: Knot Theory by Computer*
by S. Jablan & R. Sazdanovic

Vol. 20: *The Mystery of Knots — Computer Programming for Knot Tabulation*
by C. N. Aneziris

Vol. 19: *Ideal Knots*
by A. Stasiak, V. Katritch & L. H. Kauffman

Vol. 18: *The Self-Evolving Cosmos: A Phenomenological Approach to Nature's Unity-in-Diversity*
by S. M. Rosen

Vol. 17: *Hypercomplex Iterations — Distance Estimation and Higher Dimensional Fractals*
by Y. Dang, L. H. Kauffman & D. Sandin

Vol. 16: *Delta — A Paradox Logic*
by N. S. Hellerstein

Vol. 15: *Lectures at KNOTS '96*
by S. Suzuki

Vol. 14: *Diamond — A Paradox Logic*
by N. S. Hellerstein

Vol. 13: *Entropic Spacetime Theory*
by J. Armel

Vol. 12: *Relativistic Reality: A Modern View*
edited by J. D. Edmonds, Jr.

Vol. 11: *History and Science of Knots*
edited by J. C. Turner & P. van de Griend

Vol. 10: *Nonstandard Logics and Nonstandard Metrics in Physics*
by W. M. Honig

Vol. 9: *Combinatorial Physics*
by T. Bastin & C. W. Kilmister

Vol. 8: *Symmetric Bends: How to Join Two Lengths of Cord*
by R. E. Miles

Vol. 7: *Random Knotting and Linking*
edited by K. C. Millett & D. W. Sumners

Vol. 6: *Knots and Applications*
edited by L. H. Kauffman

Vol. 5: *Gems, Computers and Attractors for 3-Manifolds*
by S. Lins

Vol. 4: *Gauge Fields, Knots and Gravity*
by J. Baez & J. P. Muniain

Vol. 3: *Quantum Topology*
edited by L. H. Kauffman & R. A. Baadhio

Vol. 2: *How Surfaces Intersect in Space — An Introduction to Topology (Second Edition)*
by J. S. Carter

Vol. 1: *Knots and Physics (Third Edition)*
by L. H. Kauffman

Milton Keynes UK
Ingram Content Group UK Ltd.
UKHW050942180724
445666UK00002B/10

9 789811 292750